D1285557

Physics and Whitehead

SUNY Series in Constructive Postmodern Thought
David Ray Griffin, editor

Physics and Whitehead

Quantum, Process, and Experience

Edited by
Timothy E. Eastman
and
Hank Keeton

State University of New York Press

Cover image: This ghostly apparition is actually an interstellar cloud caught in the process of destruction by strong radiation from the nearby star, Merope. This haunting picture of Bernard's Nebula, located in the Pleiades, suggests the interconnectedness, openness, order, and creativity of the universe. Modern physics now understands the multi-scale coupling, interdependence, and pervasiveness of such stellar radiation and space plasmas.

Image Credit: NASA and The Hubble Heritage Team (STScI/AURA)

Acknowledgment: George Herbig and Theodore Simon (Institute for Astronomy, University of Hawaii)

Published by
State University of New York Press, Albany

Printed in the United States of America

For information, address State University of New York Press,
90 State Street, Suite 700, Albany, NY 12207

Production by Judith Block
Marketing by Jennifer Giovani

Library of Congress Cataloging-in-Publication Data

Physics and Whitehead : quantum, process, and experience / edited by Timothy E. Eastman and Hank Keeton.
p. cm. — (SUNY series in constructive postmodern thought)
Includes bibliographical references and index.
ISBN 0-7914-5913-6 (alk. paper) — ISBN 0-7914-5914-4 (pbk. : alk. paper)
1. Whitehead, Alfred North, 1861–1947 — Congresses. I. Eastman, Timothy E. II. Keeton, Hank. III. Series.

B1674.W354P48 2003
192—dc22

2003059022

10 9 8 7 6 5 4 3 2 1

To John B. Cobb Jr.

whose encouragement and support of the dialogue between process thought and natural science made this work possible.

And in memory of scholars whose work substantially advanced the dialogue between process thought and natural science.

David Bohm	(1917–1992)
Milič Čapek	(1909–1997)
Charles Hartshorne	(1897–2000)
Ivor Leclerc	(1915–1999)
Victor Lowe	(1907–1988)
Ilya Prigogine	(1917–2003)

Contents

Preface

The Center for Process Studies in Claremont, California held a conference on *Physics and Time* in August 1984 with David Bohm, Ilya Prigogine, Henry Stapp, and other leading scientists and philosophers. During that conference, a spontaneous conjunction of energies emerged from a conversation between two participants who also appear in these pages (Stapp and Keeton). Stapp presented a paper on Einstein and Alfred North Whitehead, after which Keeton asked Stapp if they might collaborate on reinterpreting Whitehead's gravitational theory of 1922. Stapp suggested they write a paper together and discuss the project over lunch that day. As they left the auditorium, other conference participants joined the conversation, swelling the original pair to more than 25. It was a loud and energetic session, and two more of those joining the effort that day also appear in this volume (Eastman and Tanaka). Our enthusiasm for the project of reinterpretation grew into a broad collaboration lasting nearly 10 years, including research, annual gatherings, and publications. The present volume is an extension of that collaboration. In the decades preceding 1984, efforts to explore and enhance process concepts within modern scientific research had gradually accelerated (see Bibliography, especially Internet resources).

In the 1940s one major work was published (Lillie), and in the 1950s three volumes appeared (Agar, Smith, Synge). The 1960s included a major investigation of Whitehead's philosophy of science (Palter) and a study emphasizing Whitehead's relativity theory (Schmidt). In the next three decades the pace quickened. The 1970s saw three significant works on Whitehead and science (Fitzgerald, Fowler, Plamondon), and the 1980s had 11 major volumes, including one on mathematics (Code), one on logic (Martin), one a collection of papers from the 1984 conference mentioned above (Griffin), and a collection of papers on postmodern science (Griffin). The 1990s experienced increasing productivity and diversity in 13 major works on process thought and natural science: in the fields of biology (Birch), computer science (Henry), philosophy of science (Athern), and physics (Fagg, Jungerman, Lobl, Ranke, Shimony, Stapp,

Stolz). Then a major effort was mounted by the Center for Process Studies during 1997–98 with two special focus issues in *Process Studies*, "Process Thought and Natural Science" (Eastman, ed.). In the past century, there have been more than 200 significant publications focusing on Whitehead or a process-relational perspective and the physical sciences. Of these, more than 90% have appeared since 1950, and 25% in the past decade alone. This rapid increase of interest in Whitehead's philosophy and its associated variations reflects a growing recognition of the creative possibilities a process perspective brings to contemporary science and human experience. The 1998 conference papers included in this volume are a continuation of that momentum, with a focus on Whitehead's contributions to mathematical physics, and the implications of his philosophy for contemporary physics.

Whitehead's academic career spanned more than five decades, from 1880 well into the twentieth century, covering a variety of fields from mathematics and symbolic logic to philosophy of nature and philosophy of science, to epistemology, cosmology, and metaphysics. Chronologically, his publications have been popularly grouped into three general categories parallel to this demarcation of interests, no category being completely exhaustive or mutually exclusive. A careful reading of materials published throughout his career reveals an uncanny and persistent hallmark of continuity between specific focus and broad generalization within and between each work. During his early work in mathematics and logic (1880–1912—generally the years at Trinity College, Cambridge) he clearly enlarged his specific concentration on mathematics to include applications for other more physical sciences. As he gradually generalized those early investigations into the foundations of broader science, he naturally began expressing his evolving insights using more philosophical language and categories (1912–1924—generally the years in London at University College and the Imperial College of Science and Technology). This more philosophical discourse led to even broader categorical investigations that resulted in the challenging cosmology of his mature thinking (1924–1947—generally the years at Harvard University's philosophy department, and retirement). The threads linking this complex scheme of progressively more comprehensive ideas can be found within each work, and exhibit themselves throughout his career. His ability to maintain and expand those threads within the wide range of his interests is a hallmark. Other thinkers (e.g., Alexander, Bergson, James, Peirce) track parallel paths through similar issues, and together help constitute an emerging field within philosophy focusing on *relationality* and the *process* nature of the universe. This philosophical field was described as *process philosophy* in the 1960s and found institutional support at Harvard and the University of Chicago.

Most interpreters of Whitehead's thought focus on the more philosophical works of his Harvard years, but the authors in the present volume seek an expansion of those ideas into the contemporary worlds of quantum and relativity physics. In both of these worlds, Whitehead has inspired contemporary thinkers to suggest new and challenging interpretations for current experimental data and to formulate innovative schemes expanding the foundations of physics. Although Whitehead worked actively during the formative years of quantum theory, he published no works dealing with those developments in particular. Interestingly, much of the current interpretive work focusing on his ideas demonstrates just how parallel his thinking was to that of Bohr, de Brolie, Heisenberg, and Schrödinger on matters of continuity and atomicity. In fact it is the subtly profound nature of Whitehead's quantum thinking, woven into the fabric of his emerging philosophy of nature, that stimulates several authors in this present volume.

However, from very early in his career, in numerous publications and lectures, Whitehead wrestled openly with the concepts surrounding space-time and motion, finally resulting in his own version of gravitational theory in 1922. What inspired Whitehead to formulate his own theory? As a mathematical physicist, he was aware of the success Einstein's first theory of relativity enjoyed in the scientific community. But what intrigued Whitehead most was the philosophical basis of Einstein's theory. For Whitehead's own theory, he begins with different philosophical assumptions. After 1905 his work, both mathematically and philosophically, revealed his growing dissatisfaction with the classical concepts of mass, time, and space. He had a strong intuition that Einstein, and with him the mainstream scientific community, was traveling a path that might look entirely different if the journey began with different assumptions. For Whitehead, the profound coupling of mass and energy that Einstein proposed in his special theory was a brilliant development. But something else was revealed in this mass-energy relationship. A sense of limits appeared in the concepts. Whitehead saw something beyond or behind the focus on mass-energy and realized he was looking at the coupling relations themselves, not just mass or energy. If these new relativistic theories about the interaction of mass-energy resulted in *refinement* of the basic concepts, what might happen if the concepts themselves were *radically* reconceived? How might gravity be conceived differently?

Whitehead began at a place quite foreign to most scientists during the early part of the century. Rather than focus on the *things* that were being measured and tested (whether massive objects or massless objects), Whitehead choose to focus on the *events* that constituted or included those *things* instead. What does it mean to focus on *events* rather than on things as *objects*? The papers in this volume approach this question from a variety of angles. They seek to explore the conceptual adjustments required if event-like structures replace object-like structures in physical theories, both quantum and relativistic.

The authors of the chapters in this volume have collaborated from the standpoint of a felt need. That need emerges from inquiries into the limits of mass-spacetime concepts, in the face of experimental data suggesting that foundational concepts of mass-spacetime have reached particular limits of applicability. This book emerged from three seminars offered at the Third International Whitehead Conference held in August 1998 in Claremont, California, under the sponsorship of the Center for Process Studies. These workshops were one stage in the process of uniting the variety of voices into a cohesive whole. The sequence of the workshops largely parallels the structure of the present volume, to which the editors have added introductory and concluding sections, with bibliography.

Part I consists of an introduction to process thought (Clayton), a summary of contrasts between classical, modern, and postmodern scientific-theoretic categories (Eastman), and an overview of Whitehead's work as a mathematical physicist (Keeton), culminating in suggestions for a process physics (Jungerman). Following each set of chapters is a selection of the workshop dialogue pertinent to that section.

Part II focuses on order and the phenomenon of emergence by exploring coherence in chemical systems (Earley), comparing Whitehead's *actual entities* and the collapse of quantum states (Malin), developing a foundation of physics based on the contrast between the classical *material* reality and a Whiteheadian *historical* reality (Chew), and interpreting experimental *choice* in quantum theory (Stapp), followed by pertinent workshop dialogue.

Part III focuses on fundamental processes by exploring the relationship between symmetry and asymmetry in physics (Rosen), interpreting special relativity to account for temporality (Hansen), revising quantum *individuality* using quantum logic to define *commensurability* (Tanaka), and reimaging the concept of physical *law* in terms of the its relationship with dynamics and kinematics resulting in the notion that *process* replaces *law* in very stimulating and challenging ways (Finkelstein), followed by further dialogue.

Part IV expands the focus from physics to metaphysics by comparing process metaphysics with the scientific metaphysics of Bunge (Riffert) and speculatively revisioning the role of human experience as quanta of information with extensional relations and causality (Nobo), plus dialogue. This section enlarges major themes emerging from the workshops and presents significant opportunities for further development at the interface of process and physics.

We hope this work, together with the bibliography and related Web site, will stimulate further research and constructive thinking employing the expanding resources of process thought.

We are especially grateful to the following: John B. Cobb Jr. and the codirectors of the Center for Process Studies for their unflagging support of academic advancement; Philip Clayton for special organizing; Jorge Nobo for general editing; our patient spouses, Carolyn Brown and Norma Jean Standlea, for their advice and encouragement throughout this project; Lyman Ellis for professional video recordings of the workshops; and key workshop participants beyond those directly contributing to this book, Ian Barbour, Murray Code, Lawrence Fagg, Stanley Klein, and Robert Valenza. Our thanks as well to the

Center for Process Studies, Sam Dunnam, and others who provided significant support for the workshop. We extend our gratitude to the staff of SUNY Press who created the opportunity for expressing this vision, especially Ms. Jane Bunker, Senior Acquisitions Editor, Ms. Judith Block, Senior Production Editor, and Jennifer Giovani, Marketing Manager.

Introduction to SUNY Series in Constructive Postmodern Thought[1]

The rapid spread of the term *postmodern* in recent years witnesses to a growing dissatisfaction with modernity and to an increasing sense that the modern age not only had a beginning but can have an end as well. Whereas the word *modern* was almost always used until quite recently as a word of praise and as a synonym for *contemporary*, a growing sense is now evidenced that we can and should leave modernity behind—in fact, that we *must* if we are to avoid destroying ourselves and most of the life on our planet.

Modernity, rather than being regarded as the norm for human society toward which all history has been aiming and into which all societies should be ushered—forcibly if necessary—is instead increasingly seen as an aberration. A new respect for the wisdom of traditional societies is growing as we realize that they have endured for thousands of years and that, by contrast, the existence of modern civilization for even another century seems doubtful. Likewise, *modernism* as a worldview is less and less seen as The Final Truth, in comparison with which all divergent worldviews are automatically regarded as "superstitious." The modern worldview is increasingly relativized to the status of one among many, useful for some purposes, inadequate for others.

Although there have been antimodern movements before, beginning perhaps near the outset of the nineteenth century with the Romanticists and the Luddites, the rapidity with which the term *postmodern* has become widespread in our time suggests that the antimodern sentiment is more extensive and intense than before, and also that it includes the sense that modernity can be successfully overcome only by going beyond it, not by attempting to return to a premodern form of existence. Insofar as a common element is found in the various ways in which the term is used, *postmodernism* refers to a diffuse

[1]The present version of this introduction is slightly different from the first version, which was contained in the volumes that appeared prior to 1999.

sentiment rather than to any common set of doctrines—the sentiment that humanity can and must go beyond the modern.

Beyond connoting this sentiment, the term *postmodern* is used in a confusing variety of ways, some of them contradictory to others. In artistic and literary circles, for example, postmodernism shares in this general sentiment but also involves a specific reaction against "modernism" in the narrow sense of a movement in artistic-literary circles in the late nineteenth and early twentieth centuries. Postmodern architecture is very different from postmodern literary criticism. In some circles, the term *postmodern* is used in reference to that potpourri of ideas and systems sometimes called *new age metaphysics*, although many of these ideas and systems are more premodern than postmodern. Even in philosophical and theological circles, the term *postmodern* refers to two quite different positions, one of which is reflected in this series. Each position seeks to transcend both *modernism*, in the sense of the worldview that has developed out of the seventeenth-century Galilean-Cartesian-Baconian-Newtonian science, and *modernity*, in the sense of the world order that both conditioned and was conditioned by this worldview. But the two positions seek to transcend the modern in different ways.

Closely related to literary-artistic postmodernism is a philosophical postmodernism inspired variously by physicalism, Ludwig Wittgenstein, Martin Heidegger, a cluster of French thinkers—including Jacques Derrida, Michel Foucault, Gilles Deleuze, and Julia Kristeva—and certain features of American pragmatism.[2] By the use of terms that arise out of particular segments of this movement, it can be called *deconstructive*, *relativistic*, or *eliminative* postmodernism. It overcomes the modern worldview through an antiworldview, deconstructing or even entirely eliminating various concepts that have generally been thought necessary for a worldview, such as self, purpose, meaning, a real world, givenness, reason, truth as correspondence, universally valid norms, and divinity. While motivated by ethical and emancipatory concerns, this type of postmodern thought tends to issue in relativism. Indeed, it seems to many thinkers to imply nihilism.[3] It could, paradoxically, also be called *ultramodern*-

[2]The fact that the thinkers and movements named here are said to have inspired the deconstructive type of postmodernism should not be taken, of course, to imply that they have nothing in common with constructive postmodernists. For example, Wittgenstein, Heidegger, Derrida, and Deleuze share many points and concerns with Alfred North Whitehead, the chief inspiration behind the present series. Furthermore, the actual positions of the founders of pragmatism, especially William James and Charles Peirce, are much closer to Whitehead's philosophical position—see the volume in this series titled *The Founders of Constructive Postmodern Philosophy: Peirce, James, Bergson, Whitehead, and Hartshorne*—than they are to Richard Rorty's so-called neopragmatism, which reflects many ideas from Rorty's explicitly physicalistic period.

[3]As Peter Dews points out, although Derrida's early work was "driven by profound ethical impulses," its insistence that no concepts were immune to deconstruction "drove its own ethical presuppositions into a penumbra of inarticulacy" (*The Limits of Disenchantment: Essays on Contemporary European Culture* [London: New York: Verso, 1995], 5). In his more recent thought,

ism, in that its eliminations result from carrying certain modern premises—such as the sensationist doctrine of perception, the mechanistic doctrine of nature, and the resulting denial of divine presence in the world—to their logical conclusions. Some critics see its deconstructions or eliminations as leading to self-referential inconsistencies, such as "performative self-contradictions" between what is said and what is presupposed in the saying.

The postmodernism of this series can, by contrast, be called *revisionary*, *constructive*, or—perhaps best—*reconstructive*. It seeks to overcome the modern worldview not by eliminating the possibility of worldviews (or "metanarratives") as such, but by constructing a postmodern worldview through a revision of modern premises and traditional concepts in the light of inescapable presuppositions of our various modes of practice. That is, it agrees with deconstructive postmodernists that a massive deconstruction of many received concepts is needed. But its deconstructive moment, carried out for the sake of the presuppositions of practice, does not result in self-referential inconsistency. It also is not so totalizing as to prevent reconstruction. The reconstruction carried out by this type of postmodernism involves a new unity of scientific, ethical, aesthetic, and religious intuitions (whereas poststructuralists tend to reject all such unitive projects as "totalizing modern metanarratives"). While critical of many ideas often associated with modern science, it rejects not science as such but only that *scientism* in which only the data of the modern natural sciences are allowed to contribute to the construction of our public worldview.

The reconstructive activity of this type of postmodern thought is not limited to a revised worldview. It is equally concerned with a postmodern world that will both support and be supported by the new worldview. A postmodern world will involve postmodern persons, with a postmodern spirituality, on the one hand, and a postmodern society, ultimately a postmodern global order, on the other. Going beyond the modern world will involve transcending its individualism, anthropocentrism, patriarchy, economism, consumerism, nationalism, and militarism. Reconstructive postmodern thought provides support for the ethnic, ecological, feminist, peace, and other emancipatory movements of our time, while stressing that the inclusive emancipation must be from the destructive features of modernity itself. However, the term *postmodern*, by contrast with *premodern*, is here meant to emphasize that the modern world has produced unparalleled advances, as Critical Theorists have emphasized, which must not be devalued in a general revulsion against modernity's negative features.

From the point of view of deconstructive postmodernists, this reconstructive postmodernism will seem hopelessly wedded to outdated concepts, because

Derrida has declared an "emancipatory promise" and an "idea of justice" to be "irreducible to any deconstruction." Although this "ethical turn" in deconstruction implies its pulling back from a completely disenchanted universe, it also, Dews points out (6–7), implies the need to renounce "the unconditionality of its own earlier dismantling of the unconditional."

it wishes to salvage a positive meaning not only for the notions of selfhood, historical meaning, reason, and truth as correspondence, which were central to modernity, but also for notions of divinity, cosmic meaning, and an enchanted nature, which were central to premodern modes of thought. From the point of view of its advocates, however, this revisionary postmodernism is not only more adequate to our experience but also more genuinely postmodern. It does not simply carry the premises of modernity through to their logical conclusions, but criticizes and revises those premises. By virtue of its return to organicism and its acceptance of nonsensory perception, it opens itself to the recovery of truths and values from various forms of premodern thought and practice that had been dogmatically rejected, or at least restricted to "practice," by modern thought. This reconstructive postmodernism involves a creative synthesis of modern and premodern truths and values.

This series does not seek to create a movement so much as to help shape and support an already existing movement convinced that modernity can and must be transcended. But in light of the fact that those antimodern movements that arose in the past failed to deflect or even retard the onslaught of modernity, what reasons are there for expecting the current movement to be more successful? First, the previous antimodern movements were primarily calls to return to a premodern form of life and thought rather than calls to advance, and the human spirit does not rally to calls to turn back. Second, the previous antimodern movements either rejected modern science, reduced it to a description of mere appearances, or assumed its adequacy in principle. They could, therefore, base their calls only on the negative social and spiritual effects of modernity. The current movement draws on natural science itself as a witness against the adequacy of the modern worldview. In the third place, the present movement has even more evidence than did previous movements of the ways in which modernity and its worldview *are* socially and spiritually destructive. The fourth and probably most decisive difference is that the present movement is based on the awareness that *the continuation of modernity threatens the very survival of life on our planet.* This awareness, combined with the growing knowledge of the interdependence of the modern worldview with the militarism, nuclearism, patriarchy, global apartheid, and ecological devastation of the modern world, is providing an unprecedented impetus for people to see the evidence for a postmodern worldview and to envisage postmodern ways of relating to each other, the rest of nature, and the cosmos as a whole. For these reasons, the failure of the previous antimodern movements says little about the possible success of the current movement.

Advocates of this movement do not hold the naively utopian belief that the success of this movement would bring about a global society of universal and lasting peace, harmony, and happiness, in which all spiritual problems, social conflicts, ecological destruction, and hard choices would vanish. There is, after all, surely a deep truth in the testimony of the world's religions to the presence of a transcultural proclivity to evil deep within the human heart, which no new paradigm, combined with a new economic order, new child-rearing practices, or any other social arrangements, will suddenly eliminate. Furthermore, it has correctly been said that "life is robbery": A strong element of competition is inherent within finite existence, which no social-political-economic-ecological order can overcome. These two truths, especially when contemplated together, should caution us against unrealistic hopes.

No such appeal to "universal constants," however, should reconcile us to the present order, as if it were thereby uniquely legitimated. The human proclivity to evil in general, and to conflictual competition and ecological destruction in particular, can be greatly exacerbated or greatly mitigated by a world order and its worldview. Modernity exacerbates it about as much as imaginable. We can therefore envision, without being naively utopian, a far better world order, with a far less dangerous trajectory, than the one we now have.

This series, making no pretense of neutrality, is dedicated to the success of this movement toward a postmodern world.

David Ray Griffin
Series Editor

Part I

Physics and Whitehead

1

Introduction to Process Thought

PHILIP CLAYTON

What Is Metaphysics?

Metaphysics is not new to the twentieth century. The authors in this volume who engage in metaphysical reflection or who speak of A. N. Whitehead as one of the great metaphysicians of this century are part of the heritage of a tradition running back at least 2,400 years. Various treatments of metaphysics in general, and of Whitehead's work in particular, have emphasized the connections with Plato, Aristotle, Aquinas, Kant, Leibniz and Bergson—to name just a few of the antecedents.[1] What is this deep preoccupation of which the present volume represents the newest installment?

Metaphysics must surely represent one of the oldest and most audacious pursuits of the human mind. It lies at the origins of the Western intellectual tradition and of natural philosophy, making it one of the progenitors of science itself. To engage in metaphysical reflection is to attempt to give a rational and systematic presentation, in the most rigorous manner possible, of one's responses to some of the most perplexing and profound questions that humans have posed: What is the ultimate nature of reality? Is it unified or atomistic? Is change illusory or is it fundamental? Is subjectivity pervasive within the world, or is consciousness an incidental by-product of processes that are at core fundamentally physical?

Methodological Assumptions

Alfred North Whitehead, the philosopher most often cited in the chapters that follow, was a metaphysician—a thinker interested in fundamental questions of this sort. But he was also part of (and helped to cause) a transformation in how metaphysics is perceived and pursued. In contradistinction to the German Idealists and the British Neo-Idealists who preceded him, Whitehead took the results of physics as his major starting point for philosophical reflection. In the formative years that preceded his work as a systematic philosopher, he was preoccupied with thinking through the logical foundations of mathematics and the sciences and with the implications of special and general relativity. A series of articles written at this time show him wrestling with conceptual transformations required by the effects of gravity on space, the speed of light as a limiting velocity, and the switch to space-time as the fundamental unit of physical theory. The attempt to understand the physical world fuels Whitehead's work as well as the research contained in this volume.

It is Whitehead's commitment to comprehending the natural world that makes him a model for most of the chapters that follow. Each author takes contemporary physics—empirical results, widely accepted theories, or general directions in theory development—with utmost seriousness, and most actually begin with concrete experimental results. But none of the authors writes merely as a physicist. Each detects lines of reflection and inference emanating outward from physics, creating vector spaces in which *meta*-physical questions are posed and proposals evaluated. This type of reflection is controlled by the results of the natural sciences, and some of the theories presented here are in principle open to future empirical verification or falsification. At the same time, these chapters are not reducible to the discipline of physics; they wrestle with questions that demand a type of reflection that is broader and less constrained than direct work in natural science. In our view, this is the appropriate genre for physics and philosophy after Whitehead.

Why Should Physicists Concern Themselves with Metaphysics of *Any* Kind?

At first blush, it is not obvious that physicists need to concern themselves with metaphysics of any kind. After all, the physical sciences since (at least) Newton, and especially in our century, have repeatedly defined themselves *in opposition to* philosophical reflection. A firm grounding in empirical observations and controlled experiments, the application of rigorous laws that allow for prediction, and above all the power of mathematical formalism—all of these methods have been used by physicists in this century to justify a firm "demarcation" (Karl Popper) between physics and metaphysics. "Why should we concern ourselves with inexact and untestable theories," the physicist may respond, "when physics offers the most highly attested theoretical structure the human mind has ever obtained? Why explain the precise by means of the less precise? Where the tools of mathematical and experimental physics allow for genuine scientific control, there we do physics. And when the conditions for empirical science are unsatisfied, we turn the microphone over to others of whatever persuasion—as long as they do not claim factual warrant for their speculations."[2]

This desire to draw a sharp demarcation between physics and philosophy is shared by few, if any, of the authors in the present volume. Instead, each author has found some area of metaphysical reflection (in most cases process philosophy, and in many cases the work of Whitehead in particular) to make a specific and valuable contribution to some area of physical theory. Sometimes the process-philosophical categories help one comprehend the *types* of results that physicists have achieved in recent decades; sometimes the compatibility (or even mutual entailment) between process metaphysics and a particular set of physical results (decoherence, say, or nonseparability) is emphasized; and sometimes process categories are used as the source for new hypotheses or new directions of research. In every case, the authors claim, an understanding of physical science is achieved that a sharp demarcation between science and non-science would make impossible.

It is difficult not to view this mutual interaction between directly physical work and broader philosophical categories as positive. The decades of combat between theoretical physics and speculative philosophy were characterized by a certain rigidity, a defensiveness that had more to do with turf-protection than with genuine theoretical progress. I would venture to suggest that no author in this book is confused about the standards for successful work in physics; certainly none equates physics and philosophy. Yet they also see that rigor of physical theory is not compromised by the use of broader conceptual resources; distinction *allows for* cross-fertilization. There are gains in both directions. Readers will find in these pages a more rigorous type of metaphysics than that which dominates the "metaphysics" section of their local bookstores—a metaphysics that is responsive to the complex theoretical developments in physics, that is prepared to abandon beliefs that conflict with empirical data, and that seeks to be of service in clarifying the foundations of contemporary natural science.[3] Conversely, they will also find a mode of doing science that is more open to questions at the boundaries, one in which physics enters freely into dialogue with neighboring disciplines.

What Is Process Thought in General?

A process is a structured series of events in which time (or space-time) plays a crucial role in the structuring. Thus Nicholas Rescher defines process as "a sequentially structured sequence of successive phases which themselves are types of events or occurrences (in the case of an abstract process) or definite realizations of such types (in the case of a concrete process). A structureless sequence—just one darn thing after another—is not a process."[4]

The broader commitment of the essays in this book lies in the notion of process itself. At the same time, fundamental themes of Whitehead's work continue to preoccupy the writers in this volume; many hold Whitehead's categories to be particularly valuable for making sense of results in contemporary physics. It is intriguing to see how many dimensions of Whitehead's system are still fruitful for reflection on the foundations of physics. Indeed, some will argue that developments in physics *since* the 1920s provide an even greater physical justification for Whitehead's basic principles than was available at that time—including some cases where physical theories have become established of which Whitehead could have had scant premonition. If one looks at the physical speculations of Geoffrey Chew (ch. 8) on the one hand and the philosophical speculations of Jorge Nobo (ch. 17) on the other—both of which are clearly informed by Whiteheadean categories—one recognizes how broad is the range of potential application.

At the same time, the aim has not been to canonize Whitehead but to use his work as a starting point for philosophical reflection on contemporary empirical and theoretical physics. Where modifying Whitehead's thought (even radically), or replacing it with process insights derived entirely from other thinkers, is more useful in the dialogue with contemporary physics, authors have not been shy to do so. Process thought refers, after all, to a loosely knit family of theories rather than to one particular theory. In Rescher's words, process philosophy "is not a doctrinal monolithic tendency predicated in a particular thesis or theory, but a general and programmatic approach. To see it as a unified doctrine would in fact be as much an error as it would be to identify it with the teachings and ideas of a single thinker."[5]

Key Themes in Whitehead's Thought

Still, since many of the papers owe not only their method and style of reflection but also some of their theory to Whitehead, a brief summary of his position is necessary. Our goal cannot be to provide an overall summary of Whitehead's metaphysics: many excellent secondary sources are available,[6] and others—in particular those involving the ethical, historical, and religious aspects—are not as relevant to our specific interest in the foundations of physics. Three simple tenets, however, recur in the essays that follow:

1. Experiential units, constitutive of both processes and objects, are basic elements of reality.

2. Time is not an incidental aspect of reality, added on to fundamentally static things; instead, temporal change is a fundamental feature of the physical world itself. The switch in relativity theory to space-time means that time is not just an index of change for things, which really exist only as three-dimensional objects. Duration thus becomes an intrinsically spatiotemporal notion.

3. Objects should not be taken in isolation, defined on their own, and then considered in their relations to other objects. Instead, rela-

> tions are primary, and objects are defined in terms of the network of relations of which they are part–relations between other parts of the physical world, between other temporal instances present and past, and perhaps between nonphysical moments as well. Ultimately, the entire physical cosmos represents a single system in terms of which individual objects are to be defined.

For present purposes, it will have to suffice to limit the presentation to a half dozen observations about the resulting interpretation of the physical world:

1. On this view, *process is metaphysically prior to substance.* Where Aristotle, and later Aquinas and his followers, began by defining individual substances and the essential features associated with each, Whitehead took as his starting point the process itself. (If one wished to put it less radically, one could say that Whitehead redefines substance *in terms of* process, or that he makes process and substance coterminous in every experiential unit.) A physics that begins with individual objects or that construes individual atoms as the primary building blocks of reality would, for example, be closer to Aristotle; a physics of fields of which objects are the expression (as in quantum field theory) stands much closer to Whitehead.

2. It follows from the first point that *there is no being without becoming.* The becoming/being distinction goes back to the earliest days of Western metaphysics; indeed, the focus on pervasive becoming and the transitory nature of lived reality is equally fundamental within many of the Eastern traditions. Parmenides was famous for the emphasis on the Logos or static reality ("all is one"), while Heraclitus is remembered for the brief fragments of his philosophy of becoming: "all is flux" and "you can't step into the same river twice." Whitehead's metaphysics offers a comprehensive theory of ubiquitous becoming. Each "actual occasion," which is the basic experiential unit of reality for Whitehead, exists only for a moment—just long enough to take in or "prehend" the realities of the previous events and to form a response to them (the "subjective form").[7] When an occasion has formed its subjective response or *prehension* it immediately passes out of existence as an active experiential entity with its own perspective.[8] Moving into a state of "objective immortality," it then becomes available as a datum for subsequent actual occasions. Reality thus consists of this continual flow of events, each enjoying its own duration.

3. *These basic experiential units of reality are not separable objects, but occasions-in-connection* or, more specifically, occasions-of-experience-in-connection. As a result, Whitehead advocates the doctrine of the "continuity of nature." He writes, "This doctrine balances and limits the doctrine of the absolute individuality of each occasion of experience. There is a continuity between the subjective form of the immediate past occasion and a subjective form of its primary prehension in the origination of the new occasion."[9] This limitation on individuality is reminiscent of the classical philosophical idea that "the whole is

prior to the parts," a position shared in common by most idealists, by Neoplatonists, and by most of the Eastern Vedanta traditions.

The matter here is a bit more complex, however. Whitehead is, after all, a particular type of atomist; he takes the individual actual occasion as the basic building block of his metaphysics.[10] Societies—that is, groups of occasions standing in a determinant relationship to one another over time—are merely composites of individual occasions and have subsidiary reality. Yet the beauty of Whitehead's position is that each individual occasion is nonetheless integrally related to all other occasions. It is atomism without isolation: nothing exists as an island unto itself, but all things, in the moment of their coming into existence, are linked to all previous events, and each in turn becomes the objective data for future events. It could be said that on the classical philosophical problem of the One and the Many, Whitehead places slightly more emphasis on the Many. But this Many is so intimately and organically interlinked that it produces not the reduction to individual parts for which classical atomism was notorious but rather a single unbroken web of interconnection. As George Bosworth Burch summarizes the view in his note from Whitehead's 1926–27 lectures, "every actual entity requires all other entities, actual or ideal, in order to exist."[11]

4. It follows from what we have said so far that *complexes of relations are primary, not discrete objects*. Keeping in mind Whitehead's nonsentient generalization of experience (see chapter 17) one might say that *subjectivity and objectivity are equally primordial in every experience*. The deepest level of reality consists of the flow of actual occasions or individual events of becoming. Past occasions in their "objective" form as data are merely inputs for the actual occasions in their present-tense subjectivity. Immediate events are the only real agents, responsible for making the universe what it is.

Having introduced this topic, one must immediately note that Whitehead does not advocate some form of *subjectivism*—for instance, the view that all reality is composed out of subjects that are like human subjects. Whitehead is seeking to establish some features of experiencing that are general enough to apply to all levels of reality. For this reason, his position is a *pan-experientialism*, not *pan-psychism*: there is experience at all levels of reality, but there are not *psyches* at all levels. Subjectivity at the human level is a very complex and high-level emergent capability generated through vast complexes of prehensions and interactions. When Whitehead states that "subjectivity is prior to objectivity" he is making a technical, metaphysical claim about "actual occasions" and not about human psyches.

This primacy of a network of relations over discrete objects is a perspective that is easily recognized by modern physicists who have worked with field theories. Scientists understandably wish to construe human subjectivity as the result of underlying *objective* physical processes. Consciousness, for example, many would take as an emergent property of the more fundamental *physical* level of reality. To put it differently, by disciplinary training, physicists will tend to stretch the explanatory power of objective reality "up" the ladder of complexity in the natural world toward (or to) human mental experience, rather than stretching the realm of subjective reality "down" the ladder to fundamental physical processes. Whitehead followed a different intuition: the subjective and the objective both joined together actively in every event to create greater complexity and possibility. Explaining one simply in terms of the other results in a reductio ad absurdum, which Whitehead felt was a stumbling block for modern philosophy and science. Such classical dualisms are avoided in a new duality of process (see chapter 2).

5. As a result of the previous point, Whitehead places great stress on *the phenomenon of feeling*. Breaking the normal link of feelings and consciousness, he attempts to define a notion of "physical feeling." In one classic (albeit dense) summary of the position, he writes,

> Thus in a simple physical feeling there are two actual entities concerned; one of them is the subject of that feeling, and the other is the *initial* datum of the feeling. A second feeling is also concerned, namely, the *objective* datum of the simple physical feeling. This second feeling is the "objectification" of *its* subject for the subject of the simple physical feeling. The initial datum is objectified as being the subject of the feeling which is the objective datum: the objectification is the "perspective" of the initial datum.[12]

The details of "physical" and "conceptual" feelings in this quotation need not detain us. Still, one cannot fail to notice the direction of movement. Whitehead *starts with* a moment of experiencing and moves outward: backward to the data that are experienced, and forward to the subject's reactions to its own experience.

This metaphysics stands opposed to any physics that would make the standpoint of an individual particle secondary or unimportant. There is no homogenization in Whitehead's thought: rather than conceiving all of his basic atomic units as identical, he emphasizes that each of them has its own perspective on the universe (or at least on that part of the universe that it can prehend). That particular perspective is intrinsic for defining what it is and what its subsequent effect on its environment will be. Whitehead is thus a perspectivalist par excellence.

6. Whitehead's metaphysics has been called a "philosophy of organism."[13] There is deep insight in this designation: in many ways, the biological metaphor remained fundamental to his thinking, as Charles Birch has emphasized in numerous publications.[14] All things are understood as growing and developing, as having their own perspective and striving toward their own goals, as being interlinked like various organisms within one overarching ecosystem. Many of

the features of biological organisms are thus understood to pertain to all units of reality, rather than merely emerging out of a physical-chemical substrate.

The philosophy of organism might be useful for an evolutionary cosmology such as that of Lee Smolin,[15] according to which Darwinian principles control the evolution and then survival (and even reproduction) or collapse of multiple universes. But generally physicists have resisted a biological model of physical processes, tending instead to view biological processes as ultimately reducible to their underlying physical constituents and laws.

7. I have omitted many of the features of Whitehead's thought frequently discussed in the secondary literature: questions of culture and history, initial aims, eternality, creativity, values, and the host of theological questions concerning the existence of God and values. These debates, however fascinating, are not directly relevant to this work on physics and philosophy.[16] There may be broader value in Whitehead's thought, and perhaps in many respects it will prove of interest for theologians and ethicists. Our task, however, is to probe it on one dimension only: its connection with contemporary physics, and in particular its usefulness for constructing a metaphysics that is responsive to, and helpful for, interpreting the world of experience.

Event Metaphysics

The essays in this volume are not the first to apply Whitehead's metaphysics to contemporary science; similar applications have been made in the past by (among many others) David Bohm, Karl Pribram, Ilya Prigogine, Rupert Sheldrake, and Henry Stapp.[17] In this sense the present volume stands within a long tradition of interaction between Whitehead and emerging scientific thought. It does, however, represent one of the most extensive and diverse applications yet made of process thought to the field of physics.

In particular, some very significant constructive work has been done in these pages using Whiteheadian or quasi-Whiteheadian concepts such as occasion, event, process, and field. This work is new—and difficult—enough that a few summary comments are in order. Without intending either to summarize the authors' work or to replace it with a different conception of my own, I wish to close by drawing attention to the uniqueness and importance of this speculative work for the interpretation of contemporary physics. As a shorthand, I will refer to this shared position (to the extent that there *is* one) as *event metaphysics* (EM).

Whitehead has been criticized in the past for interpreting actual occasions too heavily using the conceptual framework of organisms, thereby making these occasions into a type of mini-*thing*. By contrast, the core proposition of EM is that the basic unit of reality is not a thing at all but rather an event—an occurrence or happening. Only an ontology of events can do justice to this insight, and thing language (on this view) must be carefully subordinated to event language.

Among the various candidates for an ontology of events, the language of fields in contemporary physical theory is particularly attractive. Field language stresses the principle of continuity. In quantum field theory, for example, the field is sometimes expressed at a particular space-time point as an object; still, the microphysical object that we detect remains an expression of the field rather than being understood as a basic physical given. Indeed, it is probably more accurate and less confusing in the end to speak of the resulting phenomenon as an event than as an object.

One might say that there are essentially three ways to interpret the physics in question, that is, three ways to understand the mathematical formalism. One could interpret physical theory as about objects in the world, with fields understood as derivative from objects; it could be about the fields, with objects as merely the expression of fields; or fields and objects could be understood as equally fundamental. According to EM, the third position is the most appropriate interpretation of contemporary physics. In an event understood as the actualization or "concretization" of a field, both sides—the field and its actualization of certain space-time parameters—are equiprimordial.

Many further questions remain: How are fields and events to be understood, and what is their nature? How do these basic events produce the macrophysical world of our experience? Are they self-organizing or self-determining in any way? Are all collections of events mere aggregates, or can groups of events form larger objects that have a reality of their own?

However these questions are answered—and they are answered in rather different ways by the authors of these collected essays—it is clear that this list of questions is closely linked to some of the fundamental interpretive questions that have challenged physicists. Direct links can be drawn between the features of EM and key interpretive issues such as the measurement problem, the collapse of the wave function, the connection of relativistic and quantum physics, the relation between quantum and classical physics, and the more recent debates surrounding Bell's theorem, inseparability, and decoherence. It is our hope that the present volume will contribute toward understanding the evolution of physics during the twenty-first century.[18]

Notes

1. See William A. Christian, *An Interpretation of Whitehead's Metaphysics* (New Haven: Yale University Press, 1959); Lewis S. Ford, *The Emergence of Whitehead's Metaphysics: 1925–1929* (Albany: State University of New York Press, 1984); William W. Hammerschmidt, *Whitehead's Philosophy of Time* (New York: Russell and Russell, 1975); Lewis S. Ford and George L. Kline, eds., *Explorations in Whitehead's Philosophy* (New York: Fordham University Press, 1983); F. Bradford Wallack, *The Epochal Nature*

of Process in Whitehead's Metaphysics (Albany: State University of New York Press, 1980); A. Johnson, *Whitehead's Theory of Reality* (New York: Dover Publications, 1962); and Jorge Luis Nobo, *Whitehead's Metaphysics of Extension and Solidarity* (Albany: State University of New York Press, 1986).

2. See E. O. Wilson in *Consilience: The Unity of Knowledge* (New York: Knopf, 1998); Stephen Jay Gould, *Rocks of Ages: Science and Religion in the Fullness of Life* (New York: Ballantine, 1999).

3. Some might argue that metaphysics has received such a bad name among scientists that one should not even use the same term for both activities. But it is necessary to find some label for the sort of reflection that scientists and others engage in when they reflect more broadly on scientific results and their implications. Since *metaphysics* or *natural philosophy* have been used to designate this sort of reflection since the dawn of Western philosophy, I argue that it is better to fight for the term than to switch.

4. See Nicholas Rescher, "On Situating Process Philosophy," *Process Studies* 28/1–2 (1999): 37–42, quote p. 38, also Nicholas Rescher, *Process Metaphysics: An Introduction to Process Philosophy* (Albany: State University of New York, 1996).

5. Ibid., "On Situating Process Philosophy," 41f.

6. See, for example, the works cited in note 1.

7. Alfred North Whitehead, *Process and Reality* (New York: The Free Press, 1978), 30.

8. See chapter 17.

9. Alfred North Whitehead, *Adventures of Ideas* (New York: Macmillan, 1933), 235.

10. For an effective critique of Whitehead's atomism, see Wolfhart Pannenberg, "Atomism, Duration, Form: Difficulties with Process Philosophy," *Metaphysics and the Idea of God*, trans. P. Clayton (Grand Rapids: Eerdmans, 1990), chapter 6. On the question of groups of occasions, the effective size of actual entities, and the causal efficacy of a nexus in Whitehead, see also chapter 17.

11. Ford, *Emergence*, 312.

12. Alfred North Whitehead, *Process and Reality*, 308.

13. Dorothy Emmet, *Whitehead's Philosophy of Organism*, 2nd ed. (London: Macmillan; New York: Saint Martin's Press, 1966).

14. Charles Birch, *On Purpose* (Kensington, NSW, Australia: New South Wales University Press, 1990); *Nature and God* (London: SCM Press, 1965); with John B. Cobb Jr., *The Liberation of Life: From the Cell to the Community* (New York: Cambridge University Press, 1981); and Birch's contribution to John B. Cobb Jr. and David Ray Griffin, eds., *Mind in Nature: Essays on the Interface of Science and Philosophy* (Washington: University Press of America, 1977).

15. Lee Smolin, *The Life of the Cosmos* (New York: Oxford University Press, 1997).

16. For further details see John B. Cobb Jr. and David Ray Griffin, *Process Theology: An Introductory Exposition* (Philadelphia: Westminster Press, 1976).

17. For a good summary of some of these applications see John P. Briggs and F. David Peat, *Looking Glass Universe* (New York: Cornerstone Library, 1984).

18. I am grateful to Joseph Bracken for in-depth discussions that helped in the formulation of this section. Bracken's own work in this direction can be found in (inter alia) *Society and Spirit: A Trinitarian Cosmology* (London and Toronto: Associated University Presses, 1991); and *The Divine Matrix: Creativity as Link Between East and West* (Maryknoll, NY: Orbis Books, 1995).

2

Duality Without Dualism

TIMOTHY E. EASTMAN

It is a matter of common sense that the world is composed of a multiplicity of discrete, separable objects. Indeed, this worldview of perceptual objects is practical and essential for everyday life. As the atomic theory of matter emerged, it was rather natural to assume that atoms were small-scale counterparts of everyday objects. Indeed, such a simplistic, philosophical atomism became an accepted part of the implicit metaphysics of classical physics. Similarly commonsensical is the basic distinction between mind and matter which, with Descartes, was elevated to a philosophical first principle.

Descartes's mind-matter distinction became the prime exemplar of dualism, the notion that certain concepts related in experience are not really related but belong to different categories, and that the apparent relationship of such dual pairs derives only from secondary connections. In contrast, the process-relational approach emphasizes a "duality without dualism" that affirms the fundamental connection of such concepts as being/becoming, mind/matter, and symmetry/asymmetry, but does so in a way that avoids a simplistic symmetry of the dualities.

In particular, contrasts are presented of the form "both *A* with respect to *x* and *B* with respect to *y*" instead of simple dualisms that set up some form of absolute "*A* versus *B*" opposition. In this way, a process framework grounds the "both-and" approach described below, which is embodied in the transition from classical to modern physics. By correlating various dualities in modern physics and philosophy, we demonstrate a new tool for testing certain philosophical claims and for suggesting new hypotheses of interest in physics. The nuanced correlates discussed by Whitehead and some process philosophers are closer to complementary pairs in modern physics than traditional dualisms.

Perceptual Objects and Particularism

The worldview of perceptual objects, with its discreteness and separability, has continued its imaginative hold up to the present day, although many aspects of modern physics support a process-relational understanding of the world, as Whitehead proposed (Jungerman 2000). In contrast to this process interpretation, it has been claimed that "field theories are radically reductionistic: the whole reality of a field in a given region is contained in its parts, that is to say, its points" (Howard 1989, 235). This common view of field theories is rationalized by assuming that the discreteness and separability of mathematical points can be simply mapped onto physical systems. A more technical statement of this claim is that "by modeling a physical ontology upon the ontology of the mathematical manifold, we take over as a criterion for the individuation of physical systems and states within field theories the mathematician's criterion for the individuation of mathematical points" (236). Einstein presupposed the kind of separability indicated here and used it, along with a locality principle, as part of his strong attack on the completeness of quantum theory.

Yutaka Tanaka (ch. 13) demonstrates the construction of Bell's inequalities, applicable to certain quantum systems and their experimental testing, which have provided a definitive test of the assumptions of separability and locality. These assumptions together have now been shown conclusively as false (see chapters in Cushing and McMullin 1989). This experimental disconfirmation of Bell's inequalities confirms quantum over classical prediction. Paul Teller refers to the assumption of ontological locality, employed in setting up Bell's inequalities, as particularism (Teller 1989, 215) and points out that this assumption allows "only one kind of locality: causal locality." In discussing the correlations revealed by the Bell inequality violations, Teller states that recognizing 'relational holism' and avoiding particularism allows one to see "The correlation . . . as simply a fact about the pair. This fact . . . need not itself be decomposable in terms of . . . more basic, nonrelational facts" (222).

There are numerous historical and philosophical reasons to be skeptical of classical notions of separability and particularism, as shown by Leclerc (1986, 1986), and to be skeptical of the nonrelational, container view of space and time that is generally linked with the worldview of classical physics (Čapek 1971; Angel 1974).

Experience and Dualities

Stapp (1993) has shown how high-level consciousness or mind can be considered as an integral part of basic physical systems without assuming a simplistic reductionism or turning to mind-body dualism. One essential, nonreducible feature of quantum measurement is determining which question is posed (Stapp, ch. 9). As shown below, dualities are common in modern science, but these are dualities that need not be interpreted as simple dualisms. Whitehead used experience-like features as a basic analogue for his treatment of actual entities and systems (Clayton, ch. 1). However, the use of this analogue is not "pan-psychism" as so commonly supposed because Whitehead does not treat 'actual entities' as simply small-scale mental entities (psyches) that compose all large-scale mental entities. Instead, Whitehead's basic hypothesis is that some type of low-level experiencing (prehension) is ubiquitous and a basic metaphysical principle (Nobo, ch. 17). Griffin (1988) has shown why a better term is "pan-experientialism." The multilevel systems approach with emergence that has long been common in process approaches in considering the problem of high-level consciousness is now seen by many as a logical consequence of a nonlinear, dynamical treatment of living systems (Prigogine and Stengers 1984; Kaufmann 1993). On this basis, a Whiteheadian pan-experientialism appears as a natural hypothesis for answering how Stapp's posing of the question to quantum systems is resolved more generally.

Whitehead's philosophy is a type of general systems theory that he called a philosophy of organism.[1] Clayton (ch. 1) shows how it frames an ecological perspective that arises naturally from a comprehensive event metaphysics. Although the basic elements of his system, actual entities, are generally treated as microscopic in scale, Whitehead never associates any particular size with actual entities. One problem with a simple microscopic interpretation is revealed by Nobo (ch. 17) through detailed analysis of Whitehead's discussion of the two aspects of becoming: concrescence (microscopic in orientation) and transition (macroscopic in orientation). Tanaka (ch. 13) addresses a second problem that emerges in quantum measurement theory wherein "an individual quantum event is not necessarily microscopic. The simultaneous correspondence of the EPR experiment shows us the individuality of a quantum process at long distances . . . the region of an individual quantum process may have an arbitrary size with respect to space-time coordinates." A. H. Johnson reports that Whitehead directly acknowledged an overemphasis on the microlevel in his works and recognized the need for a category of emergence so that nexūs of actual entities can have emergent properties, both concepts being implicit in Whitehead's works and made explicit by Johnson (1983, 53). Nobo (ch. 17) and Lango (2000) both develop details concerning how the enduring objects of our everyday life are constituted from societies or nexūs of actual entities.

Relational Holism

Alfred North Whitehead attacked various types of particularism and container views and introduced a comprehensive system of thought that replaces inert substance with relations, and expands upon a philosophy of organic, relational holism.[2] Related philosophical approaches have been introduced as systems theory (Laszlo 1972; Auyang 1998), hierarchy or complexity theory (Chandler

1999), dialectical holism (Harris 1988), ecological perspectives (Nisbet 1991), evolutionary worldviews (Jantsch 1980), and varieties of holism in pragmatism and contextualism (Rescher 1999).

These various forms of relational holism are gaining in recognition for the following reasons:

1. Quantum theory is inconsistent with the classical notion of a philosophical atom (Leclerc 1986) and "requires us to renounce objects" (Finkelstein 1996, 35).

2. A detailed analysis of parts and wholes in low-energy physics provides for "the rigorous establishment of emergence; that is, the exhibition of macroscopic properties radically different from those of the constituents" (Shimony 1987, 421).

3. Field theory, evolution, and systems concepts more generally illustrate how physical systems cannot be simple classical substances, sufficient unto themselves, but are constituted by their interactions and relations with other particle and field entities.

4. Experiments show the failure of Bell's inequality and thus explicitly deny the combined assumptions of separability and locality used to create the inequality. Leading theorists in the philosophy of physics agree that 'particularism' is false and are seriously considering various types of relational holism as a viable metaphysical framework most compatible with the physics results (Kitchener 1988; Cushing and McMullin 1989).

5. Einstein's relativity is a relational theory of space and time that is incompatible with the 'container' view held in association with classical physics and strongly associated with particularist and reductionist approaches (Angel 1974; Čapek 1961, 1971).

6. The linguistic turn in philosophy is a wrong turn because "existence is not simply a matter of the satisfaction of a description" (Bradley 2002) as assumed for "weak" theories of existence. In contrast, those who engage in the business of modern science and technology generally presume a strong theory of existence. Whitehead takes account of arguments made for the linguistic turn even while maintaining critical realism.

Table 2.1 orders those concepts which are common across all three conceptual systems under discussion here (classical, quantum, process) along with those that exhibit contrasts or that illustrate complementary aspects between the classical and quantum frameworks. Here, *quantum* refers to mod-

Table 2.1. Comparison of Common, Contrasting, and Complementary Concepts in Physics

	Classical	*Quantum*	*Process*	*Contributor*
Concepts in Common	Long-range order	Long-range order	Interconnectedness	Jungerman, Tanaka
	Unpredictable evolution (praxis)	Unpredictable evolution (praxis)	Future as open	Hansen, Jungerman
	'State'-definite values	'State'-definite probability values	Probabilities and definiteness	Finkelstein, Malin
	Laws as absolute	Laws as invariant relations	Generality, coherence, consistency	Chew, Riffert
	Methodological and epistemological reduction	Methodological reduction	Methodological reduction	Riffert, Stapp
Contrasting Concepts	Space, time separation	Space-time relations	Interactive relatedness	Clayton, Hansen, Keeton
	Predictable evolution (theory)	Unpredictable evolution (theory)	Innovation/novelty	Jungerman, Malin
	State, trajectories	No state, nonunique trajectories	Creative advance	Finkelstein, Nobo
	Objects	No 'object' physics	Act, process	Finkelstein, Malin
	Laws as absolute	Laws as habits	Cosmic epochs	Chew, Finkelstein
	Absolute system	Relative systems	Interactive relatedness	Chew, Clayton, Nobo
	No experiential aspect	Experiential aspect	Experience as paradigm	Clayton, Nobo, Stapp

	No intrinsic limits	Fundamental limits	Limits, approximation	Chew, Stapp
	Global time	Local time	Local temporalism	Hansen, Tanaka
	Reductionist	Reduction without reductionism	Reduction without reductionism	Earley, Finkelstein
Complementary Concepts	Symmetry	Both symmetry/asymmetry	Dualities without dualism	Eastman, Rosen
	Space only; time spatialized	Both space/time	Becoming and being	Clayton, Hansen
	Particles only	Both particles/waves	Actual entities	Chew, Finkelstein, Nobo
	Determinism only	Both prediction/determination and indeterminism	Subjective aim	Finkelstein, Nobo
	Continuity only	Both continuity and quantization	Coordinate and genetic analyses	Tanaka, Nobo
	Order only	Both order/disorder, novelty	Innovation/novelty	Earley, Jungerman
	External relations	Both external/internal relations	Relations as fundamental	Clayton, Earley, Nobo
	Substance/objects only	Both substance and process/events	Subject/object, process	Finkelstein, Malin, Tanaka
	Global time	Both global/local time	Historical reality	Chew, Eastman

Note: Classical, quantum, and process concepts are ordered here by those concepts which are common across all three conceptual systems along with those that exhibit contrasts or that illustrate complementary aspects between classical and quantum frameworks. Contributors to this volume that address each concept are listed at the right.

ern physics broadly, exemplified especially by quantum theory and relativity theory.

Concepts in Common

Concepts that are roughly common across the classical, quantum, and process frameworks are listed in the first major row of the table. For each concept, related process phrases or terms are listed in the process column followed by contributors to this volume who provide more detail concerning that concept. Methodological (technique-based) reduction is common throughout just as it is an integral part of scientific method. Epistemological (knowledge-based) reduction fails in the quantum case because, as noted by Stapp, the physics alone does not specify which question is posed. The interconnectedness emphasized in process thought is now recognized as important in nonequilibrium, dissipative structures (Prigogine and Stengers 1984; Earley, ch. 6) and is common to all systems requiring long-range field interactions via the electromagnetic or gravitational fields, i.e., all macroscopic physical systems.

The temporal development of all complex systems, classical or quantum, is not predictable in absolute detail. Chew points out that approximation is an intrinsic part of all measurement, although classical physics suggested erroneously that precision had no limit in principle. Unlimited specification of values or of states in classical physics has a counterpart in quantum physics in the specification of wave functions but, as shown by Born (1949), wave functions are associated with probability distributions and not with simply located states of objects as suggested in classical physics. In this way, quantum physics incorporates both probabilities and definiteness.

Insofar as modern physics continues to seek a unified field theory with the presumption of one law or set of physical relationships, there is commonality in this sense between classical and modern physics. Methodologically, this emphasis on lawlike behavior in physical systems is a key part of scientific metaphysics with the criteria of generality, coherence, and consistency as discussed in detail by Riffert (ch. 16). However, Finkelstein (ch. 14) points out that this emphasis on lawlike behavior need not require ultimate reduction to a single set of physical relationships.

Contrasting Concepts

Numerous books emphasize contrasts between classical and modern physics. A pioneer in this genre is Milič Čapek (1961). He documented how the worldview of classical physics included a container view of space and time, within which space is a container for objects and a spatialized time is a container for events. In contrast, relativity theory is based on dynamic space-time relations, even though Minkowski space-time diagrams (in which time and space are plotted along X and Y coordinates) can suggest a mere spatialization of time. Spatial and temporal relations are fundamentally linked via the Lorentz transformations in a symmetric but not fully equivalent way due to an imaginary coefficient for time in the geodesics (Bunge 1967). Classical physics treated all systems within some global, absolute coordinate frame. In contrast, relativity enables the basic physical relationships or laws, including Maxwell's laws of electromagnetism, to remain invariant in form between moving frames but at the expense of considering all systems as frame dependent. Whitehead was well versed in relativity theory (Whitehead 1922) and recognized the need for such frame dependence as early as 1905, independent of Einstein (Whitehead 1906). Further, Whitehead's 1905 memoir "undertook the unification of geometry and physics by means of . . . symbolic logic. . . . Not until 1916, in the General Theory of Relativity, did Einstein express the unification of geometry and physics" (Schmidt 1967, 4).

The temporal development of all classical systems is theoretically predictable in absolute detail. Thus, classical states, trajectories, systems, and system evolution are treated as fully specifiable. As noted above, the unlimited precision of classical physics is incorrect both in principle and in practice. Modern physics has been built upon the recognition of fundamental limits that were denied in classical physics: relativity theory requires propagation velocity limits, and quantum theory embodies Planck's constant limit to the specification of conjugate variables such as velocity and momentum. Finkelstein proposes to extend these limits even to the primary physical relationships themselves in a way reminiscent of Whitehead's discussion of cosmic epochs.

Classical physics presupposes a God's-eye view of natural systems and an associated, unique global time. Relativity theory drops global time altogether and retains only local, frame-dependent times. Such local time is incorporated in Whitehead's local temporalism, as discussed in detail by Hansen (ch. 12). One way to offset this loss of global time is to introduce two times in a model of historical reality, as proposed by Chew (ch. 8). Local temporalism remains central but is then augmented by a measure of an actual occasion's total history.

Because of the particularism assumed in classical atomism, combined with unlimited specification of classical states and trajectories, a reductionist metaphysics is generally associated with classical physics. Nevertheless, Kant, Bergson, and other philosophers tried to make some opening within the classical paradigm for consciousness and other wholistic features without resorting to any simple dualism. In contrast to the apparent reductionism of classical physics, quantum physics can have composite states that are not generally reducible to states of discrete constituents (Howard 1989, 253). Indeed, if one questions the reduction of fields to mere mathematical points, collisionless space plasmas represent a classical system within which the basic dynamics is reducible only to the scale of Debye screening spheres, which is roughly one kilometer in Earth's outer magnetosphere. There is no possible contiguity in

Hume's sense in such a system, which is effectively without collisions because particle densities are only about 0.01 to 10 per cubic centimeter. The exchange of forces in such plasmas is without contact and mediated via electric and magnetic fields (Eastman 1993).

Meaningful reference to objects at all scales is presumed in classical physics. In contrast, our everyday world of perceptual objects has no counterpart in quantum microphysics. Discrete self-identical objects, the essence of the classical notion of substance, are replaced with interacting wave-particle entities. Finkelstein points out that Bohr and Heisenberg talked about "no object" physics. In his major work, *Quantum Relativity*, Finkelstein (1996) systematically develops a new conception of act and process to supplant objects. Relativity theory makes reference to events and clocks but contains no fundamental reference to objects (Schmidt 1967, 30). Objects and substances may be considered as derivative notions in modern physics and need not be treated as primary concepts as they were in classical physics.

Complementary Dualities in Both Physics and Philosophy

Modern physics comprises many complementary pairs or dualities, and a number of these are correlated with dualities in philosophy. Further, the transition from classical to modern physics can be illustrated as the movement from one to both poles of various dualities, as shown in Table 2.2. The basic form of this transition is from *A* only to *A* plus *B* where *A* and *B* are complementary pairs. The worldview of classical physics tended to make *A* terms exclusive or ultimate in some way that is now recognized as misleading or incomplete. Just as classical physics can be considered as a limiting case of quantum and relativity physics, its *A* characterizations can be considered as a limit or subset of more complete *A-B* complementary pairs. The quantum and process views construct a more inclusive duality in each case that treats *A* only as a type of classical limit analogous to how Newton's equations can be retrieved as a classical limit to equations for quantum dynamics.

This explicit recognition of dualities is becoming increasingly understood as an integral part of modern physics (Witten 1997). The corresponding both-and approach cuts through most of the alleged puzzles that are so much a part of the current genre of physics popularizations. Many of these "puzzles" depend on the comparison of a confirmed physics result with a presupposition linked to the worldview of perceptual objects or, effectively, of classical physics. For example, the insistence on identifying a simply-located particle going through one slit or the other in a two-slit quantum experiment depends on treating particles only as traditional philosophical atoms and not as quantum wave-particle entities. When properly constructed, the puzzle dissolves.

Although the both-and and duality themes of process thought are emphasized here, we wish to avoid another simplistic reduction. Dualities are often

Table 2.2. Movement from One to Both Poles Simultaneously of Various Dualities in the Transition from Classical to Modern Physics

Classical Physics	*Modern Physics*
Substance only; materialism	Both substance and event-oriented descriptions
External relations only	Both external and internal relations
Continuity only; no ultimate discreteness	Both continuity and quantization
Symmetry only	Both symmetry and asymmetry
Space only; time spatialized	Both space and time; coupled space-time metric
Determinism only	Both predictability/determination and indetermination[a]
Particles only	Both particles and waves
Parts only	Both parts and wholes[b]
External only (source for order)	Both external and internal sources of order; self-organization[c]
Efficient cause only	Both efficient cause and other types[d]

[a]The term *determination* is used here to denote the predictability of causal order as practiced in science, whereas *determinism* is a metaphysical claim requiring philosophical argument. Science is neutral with respect to the philosophical question of determinism.

[b]Macroscopic processes emerge in collisionless space plasmas where the "large-scale dynamics are immune from the details of microphysics." In turn, some systems such as superfluids exhibiting Bose-Einstein condensation have a close coupling of micro- to macroscale, and there are other systems that fill in between these two extremes (E. Siregar, S. Ghosh, and M. L. Goldstein. "Nonlinear entropy production operators for magnetohydrodynamic plasmas." *Phys. Plasmas* 2, no. 5, 1481 (May 1995); see also T. E. Eastman. "Micro- to macroscale perspectives on space plasmas." *Physics of Fluids B (Plasma Physics)* 5, 2671 (1993).

[c]For an in-depth study of nonlinear systems, self-organization, and their application to biological systems and evolution, see Stuart A. Kauffman, *At Home in the Universe: The Search for Laws of Self-Organization and Complexity* (Oxford: Oxford University Press, 1995).

[d]Mario Bunge, *Causality: The Place of the Causal Principle in Modern Science* (Cambridge: Harvard University Press, 1959).

enclosed in triadicities, encouraging us, with Peirce, to "think in trichotomies not mere dichotomies, the latter being crude and misleading by themselves."[3] Similarly, the complementary pairs of modern physics often point beyond themselves to higher levels of abstraction or other (meta)physically constructed solutions.

Dualities of various types permeate philosophy and undergird Plato's dialogical method. Table 2.3 lists many such dualities adapted from a list by Charles Hartshorne, who discusses a technique for relating the terms of each dual pair.[4] This list of philosophical dualities shows how many important topics lend themselves to a dual or complementary construction and that these dualities are often not merely symmetric in character.

The Western intellectual bias is very different from that of Chinese culture. Chinese scholars David Hall and Roger Ames distinguish two modes of

Table 2.3. Dualities in Philosophy

Relative	Absolute
Dependent	Independent
Internally related	Externally related
Experience, subject	Things experienced, object
Whole, inclusive	Constituent, included
Effect, conditioned	Cause, condition
Later, successor	Earlier, predecessor
Becoming	Being
Temporal	Atemporal
Concrete	Abstract
Particular	Universal
Actual	Possible
Contingent	Necessary
Finite	Infinite
Discrete	Continuous
Complex	Simple
Asymmetry	Symmetry

thinking. "Correlative" or "analogical" thinking, the dominant mode in classical Chinese culture, "seeks to account for states of affairs by appeal to correlative procedures rather than by determining agencies or principles" (Hall and Ames 1995, xviii). It accepts no priority of permanence over process and presumes no ultimate agency responsible for the general order of things. Causal thinking, the dominant Western mode, tends to assert a priority of being over becoming and to see the cosmos as a single-ordered world and as the consequence of some primordial agency. Hall and Ames argue that correlative thinking is dominant in Chinese culture, whereas causal thinking is recessive. In turn, causal thinking is dominant for the West, whereas correlative thinking is recessive. The Western tendency to convert certain dualities into either a dualism or a univocal prioritization of one pole over the other is also offset by correlative thinking. "In a correlative sensibility such as we find within the Chinese tradition, terms are clustered with opposing or complementary alter-terms. Classical Chinese may be uncongenial to the development of univocal propositions for this reason" (230).

With the philosophical dualities of Table 2.3 as both motivation and a basis of comparison, I now consider in Table 2.4 various dualities in modern science as exemplified by quantum and relativity theory.

Some of the physics dualities are direct correlates to those in the philosophy list (discreteness-continuity; actuality-possibility; synthetic-analytic; asymmetry-symmetry; and final act-initial act, associated with successor-predecessor). These dualities have a structure that is similar in both physics and

Table 2.4. Dualities in Physics

Field	Source/matter	Field theories
Wave	Particle	Bohr's complementarity
Momentum	Position	principle (Wilkins 1987)
Magnetic	Electric	Gauge theories (Witten 1997)
Discreteness	Continuity	Topology (Geroch 1985, 142)
Actuality	Possibility	Quantum properties (Bub 1997)
Nonlinear	Linear	Both classical and quantum
Final act/absorption	Initial act/emission	Quantum actions
Synthetic	Analytic	Geometry[a]
Asymmetry	Symmetry	Linear mappings (Geroch 1985, 116)
Episystem	System	Quantum system cut
Nonlocal	Local	Quantum levels

[a]"The qualitative (coordinate-free) and quantitative (coordinate-based) formulations of geometry are traditionally called synthetic and analytic geometry. Analogously, one may speak of synthetic and analytic quantum theory" (Finkelstein 1996, p. 186).

philosophy. For example, an analysis of asymmetry and symmetry in physics yields the same result as applying Hartshorne's interpretive rules for dual pairs in philosophy. Rosen (ch. 11) states this result as follows: "Asymmetry is a necessary condition for symmetry. For every symmetry there is an asymmetry tucked away somewhere in the world."

Shimon Malin (ch. 7) argues that the collapse of quantum states in quantum physics enforces a balance between actuality and potentiality, and between complexity and simplicity. The actuality-possibility duality is often debated in interpretations of modern science. One apparently explicit way in which this duality enters is in an analysis of the uniqueness theorem for quantum measurement theory. As stated by Bub (1997, 239), "Classically, only the actual properties are time-indexed; quantum-mechanically, both the actual properties and the possible properties are time-indexed . . . there is nothing inherently strange about the notions of possibility or actuality in quantum mechanics."

Classical physics is often thought as superior to quantum physics at the macroscopic scale, which is often regarded as simply linear. However, Finkelstein (1996, 388) points out that "classical non-linearity is a simplification of the quantum non-linearity inherent in the many-system kinematics of the composite system whose classical limit we take."

The final act-initial act pair in Table 2.4 is discussed in depth by Finkelstein (1996).

> Initial and final actions taken together are collectively called external (or terminal) actions. . . . The duality between initial and final modes, between before and after the fact, is the most important symmetry of quantum theory. . . . Often we call initial actions "creation operations" and final actions "annihilation operations."

> There is no need, however, to imagine creation from nothing or annihilation to nothing. These are acts of an experimenter with a large reservoir of quanta from which to draw and in which to deposit. Some use the terms "emission and "absorption" which have more appropriate associations. (14; see also 17, 40, 47, 48)

The philosophical pairing of whole-constituent has two related physics entries, episystem-system and nonlocal-local. In quantum physics, episystem is defined as follows:

> What acts on the system we will call the episystem. The episystem consists of everything playing a significant part in the experiment that is not part of the system, including the experimenter, the apparatus, the recording system, and an entropy dump. We call this division of an experiment into system and episystem, the system cut. . . . An action vector does not describe a state of being of the system but an action of the episystem on the system. (Finkelstein 1996, 16)

The system cut "is permeable and movable" (395).

Applying Hartshorne's interpretive rules for dual pairs to episystem-system results in the proposition that an episystem necessitates some system; whereas a system necessitates only that there be some episystem or other, yet which particular episystem is contingent. The unavoidable yet contingent presence of an episystem is widely recognized in quantum physics and is a central part of the argument by Stapp (ch. 9) that necessarily a question must be posed to quantum systems, although there is some contingency in both details about the episystem and the answer that nature gives to any particular question.

Nonlocal-local pairs naturally emerge in quantum physics and resolve the meaning of action. The action principle is the key variational principle of physics. However, in classical physics, there is no reason why a particle's motion should be affected by values of the action on paths it does not take. In contrast, "Quantum physics is kinematically nonlocal though dynamically local. Quanta only act where they are, but most initial and final actions, even sharp ones, do not determine where they are" (Finkelstein 1996, 372).

Dualities Without Dualism

As we have seen, there is substantial correspondence between many duality pairs in philosophy and physics. In those cases where the correspondence is most clear, the physics results may have important philosophical implications. Where the physics pairings have less clear counterparts in the philosophy list, for example with gauge theory dualities, those cases should be fruitful areas of study. For example, the magnetic-electric pair and other such pairings in gauge theory are deeply linked to symmetry principles and are closely related to the asymmetry-symmetry pairing. Philosophical analysis of these cases, such as the application of Hartshorne's interpretive rules, points to the need for transcending simple symmetry and recognizing symmetry breaking, a key result of modern physics (Witten 1997).

Finding a Balance

The process-relational tradition has occasionally opted for the Heraclitean extreme that "all is change." However, Whitehead and most recent process philosophers have worked toward a middle ground in which "being and becoming, permanence and change must claim coequal footing in any metaphysical interpretation of the real, because both are equally insistent aspects of experience" (Kraus 1979, 1).

Notes

1. The famous general systems theory of Ervin Laszlo was inspired by Whitehead's philosophy. "I found . . . that the organic synthesis of Whitehead can be updated by the synthesis of a general systems theory, replacing the notion of 'organism' and its Platonic correlates with the concept of a dynamic, self-sustaining 'system' discriminated against the background of a changing natural environment" [Ervin Laszlo, *Introduction to Systems Philosophy: Toward a New Paradigm of Contemporary Thought* (New York: Gordon and Breach, 1972), viii]. However, his systems theory is classical whereas Whitehead's philosophy of organism has clear quantum aspects [private communication, David Finkelstein, March 29, 2002].

2. The term *holism* is taken broadly here to suggest systems with significant interdependence and emergence, yet with hierarchical organization that provides for relative independence of components. A view that emphasizes discrete, independent elements without relations is here called a "particularism."

3. Charles S. Peirce, the great American philosopher and originator of pragmatism, considered triads as much more adequate than dyads or tetrads as intellectual instruments (Hartshorne 1970, 100).

4. The dualities in philosophy listed in Table 2.3 are adapted from the table of Metaphysical Contraries in Charles Hartshorne, *Creative Synthesis* (1970, 100–101). Hartshorne's interpretive rules for dual pairs are located on these same pages. In Table 2.3, the dual pair asymmetry/symmetry is added based on Rosen's analysis. Hartshorne associates his r-terms (left column of Table 2.3) with Peirce's Seconds and Thirds and his a-terms (right column) with Peirce's Firsts.

References

Angel, Roger B. 1974 1980. *Relativity, the Theory and Its Philosophy*. Oxford: Oxford University Press.

Auyang, Sunny Y. 1998. *Foundations of Complex Systems Theories*. Cambridge: Cambridge University Press.

Barbour, Ian. 1990. *Religion in an Age of Science: The Gifford Lectures 1989–1991, Volume 1.* San Francisco: Harper.

Born, Max. 1949. *Natural Philosophy of Cause and Chance.* Oxford: Oxford University Press.

Bradley, James. 2002. "The speculative generalization of the function: A key to Whitehead." *Tijdschrift voor Filosofie* 64, 253–271.

Bub, Jeffrey. 1997. *Interpreting the Quantum World.* Cambridge: Cambridge University Press.

Bunge, Mario. 1967. *Foundations of Physics.* New York: Springer-Verlag.

Čapek, Milič. 1961. *Philosophical Impact of Contemporary Physics.* Princeton: D. Van Nostrand.

Čapek, Milič. 1971. *Bergson and Modern Physics. A Reinterpretation and Re-evaluation.* Dordrecht: Reidel.

Chandler, Jerry L.R. 1999. "Semiotics of complex systems and emergence within a simple cell." *Semiotica* 125, no. 1/3, 87–105.

Cushing, James T. and Ernan McMullin, eds. 1989. *Philosophical Consequences of Quantum Theory: Reflections on Bell's Theorem.* Notre Dame: University of Notre Dame Press.

Eastman, Timothy E. 1993. "Micro- to macroscale perspectives on space plasmas." *Phys. Fluids B* 5, no. 7, 2671–2675.

Finkelstein, David R. 1996. *Quantum Relativity: A Synthesis of the Ideas of Einstein and Heisenberg.* Berlin: Springer-Verlag.

Geroch, Robert. 1985. *Mathematical Physics.* Chicago: University of Chicago Press.

Griffin, David Ray, ed. 1988. *The Reenchantment of Science: Postmodern Proposals.* Albany: State University of New York Press.

Hall, David L. and Roger T. Ames. 1995. *Anticipating China: Thinking Through the Narratives of Chinese and Western Culture.* Albany: State University of New York Press.

Harris, Errol E. 1988. "Contemporary physics and dialectical holism." In *The World View of Contemporary Physics: Does It Need a New Metaphysics?*, edited by R. F. Kitchener. Albany: State University of New York Press.

Hartshorne, Charles. 1970. *Creative Synthesis and Philosophic Method.* La Salle: Open Court.

Hartshorne, Charles. 1997. *The Zero Fallacy and Other Essays in Neoclassical Philosophy.* La Salle: Open Court.

Howard, Don. 1989. "Holism, separability, and the metaphysical implications of the Bell experiments." In *Philosophical Consequences of Quantum Theory: Reflections on Bell's Theorem*, edited by J. T. Cushing and E. McMullin. Notre Dame: University of Notre Dame Press.

Jantsch, Erich. 1980. *The Self-Organizing Universe: Scientific and Human Implications of the Emerging Paradigm of Evolution.* Oxford: Pergamon Press.

Johnson, A. H. 1983. *Whitehead and His Philosophy.* Lanham: University Press of America.

Jungerman, John A. 2000. *World in Process: Creativity and Interconnection in the New Physics.* Albany: State University of New York Press.

Kauffman, Stuart A. 1993. *The Origin of Order: Self-Organization and Selection in Evolution.* Oxford: Oxford University Press.

Kitchener, Richard F., ed. 1988. *The World View of Contemporary Physics: Does It Need a New Metaphysics?* Albany: State University of New York Press.

Kraus, Elizabeth M. 1979. *The Metaphysics of Experience: A Companion to Whitehead's Process and Reality.* New York: Fordham University Press.

Lango, John W. 2000. "Whitehead's category of nexus of actual entities." *Process Studies* 29, no. 1, 16–42.

Laszlo, Ervini. 1972. *Introduction to Systems Philosophy: Toward a New Paradigm of Contemporary Thought.* New York: Gordon and Breach.

Lawrie, Ian D. 1990. *A Unified Grand Tour of Theoretical Physics.* Bristol: Adam Hilger.

Leclerc, Ivor. 1986. *The Nature of Physical Existence.* Lanham: University Press of America (originally published by George Allen & Unwin Ltd, London, 1972).

Leclerc, Ivor. 1986. *The Philosophy of Nature.* Washington, DC: The Catholic University of America Press.

Nisbet, E. G. 1991. *Leaving Eden: To Protect and Manage the Earth.* Cambridge: Cambridge University Press.

Nobo, Jorge Luis. 1986. *Whitehead's Metaphysics of Extension and Solidarity.* Albany: State University of New York Press.

Prigogine, Ilya, and Isabelle Stengers. 1984. *Order Out of Chaos: Man's New Dialogue with Nature.* New York: Bantam Books.

Rescher, Nicholas. 1999. "On situating process philosophy." *Process Studies* 28, nos. 1–2, 37–42.

Rosen, Joe. 1995. *Symmetry in Science: An Introduction to the General Theory.* Berlin: Springer-Verlag.

Schmidt, Paul F. 1967. *Perception and Cosmology in Whitehead's Philosophy.* New Brunswick: Rutgers University Press.

Shimony, 1987. "The methodology of synthesis: Parts and wholes in low-energy physics." In *Kelvin's Baltimore Lectures and Modern Theoretical Physics: Historical and Philosophical Perspectives*, edited by R. Kargon and P. Achinstein. Cambridge: The MIT Press, 399–424.

Stapp, Henry P. 1993. *Mind, Matter, and Quantum Mechanics.* Berlin: Springer-Verlag.

Teller, Paul. 1989. "Relativity, relational holism, and the Bell inequalities." In *Philosophical Consequences of Quantum Theory*, edited by J. T. Cushing and E. McMullin. Notre Dame: University of Notre Dame Press, 208–223.

Whitehead, Alfred North. 1906. "On mathematical concepts of the material world." *Philosophical Trans. of the Royal Society of London A* 205, 465–525.

Whitehead, Alfred North. 1922. *The Principle of Relativity, with Applications to Physical Science*. Cambridge: Cambridge University Press.

Wilkins, M. H. F. 1987. "Complementarity and the union of opposities." In *Quantum Implications: Essays in Honour of David Bohm*, edited by B. J. Hiley and F. D. Peat. London: Routledge & Kegan Paul Ltd., 338–360.

Wilson, Edward O. 1998. *Consilience: The Unity of Knowledge*. New York: Knopf.

Witten, Edward. 1997. "Duality, spacetime and quantum mechanics." *Physics Today* May, 28–33.

3

Whitehead as Mathematical Physicist

HANK KEETON

To appreciate the depth of Whitehead's engagement of mathematics and physics, we need to set a detailed context for his work. In 1880 Whitehead was enrolled at Trinity College, Cambridge, and continued his association there for the next 30 years. His biographer, Victor Lowe, has published a comprehensive history of Whitehead's life in *Whitehead: The Man and His Work* (*WM&W*) and I am indebted to his thoroughness for many biographical details not recorded elsewhere. In the early years of his tenure at Trinity, Whitehead learned mathematical physics from W. D. Niven, a former student of Clerk Maxwell, after Maxwell came to England in 1871 and published his *Treatise on Electricity and Magnetism* in 1873. It is probably at the influence of Niven that Whitehead chose Maxwell's *Treatise* for his fellowship dissertation in 1884. Whitehead was hired as a special math tutor that same year, giving personalized instruction to other students—an appointment signifying not only his stature in mathematics, but also an honor much sought after. That tutoring post led to an appointment as assistant lecturer for the school year 1885–86, and three short years later he received a full lectureship, which he held until 1910 when he left for London.

It is these years at Trinity that shaped the course of Whitehead's scholarship, and they are marked by several distinct features. First, as a student, he was "coached" by Dr. E. J. Routh, FRS, one of the great mathematical coaches for "a whole generation of Cambridge mathematicians" (*WM&W*, 98). Whitehead also had the opportunity to meet and discourse with Arthur Cayley, one of the great English mathematicians of the 1880s. It is not clear how Whitehead first learned about the work of Hermann Grassmann, but in 1888 Whitehead offered the first Cambridge course on Grassmann's *Ausdehnungslehre* (The Calculus of Extension), and it became a foundation for his work in mathematics and symbolic logic. "The idea that a consistent theory of n-dimensional space, like Grassmann's, has uses beyond those of three-dimensional geometry, had not arrived. Whitehead's lectures in 1887 [academic year 1887–88] and 1890 on the *Ausdehnungslehre* paid special attention to its *applications*" (*WM&W*, 154, emphasis added). It is important to note that Whitehead used the advancements of Grassmann's mathematics directed toward applications in the physical world. By the time Whitehead published his first major work in 1898, *A Treatise on Universal Algebra, with Applications* (*UA*), he credited the influence of the two decades of intense development in mathematics and logic from 1844 to 1862 with providing the foundation for his own work: Grassmann's works (1844, 1862), William R. Hamilton's *Quaternions* (1853), and George Boole's *The Mathematical Analysis of Logic* (1847), and *Symbolic Logic* (1859). As Whitehead recounted, "My whole subsequent work on Mathematical Logic is derived from these sources."[1]

The application of these developments in symbolic logic was hitched to Maxwell's star:

> We may speculate that it was either while Whitehead was writing his fellowship dissertation on Maxwell's treatise or soon thereafter that there grew in his mind a strong desire to work out eventually a theory of the *relation between matter* . . . and *space*—in other words, between *electromagnetism* and *geometry*. But the first task in Whitehead's large scheme of work was to make a detailed *investigation of systems* of *abstract mathematical ideas* that were being symbolized in various ways and *applied to physics*." (*WM&W*, 156, emphasis added)

This "first task" of the larger scheme of work became public through the treatise of 1898.

The importance of *UA* cannot be overstated: Whitehead worked for seven years, from 1891 to 1898, on the first volume of the originally envisioned two-volume work, investigating "the various systems of Symbolic Reasoning allied to ordinary algebra" (*WM&W*, 190). One of the keys to understanding his intentions lies in that portion of the title—"*with Applications*"—often eclipsed in reference. Whitehead intended to systematize, expand, and *apply* the resultant systems of logic to areas where the formulations might suggest or reveal new insights. His orientation toward spatial relations is the most direct example of application, and the generalizations he sought in the new parabolic and hyperbolic geometries became more clearly evident in the years immediately following this publication. Whitehead's organization of the volume reveals much of his underlying intent. The volume is divided into "books" or parts, the first being introductory, and the second devoted to "The Algebra of Symbolic Logic" where he explores the application of Boole's algebra to *regions of space*. Part III, "Positional Manifolds," explores non-metrical *n*-dimensional spatiality. The fourth part examines Grassmann's ideas, with the intent to apply them to the notion of *forces* (again non-metrical) in Part V. Part VI expands the calculus of extension, and applies it to elliptical, parabolic, and hyperbolic geometries. Part VII applies them to the Euclidian 3-space, with an examination of curved surfaces, and generalizations that hint of *topological transformations*.

This was his first major published work, and Whitehead had already set the tone for his intellectual life's work. The symbolic logic of *UA* is understood not as symbols for numbers, but representing *processes* of connecting elements within a consistent and comprehensive scheme. The scheme is generalized beyond its immediate focus on algebraic systems, with the intent that it will apply to a broader range of experience. The first volume of *UA* was the basis for his election as a Fellow of the Royal Society in 1903. The advancements represented by this work, plus the work in two papers submitted to the *American Journal of Mathematics*, led to his being awarded the Doctor of Science degree by Trinity in 1905. Upon completion of Volume 1, Whitehead immediately began work on Volume 2 and proceeded with that comparative study of all the algebras, new and traditional, until 1903, when for reasons partly linked to joint efforts with Bertrand Russell, he left the work unfinished. It is currently unavailable and presumed lost. During this time Russell and Whitehead began their decade-long collaboration on the logical foundations of mathematics, *Principia Mathematica* (*PM*), which was an effort to axiomatize all of then-known mathematics using Peano and Russell's logical notation. In 1931, Kurt Gödel derived a famous proof showing the basic incompleteness of any axiomatic system based on number theory, using *PM* as his primary example. Contrary to many popular characterizations, Gödel's proof does not *disprove* any of Whitehead and Russell's work, but in effect shows only that their stated goal was unachievable using the strict axiomatic method they employed.[2] The fourth volume of *PM* was intended to focus on geometry, written principally by Whitehead, but like *UA* Volume 2 was not completed.

In *UA* it was clear from the outset that the pure nature of the new algebras was not the limit of Whitehead's interest. He wanted to explore their application to the physical world, and for Whitehead that meant to geometry and physics. He employed symbolic logic as a tool for generalization and interpretation, seeking the broad connections that obtained between the algebras and the physical world. Later, he would use these same methods to investigate the elements of physics and philosophy, always seeking the most general foundational level within and among these aspects of the world of experience. More than twenty years after embarking on this series of projects, in 1912, Whitehead applied for the Chair of Applied Mathematics at University College, London stating that, "[d]uring the last twenty-two years I have been engaged in a large scheme of work, involving the logical scrutiny of mathematical symbolism and mathematical ideas. This work had its origin in the study of the mathematical theory of Electromagnetism, and has always had as its *ultimate aim* the general scrutiny of the *relation of matter and space*" (*WM&W*, 155–156, emphasis added). Although *UA* Volume 2 did not appear in print, the work behind it, and the ideas within it, most assuredly did not disappear. Instead, Whitehead read a paper in 1905 (published in 1906) before the Royal Society of London that bears a striking resemblance to the stated goals of that unpublished second volume of *UA*, but applied to geometry instead. The paper was titled "On Mathematical Concepts of the Material World" (*MC*), and in it Whitehead proposes a conceptual scheme, with specific suggestions for extending and applying the scheme to electromagnetism and gravitational laws, utilizing the linear algebras and Hamilton's quaternions, especially the notion that the relation of multiplication between symbols could be associative, and did not have to be commutative. There is a sense that this paper was really a shadow of *UA* Volume 2, not just because its content was so similar to that project and its appearance so contiguous, but also its natural progression from Whitehead's fellowship dissertation to *UA* Volume 1 is further evidenced in the steps he takes in this very complex work. More than 30 years later, after publishing his major philosophical volumes, Whitehead confided to Victor Lowe that he considered *MC* one of his best pieces of work.[3]

The seminal place of *MC* in an overview of Whitehead's thought is significant. In this relatively short memoir, he planted seeds in mathematics, physics, and philosophy that would bear mature fruit throughout the rest of his life. Those seeds were firmly rooted in symbolic logic, which was his primary mode of investigation, and *MC* clearly showed the evolving nature of his ideas. Although in this work he employed *instants* of time, which almost seem classical in their mathematical expression, it is quite obvious when he developed *Concept V* that he understood the *relationship between space and time* in a much more complex fashion; not confined by the absolute nature imparted to them in Newton's theories. The year 1905 was propitious in other ways, for Einstein also published two papers on the electrodynamics of moving bodies, inspired by Maxwell's work in electromagnetism. These papers formed the basis for Einstein's special theory of relativity, which Whitehead interpreted in his own way, starting from similar inspiration, but following a different path. For now, as we unpack Whitehead's role in mathematical physics, we will focus on his memoir of 1905, and highlight aspects of it that play crucial roles in his later work.[4]

> The object of this memoir is to initiate the *mathematical investigation* of various possible ways of conceiving the *nature of the material world.* In so far as its results are worked out in *precise mathematical detail*, the memoir is concerned with the possible *relations to space of the ultimate entities* which (in ordinary language) constitute the "stuff" in space. . . . In view of the *existence of change in the material world*, the investigation has to be so conducted as to introduce, *in its abstract form, the idea of time*, and to provide for the definition of *velocity and acceleration.* The general problem is here discussed purely for the sake of its *logical* (*i.e., mathematical*) *interest.* (*MC*, 11, emphasis added)

At the outset of this stage in Whitehead's work he tells us some interesting details about the progress of his "larger scheme" of work. He is now extending the scope of that project specifically into the "*nature of the material world*," using the logic of *relations* between space and the "ultimate entities." It is clear he is speaking practically about *matter*, but at the same time it must be understood that for him, the notations used in symbolic logic do not *mean* only *matter*. He intends to follow the symbolic logic of spatial *relations* between *stuff*, whatever that *stuff* may be. This is crucial to his later work, because we see in *MC* the logical progression from points as *stuff*, to "linear reals" as *stuff*, and finally arrive at the end of the memoir with "linear objective reals" (vectors), verging on definitions of motion, velocity, and acceleration that will issue in newly formulated laws of electromagnetism and gravitation. Not bad for an investigation into the physical world using symbolic logic as the primary tool! Mass and energy are not assumed, but are derived from the *relations* between the linear objective reals. (It is important to avoid confusion in terminology by not transposing the common philosophical definition of *objective* to Whitehead's intent. Here *objective* means "experienceable" and will later be applied to "subjective" forms of experience through the extensive-topological scheme of *Process and Reality* [*PR*] Part IV).

When Whitehead extends the work of this 1905 memoir, and in 1922 derives his own relativistic laws in *The Principle of Relativity, with Applications to Physical Science* (*R*), he states,

> The metrical formulae finally arrived at are those of [Einstein's] earlier theory, but *the meanings ascribed to the algebraic symbols are entirely different*. . . . I deduce that our experience requires and exhibits a *basis of uniformity*, and that in the case of nature, this basis exhibits itself as the *uniformity of spatio-temporal relations*. This conclusion entirely cuts away the causal heterogeneity of these relations which is essential to Einstein's later theory. (*R*, v, emphasis added)

This does not mean that space-time is *uniform*, but that the *relations* constituting space-time are uniform. Much will be said about this throughout the years following 1922, and in some chapters of this present volume.

This flash-forward to 1922 is intended to help set the stage for understanding Whitehead's evolving orientation toward the physical world, and the particular interest such orientation might have for present-day physicists and philosophers. In *MC* (1905) he takes his investigative enterprise into a broader orbit from the algebras, and specifically engages geometry, which is one of his chief intellectual interests. At the same time, his scope is broader than even the geometries can contain, and he asserts his intention to account for the existence of *change in the material world*, providing definitions for velocity and acceleration. It is essential to understand that although his notation using symbolic logic does not evidence *time* in a form other than *instants*, it is an incomplete grasp of his overall project to conclude that his orientation toward time was strictly classical and not relational.

The outline of the 1905 memoir has a simplicity that belies its content. Whitehead identifies five concepts he intends to investigate, the first two being representations of the classical conception of the physical world, and the remaining three progressively engaging the actual world of change. He warns the reader early that his intentions are not limited by particular meanings commonly afforded terms in such investigations. For instance, he invokes Leibniz's theory of the Relativity of Space, noting that it

> . . . is not a concept of the material world, according to the narrow definition here given. It is merely an indication of a *possible type of concepts* alternative to the classical concept . . . a wider view suggests that it is a protest against dividing the class of objective reals into two parts, one part (the *space* of the classical concept) being the field of *fundamental relations* which do *not* include *instants of time* in their fields, and the other part (the particles) only occurring in the fields of *fundamental relations* which *do include instants of time.* In this sense it is a protest against *exempting any part of the universe from change* . . . this theory . . . has never been worked out in the form of a precise mathematical concept. . . . Our sole purpose is to exhibit concepts not inconsistent with some, if not all, of the limited number of propositions at present believed to be true concerning our *sense-perception.* (*MC*, 14, emphasis added)

He intends to follow Leibniz's lead, find a class of objective reals that is not separated from time, derive the fundamental relations obtaining upon that class, and express these objective reals and their relations in a concise mathematical formalism that accurately accounts for change in the world of sense perception.

Concept I represents the classical concepts whose objective reals are *points of space* and *particles of matter*. The essential relations (R) of this concept have as their field the classical points of space, and are triadic between a minimum of three points in a unique order. But we must account for change in this space, so a second division of the objective reals is created, comprising *particles of matter* that are capable of motion. Then an *extraneous relation* is deduced whereby a single point of space, a single particle of matter, and a single instant of time are related, and these relations are unique. "[T]he impenetrability of matter is secured by the axiom that two different extraneous relations cannot both relate the same instant of time to the same point. The [classical] *general laws of dynamics*, and *all the special physical laws*, are axioms concerning the properties of this class of *extraneous relations.* Thus the classical concept is not only dualistic, but has to admit a class of as many extraneous relations as there are members of the class of particles" (*MC*, 29, emphasis added). Even if all the extraneous relations were to be grouped as a class, as relations of particles *occupying* points of space, the resulting axiom for the class would be a single axiom of occupation and "the laws of physics are the properties of this single extraneous relation O [occupation]" (*MC*, 29). The structure of his argument, even at the outset of the memoir, clearly evidences his intuition that *relations* constitute the class of physical experience from which the laws of physics are derived.

Concept II is a monistic variant of Concept I, where the concept of *particles* is abolished, and we "transform the extraneous relations into dyadic relations between points of space and instants of time. . . . The reason for the original introduction of 'matter' was, without doubt *to give the senses something to perceive.* If a *relation can be perceived*, this Concept II has every advantage over the classical concept" (*MC*, 29). As Whitehead progresses in his investigation, some very subtle steps occur. It is here that he introduces the notion that *relations* may be the ultimate *objects of perception*, but as this investigation is strictly mathematical-geometrical, he does not pause to examine the consequences of such a suggestion. Such a full examination does not in fact appear in a systematic way until *PR*, more than twenty-five years later.

Concept III is an effort to couple dynamics with the previous concepts and abandon the prejudice against moving points. The essential relation (R) of this concept becomes tetradic, between three objective reals and an instant of time. Since Whitehead's investigation is focused on geometric relations, these reals may be considered related points or related lines, so he can derive from three intersecting straight lines the kinetic axes, and thereby define velocity and acceleration. He has not used any classical formulae to accomplish these results, limiting himself to logical formalism and geometry. "This concept has the advantage over Concepts I and II that it has reduced the class of extraneous relations to one member only. . . . The concept pledges itself to *explain the physical world* by the aid of motion only" (*MC*, 31). Whitehead derives two variants on this concept to account for persistence of the "*same objective reals* in motion," as well as the "*continuity of motion*" of single corpuscles, creating the definition of identity for the objective reals.

So far, what Whitehead has accomplished is the formalization of the classical concepts of space, matter, and time, using only symbolic logic and Euclidian geometries. But his intent is to examine possible relations between matter and space in a changing world, and he is not satisfied with the limitations of the classical formulations. In Concept IV he introduces a complex scheme for defining entities in terms of their *intersections within a region*, and the order of that intersection. He calls these "points" *intersection-points* (shortened to *interpoints*), not to be confused with the simplistic concept of a common "point" of two intersecting lines in Euclidian geometry. "The theory of interpoints depends on that of 'similarity of position' in a relation" (*MC*, 35). The essential relations in this Concept are pentadic, with an instant of time, and one objective real intersecting three other objective reals in a unique order. In the point-ordering relation, Whitehead makes his shift from plain to projective geometry, and defines classes of interpoints by their ordering relations. He derives two versions of this Concept, IV-A being dualistic, and IV-B a monistic variant. For the purposes of setting the stage of Whitehead's context in physics, it is important to note that this Concept establishes relations between entities (interpoints) which will serve his purposes later in the paper he delivers to the First Congress on Mathematical Philosophy in Paris, titled "The Relational Theory of Space" (*RTS*), where the notion of interpoints is generalized into a notion of *inclusion*. For the 1905 concepts, what is most important is their logical consistency and applicability to the world of sense perception. "It is necessary to assume that the points in this concept [IV] disintegrate, and do not, in general, *persist from instant to instant*. For otherwise the only continuous motion possible would be representable by *linear transformations of coordinates*; and it seems unlikely that *sense-perceptions* could be explained by such restricted type of motions" (*MC*, 43, emphasis added).

Concept V introduces a crucial notion of *dimension*, which is defined by a property of classes of linear objective reals. These properties (or dimensions) are derived from geometrical relations, and exist in ways not commonly used in ordinary language. For example, Whitehead derives "flatness" in terms of classes and sub-classes of classes with a particular property. His definition of *flatness* does in fact apply to three-dimensional space in a classical sense, but its logical structure translates differently when applied to properties other than flatness. "It is this peculiar property of flatness which has masked the importance in geometry of the theory of common φ-subregions" (*MC*, 45). This Concept is an effort to define regions, and actions within those regions, again based upon geometrical relations. Insights from projective geometry inform his arguments, and the *ordering of relations* evidences an intuition of topological relations not yet formalized systematically by either Whitehead or others. The culminating concept of this memoir utilizes the "Theory of Interpoints," "Theory of Dimensions," and the unique property of *homaloty* (similar to *flatness* of a class of punctual lines), to define linear objective reals, which are the precursors of *actual entities* and the events of his mature cosmology. The "geometry in this Concept V includes more than does the geometry of Concept I. For in Concept I geometry has only to do with points, punctual lines, and punctual planes; but in Concept V geometry has, in addition, to consider the *relation of the objective reals* (which are all 'linear') and of *interpoints* to the above entities. In this respect, geometry in Concept V merges into physics more than does geometry in Concept I" (*MC*, 69, emphasis added).

It is clear that Whitehead made significant progress on his "larger scheme" with this memoir of 1905. If it is in fact a "shadow" of *UA* Volume 2, then it reveals a depth of innovation unparalleled at the time, linking the new algebras with non-Euclidian geometries in ways not previously explored. It is one step in a much larger vision, which continues to unfold through his interpretive process using symbolic logic to investigate geometry, and deducing physical laws without relying on experimental data. He concludes this effort by prescribing three cases for understanding the characteristics of the electron, and the forces conjectured to be present with it. "What is wanted at this stage is some simple hypothesis concerning the *motion* of objective reals and correlating it with the motion of electric points and electrons. From such a hypothesis *the whole electromagnetic and gravitational laws* might follow with the utmost simplicity. The complete concept involves the assumption of only *one class of entities* forming the universe. Properties of 'space' and of *physical phenomena* in space become simply the properties of this single class of entities" (*MC*, 82, emphasis added).

In 1906 and 1907 Whitehead published two volumes on geometry, *The Axioms of Projective Geometry* (*APG*) and *The Axioms of Descriptive Geometry* (*ADG*), wherein he formalizes many of the definitions, axioms, and propositions investigated in *MC*. These two volumes are designed to be read together as two parts of a classification scheme extending the details of the *point-ordering* relations explored in the 1905 memoir. Projective geometry yields ordering relations in closed series, while descriptive geometry yields ordering relations in open series. These two tracts are more mathematical in nature, and less philosophical in scope, but they represent development of the earlier notions in a systematic fashion. It should be remembered that during this same period, Whitehead and Russell were collaborating on their decade-long project of formalizing mathematics by using symbolic logic to investigate number theory. Gödel's 1931 incompleteness theorem notwithstanding, their work on *PM* in terms of classes and properties of classes continued to influence their subsequent work.

Whitehead's progression of ideas takes another form in *RTS* (1914), extending the link between *UA* and *MC* that creates a powerful web of ideas heralding Whitehead's distinctly philosophical period. In *UA* Whitehead uses symbolic logic to investigate the foundations of Euclidian and non-Euclidian algebras, applied to extended regions of space. In *MC* he continues that project by linking symbolic logic and algebra to investigate the foundations of plain, elliptical, hyperbolic, and parabolic geometries in terms of relations between extended volumes of space, culminating in axioms of projective geometry, and definition of the cogredient point (at infinity). He also opens the door to topology with his interpretation of the point-ordering relation for extended volumes. *MC* represents an advancement of his theory of extension (further developed in *RTS*), and also an effort to derive adequate definitions of velocity and acceleration, in order to account for the existence of *change* in the material world. Considering that Whitehead has included *time* in his investigations with representations only of *instants* and differences between *instants*, it seems quite natural in the overall progression that he begins to see time as an element of *relation* with *extension*, similar to that which has been his focus for space. Whitehead's natural propensity for *relation* carried every expectation that at some stage in his development, the *extended nature of time* would somehow be included within the scope of his inquiries. And indeed, in 1914 *RTS* reveals his thinking on that matter by introducing time as a *lapse*. "According to ordinary physics, the perceptions resulting from *changing relations* between physical objects occur during a *certain lapse of time*" (*RTS*, 3, emphasis added).

The extendedness of time is not so much a problem as are apparent *differences between the lapses of time* experienced by different subjects (the question of simultaneity). "[T]he real problem of time is the formation of a common time for the complete apparent world *independent of the different times* of the immediate apparent worlds of the various perceiving subjects" (*RTS*, 5, emphasis added). To begin accommodating the existence of *change,* he considers the *lapse* of extended time in conjunction with the continuity of spatial volumes and addresses the concept of limits and boundaries. This type of *extendedness* and *connection* leads him to consider *surfaces as regions of intersection* (interpoints from *MC*) and derive the concept of *inclusion*, using more overtly philosophical language, still accompanied by symbolic notation. This notion, together with distinctions of *part-whole*, *divisibility*, *continuity*, and *limits*, allows Whitehead to derive distinctly *topological* concepts in terms of *inclusion classes.* He derives the relations of *covering* and *T-incident* to expand the notion of his new *T-space*. It is unknown why he chooses the variable *T* in this work, applying it in a manner different from his other works, but its suggestiveness may not be strictly coincidental. Whitehead closes this paper with the announcement that he hopes to develop and expand these concepts in a work soon to be published, presumably *An Enquiry Concerning The Principles of Natural Knowledge* (*PNK*) in 1919.

It is clear that Whitehead's orientation toward physics remains firmly rooted in the operations of symbolic logic, and in the fundamental concepts of space, time, and matter, variously conceived. It is equally important to acknowledge the methodological consistency throughout his investigations. He continues to use symbolic logic, not experimental data or metrical-dimensional values, as the basis for his arguments. We will see that this methodological distinction forms the basis for his divergence from Einstein's formulations of relativity. In 1915, Whitehead presents a paper titled "Space, Time, and Relativity" (*STR*), wherein he expands his growing notion of the *logical relativity* of extended space and extended time, develops further the concept of durations and *temporal extension*, and in general continues to refine the *theory of extension*.

A thorough review of his logical notation might reveal when in his scheme of extension he links the persistence of motion with continuity, resulting in temporal extension. It happens sometime between 1914 and 1919, because by *PNK* he speaks very clearly about how "the spatial relations must now stretch across time" (*PNK*, 6). It is not the *nature of space*, but rather the *nature of extension* applied to space-time that motivates Whitehead as his investigations become more philosophical. In an integral development of the ideas in *PNK*,

Whitehead participated in a symposium in 1919 titled "Time, Space and Material: Are They, and If So in What Sense, the Ultimate Data of Science" (*TSM*) before publishing *The Concept of Nature* (*CN*) in 1920. His participation in this symposium gave him the opportunity to coin the term "creative advance," expand the projective-topological nature of *cogredience*, and begin speaking in terms of "event-particles." The theory of extension is also shown to include many aspects of space and time that had earlier been not so attributed.

In *PNK-CN*[5] Whitehead continues the work begun in *MC* and derives *congruence* as a class of *relations of extension* (*PNK*, 51), and he expands his emerging *theory of events* to include projective-topological relations described philosophically (*PNK*, 59). At each step his theory of extension continues to evolve, and objects are described as a *class of relations* between events. He begins to develop the formalization of the *method of extensive abstraction*, which is derived directly from the fundamental axioms of extension (*extensive connection*) (*PNK*, 101ff). He accents the existence of multiple time systems and durations, which emerge from the theory of extension when applied to motion, continuity (point ordering), and permanence (*PNK*, 110ff). Whitehead derives a complex geometry, and finally completes his goal (stated in *MC*, 82) of formally deriving laws of motion suitable to the extensional scheme in its present form. Objects and events are further differentiated, and the class of causal relations is enhanced and expanded (*PNK*, 121). Together, *PNK* and *CN* serve as introductory chapters for Whitehead's formal theory of relativity presented in *R*. Whitehead states very clearly what are his differences from Einstein's theory and the Maxwell-Lorentz assumptions on which Einstein relied: ". . . we must not conceive of *events* as in a given Time, a given Space, and consisting of *changes* in given persistent material. Time, Space, and Material are *adjunct of events*. On the old theory of relativity, Time and Space are *relations between materials*; on our theory they are *relations between events*" (*PNK*, 26, emphasis added). "The divergence chiefly arises from the fact that I do not accept [Einstein's] theory of *non-uniform space* or his assumptions as to the *peculiar fundamental character of light-signals*" (*CN*, vii, emphasis added). "The metrical formulae finally arrived at are those of [Einstein's] earlier theory, but the *meanings ascribed to the algebraic symbols are entirely different*" (*R*, iii, emphasis added). Whitehead's theory of relativity has been the focus of much attention in the scholarship, to the extent that other similar theories have been classified as "Whiteheadian-type" theories. In the effort to set his context in physics, that portion of his history speaks for itself.

Scholars have been inspired by Whitehead's work in areas beyond strictly mathematics and physics. His fundamental contribution to mathematical logic, however the formal outcome of *PM* is characterized, is incontrovertible.[6] Very recent work, requiring thorough consideration of logical notation amended by Russell, has yielded new insights into *PM* *89.16, thereby recovering mathematical induction and extending the applicability of Whitehead and Russell's work.[7] In a concise summary of many of the issues outlined in this historical review, Paul Schmidt described the range of Whitehead's significance for the natural sciences.

> In the realm of science the [philosophy of organism] provides
> (1) an ontological basis for causal connection,
> (2) a basis for vectors [and a new definition of tensors] in physics,
> (3) a foundation for the notion of energy,
> (4) an interpretation of the quantum character of nature [see chapters 7, 9, 13, 17],
> (5) an interpretation of vibration and frequency,
> (6) a foundation for induction,
> (7) a relational theory of space-time [see chapter 12],
> (8) a basis for the notions of rest, motion, acceleration, velocity, and simultaneity [see chapters 12, 13],
> (9) an explanation of the relation between geometry to experience, and
> (10) a foundation for the biological notion of an organism and the sociological notion of a society.[8]

Many of these themes have been explored in depth by interpreters of Whitehead, and his work continues to inspire explorations into understanding them from new perspectives.

Following the lines of Schmidt's point (4) above, Henry J. Folse wrote in 1974, "we might ask what is the prognosis for any future rapport between process philosophy and quantum theory. The Copenhagen interpretation has come under considerable criticism in recent years, much of which draws its strength upon an appeal to the classical ontology of mechanistic materialism. It would seem that the Copenhagen interpretation and process philosophy would make good allies in any battle against resurgent substantival materialism."[9] Readers of the present volume will be informed from several perspectives on this estimation of Whitehead's relevance for quantum theory. His relativistic-gravitational concepts still provide current stimulation for theorists. In working on Whitehead's theory, Eddington stumbled upon a previously unknown form of the Schwarzschild metric, a form thus *discovered a full thirty-six years before Kruskal's* famous singularity-free systems of 1960. Writing in 1989, A. T. Hyman describes the impact of this development by saying that "[t]hus, far from being a useless relic of an obscure philosopher, Whitehead's theory has actually contributed significantly to the progress of gravitational research during the past thirty years."[10] Even more recently, Joachim Stolz wrote *Whitehead und Einstein* in 1995,[11] which was reviewed by Robert Valenza in the journal *Process Studies.*[12] The review highlighted the significant contribution Stolz made to the reconstruction of Whitehead's philosophy. Valenza, who (together with Granville Henry) is sometimes a critic of logicism,[13] especially the variety manifest in *UA* and *PM*, modifies some nuances of his Whiteheadian critique and lauds Stolz's thesis that there exists a *deep structure* in the development of Whitehead's thought evidencing a *general covariance principle.* That interpretation parallels the historical review comprising the bulk of this summary of Whitehead as a mathematical physicist. The foundation of Whitehead's mature event ontology is the functoriality of symbolic logic applied to regions of space-time. Stolz's proposed reconstruction of Whitehead's philosophy in light of this insight may serve as an inspiration to many who explore the demanding environment of Whitehead's complexity.

It should come as no surprise that *topological invariance* and other properties[14] such as mapping[15] are now seen under new illumination by developments Whitehead uncovered during his career. Information theory, field theories of cognition,[16] and other allied disciplines, are also beginning to incorporate input from his work.

Whitehead began his career with symbolic logic and mathematics, and when he applied those tools to the material world, he discovered fields of relations of events, classes of relations of events, and an emerging topology of forms of relations. Rather than focus on objects existing in the fields of space-time, Whitehead focused on the events that constituted, or included, those objects instead. His contribution to the nineteenth–twentieth century conceptual revolution in physics, (propagated by Maxwell, Lamor, Lorentz, Einstein, and Minkowski), developed in ways that reinforced a new balance between mass-energy objects and the events or fields that embed them. Whitehead ultimately developed a cosmology where *object-being* and *event-becoming* were profoundly and *extensively* connected. Today contemporary science is embracing living systems and ecological models, while modern physics is incorporating quantum-relativistic phenomena and field theories. These movements clearly illustrate new fundamental understandings of both *change* and *permanence*, suggesting the continued fruitfulness of Whitehead's *process* vision.

Abbreviations

Works by Alfred North Whitehead

ADG — *"The Axioms of Descriptive Geometry." In Cambridge Tracts in Mathematics and Mathematical Physics*, no. 5, ed. J.G. Leathem and E.T. Whittaker. Cambridge: Cambridge University Press, 1907.

APG — "The Axioms of Projective Geometry." In *Cambridge Tracts in Mathematics and Mathematical Physics*, no. 4, ed. J.G. Leathem and E.T. Whittaker. Cambridge: Cambridge University Press, 1906.

CN — *The Concept of Nature*. Cambridge: Cambridge University Press, 1920; reprint, Cambridge: Cambridge University Press, 1955.

MC — "On Mathematical Concepts of the Material World." In *Alfred North Whitehead, an Anthology*, ed. F.S.C. Northrup and M.W. Gross, 7–82. New York: The Macmillan Press, 1953.

PM *Principia Mathematica*, with Bertrand Russell. Cambridge: Cambridge University Press, 1910–1913; v. I, 1910; v. II, 1912; v. III, 1913.

PNK *An Enquiry Concerning the Principles of Natural Knowledge.* 2nd ed. Cambridge: Cambridge University Press, 1925; reprint, Cambridge: Cambridge University Press, 1955.

PR *Process and Reality*, corrected edition, ed. D.G. Griffin and D.W. Sherburne. New York: The Free Press, 1978.

R *The Principle of Relativity with Applications to Physical Science.* Cambridge: Cambridge University Press, 1922.

RTS "The Relational Theory of Space," trans. by P.J. Hurley. University of San Diego, 1976.

STR "Space, Time and Relativity." In *The Interpretation of Science*, ed. with intro. A.H. Johnson, 90–107. New York-Indianapolis: Bobbs-Merrill, 1961.

TSM "Time, Space, and Material: Are They, and If So in What Sense, the Ultimate Data of Science." In *The Interpretation of Science*, ed. with intro. A.H. Johnson, 56–68. New York-Indianapolis: Bobbs-Merrill, 1961.

UA *A Treatise on Universal Algebra, with Applications*, v. I (no other published). Cambridge: Cambridge University Press, 1898.

Works by Other Authors

WM&W Lowe, Victor. *Alfred North Whitehead: The Man and His Work.* Baltimore: Johns Hopkins University Press, 1985.

Notes

1. Alfred North Whitehead, "Autobiographical Notes," in *The Library of Living Philosophers: The Philosophy of Alfred North Whitehead*, ed. Paul A. Schilpp (New York: Tudor Publishing Company, 1951), 9.

2. Morris Kline, *Mathematical Thought from Ancient to Modern Times*, 3 vols. (New York: Oxford University Press, 1972), 1206.

3. Victor Lowe, *Understanding Whitehead* (Baltimore: John Hopkins University Press, 1962; paper, 1966), 157.

4. I hold this position contra Lowe and others. See note 16, *WM&W*, 301. Lowe references a letter from Whitehead to Russell in September 1911, wherein Whitehead specifically states that "the idea suddenly flashed on me that *time* could be treated exactly the same way as I have now got space . . . the result is a *relational theory of time*. . . . As far as I can see, it gets over all the old difficulties, and above all *abolishes the instant in time* . . ." But Lowe follows this with a very curious report of a conversation with Russell in 1965, where "I asked whether Whitehead, like himself, favored the absolute theories of space and time early in their collaboration. 'No,' he answered, and added, 'I think he was *born* a relativist'(*WM&W*, 299, emphasis added). That conviction is shared in full by this author. Whitehead spoke, thought, and wrote in symbols. The meaning associated with those symbols was not numerical; neither was it simple. If he does not state the full meaning of the symbols, *it is our own projection* that creates those

meanings. It is almost impossible to conceive that such a mind as his, so thoroughly immersed in the generalizing of concepts relating to logical, geometrical, and spatial relations, would isolate time as an exception to those generalizations. Whitehead, in remarking about the differences (in *MC*) between the *punctual* concepts (I and II) and the *linear* concepts (III, IV, and V), states about Concept III (leading to Concept IV) that "The concept pledges itself to explain the physical world by the aid of *motion* only. It was indeed a dictum of some eminent physicists of the nineteenth century that 'motion is the essence of matter.' But this concept takes them rather sharply at their word. There is absolutely nothing to distinguish one part of the *objective reals* from another part except differences of motion. The '*corpuscle*' will be a volume in which some peculiarity of the *motion* of the *objective reals* exists and persists" (*MC*, 31). "In fact, when *motion* is considered, it will be found that the *points of one instant* are, in general, different from the *points of another instant*, not in the sense of Concept III that they are the same entities with different relations, but in the sense that *they are different entities*" (*MC*, 33, emphasis added). Whitehead was working out the details of a radically generalized concept of matter in relation to space and time, and his formalism, as well as his descriptions, give clear evidence that the logic of differences between these complex relations is the basis for his physical theory. The fact that he used *instants* of time in the formalism does not, or should not, shroud the distinction of the *corpuscles* that provide the basis for the definition of velocity and acceleration. The *corpuscles* are precursors of *durations* in *PNK* (1919). The seeds are planted in the logical analysis, and it is simply the interpretive scheme that requires further elucidation; like that recounted to Russell in the letter of September 1911, a mere six years after presentation of the memoir. To say the *relational theory of time* is not present in *MC* is to miss the structure and direction of the work. No scholarship is gained by maintaining either position in isolation. Acknowledgement of the process is enough.

5. (*CN*, vi) Whitehead is quite clear that the two volumes *PNK* and *CN* should be read and considered together. His writing at this stage has become predominantly philosophical, with adequate formalism accompanying sections he deemed to require such elucidation.

6. Jan Lukasiewicz, *Aristotle's Syllogistic from the Standpoint of Modern Formal Logic* (Oxford: Clarendon Press, 1951).

7. Gregory Landini, "The 'Definability' of the Set of Natural Numbers in the 1925 *Principia Mathematica*," *Journal of Philosophical Logic* 25, no. 6 (1996): 597–615.

8. Paul F. Schmidt, *Perception and Cosmology in Whitehead's Philosophy* (New Brunswick: Rutgers University Press, 1967), 173–74.

9. Henry J. Folse Jr., "The Copenhagen Interpretation of Quantum Theory and Whitehead's Philosophy of Organism," in *Tulane Studies in Philosophy; Studies in Process Philosophy I*, vol. xxiii (1974): 46–47.

10. A. T. Hyman, "A New Interpretation of Whitehead's Theory," in *Il Nuovo Cimento* 104B, no. 4 (1989): 389.

11. Joachim Stolz, *Whitehead und Einstein: Wissenschaftsgeschichtliche Studien in Naturphilosophischer Absicht* (Frankfurt am Main: Peter Lang GmbH, 1995).

12. Robert J. Valenza, review of *Whitehead und Einstein: Wissenschaftsgeschichtliche Studien in Naturphilosophischer Absicht*, by Joachim Stolz, *Process Studies* 29, no. 2 (2000): 378–81.

13. Granville C. Henry and Robert J. Valenza, "Whitehead's Early Philosophy of Mathematics," *Process Studies* 22, no. 1 (1993): 21–36.

14. Newton C.A. DaCosta, Otavio Bueno, and Steven French, "Suppes Predicates for Space-Time," *Synthese* (1997): 271–79.

15. James Bradley, "The Speculative Generalization of the Function," *Tijdschrift voor Filosofie* 64 (2002): 253–271.

16. Joachim Stolz, "The Research Program in Natural Philosophy from Gauss and Riemann to Einstein and Whitehead: From the Theory of Invariants to the Idea of a Field Theory of Cognition," *Prima Philosophia* 10, no. 4 (1997): 157–64.

4

Evidence for Process in the Physical World

JOHN A. JUNGERMAN

Introduction

From the world of the extremely small (atoms, nuclei, and their constituents), to nonlinear self-organizing systems on a human scale, and finally to the vast regions of the cosmos itself, at every level we find interconnectedness, openness, creativity, and increasing order.

These are fundamental tenets of process thought through its description of events, or occasions of experience, which the process philosopher Alfred North Whitehead assumes are its basic entity. Process philosophy asserts that an event, or society of events, is influenced by or connected to previous events. The event harmonizes this information with its goals and makes a choice among alternatives, an openness that may lead to creativity and novelty.

It is impressive that process thought is often in agreement with discoveries in physics that occurred decades after its formulation. I would now like to discuss in turn interconnectedness, openness, creativity, and increasing order in more detail (see also J. Jungerman, *World in Process: Creativity and Interconnection in the New Physics*, Albany: State University of New York Press, 2000).

Interconnectedness

According to the general theory of relativity, there is an interdependence of time, space, mass-energy, and gravity. They are all interlinked. Time is itself part of a process. The general theory also predicts that accelerating masses will give off gravitational radiation—a prediction recently verified with great precision by long-term observation of a binary pulsar.[1] The special and general theories of relativity give us a connection between concepts that were previously thought to be separate. Such interrelatedness is in accord with process thought—it shows a dynamic and interconnected universe.

Again, twentieth-century physics has shown us new connections between concepts originally thought to be separate: light is both a wave and a particle, and matter itself is not only a particle, but also has a wave aspect. We are invited to a transcendent conception of a wavicle for both light, or more generally electromagnetic radiation, and matter. Wave-particle duality is a fact of the physical universe. These discoveries are illustrations of the fundamental tenet of process thought that the world is formed of interconnections—in this instance conceptual ones.

Quantum mechanics is in accord with a holistic conception of the universe in that interconnectedness lies at its heart. This was another of Einstein's objections to quantum mechanics. He foresaw "spooky interactions at a distance," which he rejected. Yet in the late twentieth century there has been ample experimental evidence that particles once formed in a quantum state remain in this state and are connected to each other even though separated at great distances. In 1997 a group of Swiss physicists demonstrated that the quantum mechanical correlation was intact over a distance of 10 kilometers using fiber optic telephone lines.[2] Thus, the connection observed is independent of distance.

Particles formed from three quarks, such as the proton and the neutron, are called baryons. In addition to the neutron and proton, other baryons have been found, and these contain new types of quarks. In fact, four quarks in addition to the up and down quarks that form neutrons and protons have been discovered and given the quaint names "charm," "strange," "bottom," and "top." The last, the top quark, was found in 1995.

Atoms in an excited state decay to their ground state by emission of a photon. In an analogous manner, protons and neutrons have been found to have excited states that are formed from baryons made from up, down, charm, and strange quarks. They decay to protons and neutrons by emission of photons as well as other particles, which may themselves be structured. In this way these baryons form a web of connections, in agreement with process thought.

Self-organizing systems at the human scale form connections among trillions of molecules. A thin layer of heated oil with a thermal gradient and far from equilibrium forms an exquisite pattern of Bénard cells—an intimate connection among the myriad of molecules involved. The gravity of our sun makes it possible for the earth to be at just the right distance to be hospitable to life. The low-entropy electromagnetic radiation from the sun makes life possible—an essential connection for all of us. We humans are connected by necessity and dependent on plants and animals for our survival and depend on each other through our cultures to provide not only the necessities of life, but also the means to expand our creativity.

Within the cosmos there are many interrelationships. Gravity forms galaxies from cosmic dust and, with the passage of time, it creates stars by compressing them so that nuclear reactions can begin. We again share a common connection in being stardust together from a supernova explosion some six billion years ago. This time can be estimated by noting that the abundance of U-235 now is about 0.7 percent of that of U-238. Assuming that in the supernova event these isotopes were produced in approximately equal amounts, we can ask how long it would take for the shorter-lived U-235 to decay to its present abundance—giving the six-billion-year figure.

Openness

Process philosophy asserts that when an individual event makes a selection, there is an openness, a selection among alternatives. The entity takes into account the past, as well as the goals and possibilities for the future, but the final action is unique and unpredictable. There is corroboration of this idea in the physical world.

Heisenberg's uncertainty principle shows us that there is a lack of determinism at the microlevel. This implies that the universe is not preordained but has an inherent openness in its evolution. For example, if an electron passes through a double slit to form an interference pattern, it is not possible to predict at which fringe the electron will be found by a detector. It is as if the electron makes an individual selection among alternative fringes, subject to the overall statistical demands of quantum theory.

Neither can we predict when a virtual particle pair will be created ex nihilo in a vacuum, or when virtual photons and gluons will be exchanged to bind atoms and nuclei. We can only predict the aggregate behavior of thousands of similar particles—that is, the probability that an event will occur.

Einstein disliked quantum mechanics precisely because its view of nature is of a statistical character with nothing to say about individual events. His statement that "God does not play dice with the universe" shows the deep religious conviction that drew Einstein away from quantum mechanics, which he regarded as an "incomplete theory." Yet the possibility that each particle in some sense determines its own destiny is in accord with the process philosophy of Whitehead and his successors, such as Charles Hartshorne. It should be noted that, although there are some impressive confirmations of Whitehead's philosophy by quantum mechanics, they are only partial.[3]

Again in nonlinear self-organizing systems we find that often we can never specify their initial conditions accurately enough to permit a long-term prediction of their future course, if at all. Predictions of the weather can only be made for a few days at most because we simply cannot define sufficiently all the conditions in the atmosphere at any given moment in order to calculate its future behavior.

The cosmos is a dynamic place with the birthing and dying of stars. We can observe these events, but we are unable to predict when or where they will happen. A supernova was observed near us (remnants are the Crab nebula) in

1054 A.D., another about 400 years ago, and again in 1987 in the Magellanic cloud, about 170,000 light years away. We now observe dozens of new ones every few months by computer surveys of the far reaches of space.[4] But we have no idea when or where another will appear.

Creation of Order and Novelty

It is a reasonable assumption that the cosmos has created familiar order at times billions of years in the past and to the edges of the known universe—that physical laws seem to be the same even under these extreme conditions of space and time. We do know that astronomical objects hundreds of millions of light years away, and therefore hundreds of millions of years in the past, display the same physical constants as we observe on earth at present.

For example, the fine structure constant derived from the hyperfine structure of stellar hydrogen has the same value to less than a percent as found in a laboratory on earth.[5] Again, long-term observation of the binary pulsar 1913 + 16 shows that its gravitational constant is either the same as on earth or could differ by 10% over a time equal to the age of the universe.[6] Why is there such lawfulness? It could have been otherwise: randomness and chaos or endless repetition. The answer of process theology is that the divine lure guides the universe toward maximum enjoyment and creativity for all its entities: hence the evolution of the universe toward more complexity and organization. Without lawfulness and order this would have been impossible.

The vacuum—the space where there are no atoms or the space within an atom—is not empty. It is filled with continuous creativity—the birthing and dying of a myriad of particle pairs. We have experimental evidence confirmed by a quantum electrodynamics that these transient particle pairs are in fact part of the world we live in. Indeed, they are part of us—within the atoms of our bodies.

One of the experimental confirmations is the Lamb shift of the first excited state of the hydrogen atom. The energy level is slightly shifted due to the "polarization of the vacuum" produced by the virtual particle pairs. That is, the virtual particle pairs in their transient existence act like a dielectric within the atom, reducing the effective electric field between the proton and electron by a small, but detectable degree. Theory and experiment agree to one part in 10 billion—a truly impressive result. If virtual particle pairs are not considered, the agreement is reduced to one part in a million (Dirac's relativistic theory of the electron).

What holds the quarks together to form nuclei of atoms, or other baryons? What holds atoms themselves together? Modern field theory posits force carriers that accomplish this—virtual particles that are born and die by the billions of trillions each second. In other words, forces are created by events. Here, at the most elementary level of matter, we find corroboration of process thought—events are primary in the way the world is held together.

There are four known forces in nature: the strong force, the electromagnetic force, the weak force, and the force of gravity (in order of decreasing strength). The strong and weak forces operate only at the nuclear level, whereas the electromagnetic and gravitational forces are long range and act in our everyday world. All these forces can be described as resulting from an exchange of mediating particles that are born on one particle and die on another.

The mediating particles within the atom are virtual photons, whose creation and absorption provide the electromagnetic force that holds the electrons to the nucleus. Within the nucleus of the atom the mediating virtual particles are called gluons; their creation and absorption provide the strong force that holds the quarks together to form the nucleus. This is the view of modern field theory (called a field theory to emphasize that a field of particles is producing the force).

Again, in complex systems far from equilibrium, there is self-organizing creativity. In the example of heated oil referred to earlier, when the thermal gradient in the oil is sufficient, it breaks up spontaneously into hexagonal Bénard cells. Rotation of the liquid is in opposite directions in adjacent cells. At the critical temperature difference between the upper and lower plates, when the system changes from conductive to convective flow, the sense of a cell's rotation is established. No matter how carefully the system is controlled, the rotation sense—right- or left-handed—is unpredictable. A huge number of particles, about a trillion billion in a typical Bénard cell, are organized in a coherent fashion and each of the cells is further organized with respect to all the other cells in the system. Typical intermolecular forces are about a hundred millionth of a centimeter, whereas the cell is millions of times larger. Here we have self-organization and creativity with complex behavior of a new type: the dissipative structure.

Much more complex creativity occurs in life forms. Highly organized life forms, such as humans, are capable of a great deal more enjoyment and creativity than, say, a slime mold. Yet, even at that basic level of life there is rhythmic pulsation of creativity, a search for nourishment, and reproduction.

The big bang was a period of magnificent creativity: the space in our universe was created in an expanding bubble of intense radiation. As the radiation cooled, matter and eventually atoms were formed. Billions of years of creativity followed: galaxies formed, stars ignited. Supernovae produced heavy elements and spewed them into the cosmos. Finally, five billion years ago, our solar system formed and conditions for life as we know it were produced on our planet. The universe is not static but continuously evolving—in the past its composition was very different from its composition at present.

Our knowledge of and justification for the big bang and subsequent evolution of the cosmos are intimately connected with our knowledge of the nuclear and subnuclear realm. There again we find an order among the subnuclear particles: in the standard model, six quarks and six leptons grouped into three families organize the complexity of hundreds of subnuclear particles.

Hartshorne argues that asymmetry is essential as a metaphysical principle so that order can be created: "This pattern, symmetry within an overall asymmetry, we meet again and again. I see in it a paradigm for metaphysics."[7] In this reference Hartshorne refers to appeals made in physics to symmetry "as a sufficient argument" and regrets not being able to respond to this assumption "in my ignorance." Joe Rosen, in a recent paper in *Process Studies* titled "Response to Hartshorne Concerning Symmetry and Asymmetry in Physics" and in chapter 11 of this volume, reassures Hartshorne that "indeed, asymmetry is basic in physics too."[8] He gives as an example the asymmetry in mirror reflection in weak interactions. To this could be added the excess of matter over antimatter in the universe (which may or may not be linked to the asymmetry found in K-mesonic decay) and the flow of time from the past to the future in the macroworld.

Process, Evidenced by Events, Is Primary

At a fundamental level, birthing and dying of particle pairs fill the vacuum—a series of events. The electrical forces that hold atoms together are thought of in the field theory of physics as arising from a series of exchanges of photons between the electrons and the nucleus—again a series of events. Similarly, quarks that constitute the protons and neutrons of the atomic nucleus are held to each other by a series of exchanges of gluon events. It is remarkable that Whitehead's process metaphysics, formulated in the 1920s, is so compatible with these ideas, some of which did not emerge from physics until the 1940s and later.

Matter, such as a table or a chair, seems so solid to us, but this is a limitation of our senses. In fact, so-called solid matter is just emptiness, except for a dance of virtual particles. The mass in matter occupies an extremely miniscule volume. Since we are now more familiar with the constitution of matter, we are in a position to understand just how incredibly empty matter really is.

First, electrons have very little atomic mass and are points to the limit of our measurements. Second, more than 99.9% of the atomic mass is in its nucleus. Finally, that mass is found in pointlike particles, quarks. Recent experiments performed with protons that had an energy of a trillion electron volts (at the Fermi National Accelerator Laboratory near Chicago) show that the quarks are pointlike to 10^{-17} centimeters or less. Since atomic diameters are about 10^{-8} centimeters, the ratio of a quark diameter to an atomic diameter is at most one in 10^{9}, or one in one billion or less.

The fraction of the atomic volume in which the mass of an atom is located is less than the cube of 10^{-9}. This is less than 10^{-27} of the atomic volume itself—less than one part in a thousand trillion trillion, 10^{27}! A thousand trillion trillion is a very large number. If we estimate all the grains of sand in all the beaches of earth, the number is about 10^{20} or ten billion trillion.[9] The number 10^{27} is ten million times larger. Matter is really empty space to an astonishing degree.

However, the "empty" atom is filled with virtual pairs and also with photon and gluon force carriers that are continually being birthed and dying—a dynamic, creative process at the most elementary level of matter. So solid matter is not really solid at all, but empty; at the same time it is filled with dynamic radiation, or a "complex of events" in the language of process philosophy. Substance is really an effect of our macroscopic human senses; at the microscopic level the world is a series of events—a process.

From this we can appreciate that apparently solid matter is really space filled with an unimaginable number of events—not only those from spontaneous particle pairs but also from the virtual photons, gluons, bosons, and gravitons that are the force carriers. As mentioned previously, the hydrogen atom is held together by a trillion billion virtual photon exchanges every second. To hold together the quarks that constitute that atom's proton requires the exchange, the birthing and dying, of a trillion trillion gluons every second. Thus, the world at its most elementary level is really the realm of events, not substances. Hence, the fallacy of "substance thinking."

Since our memories permit us to recall the past, there must be a record of it somewhere in our brains. A record means that there was an event, so every conscious experience is associated with an event.[10]

Substance thinking sees connections as external to substances. Connections are secondary—an afterthought. It is like the collision of two billiard balls. The connection in the moment of collision is momentary and unimportant. The substances, the billiard balls, are what is important. On the other hand, events are constituted by their interconnections with other events—a web of connections—as evidenced by a myriad of virtual photons and gluons that are birthing and dying within matter. Their transitory existence gives rise to the forces that hold atoms and their nuclei together—so that we ourselves and the world about us can exist.

With our new knowledge of cosmology and the primacy of process, what emerges is a new mythology only decades old—a creation story appropriate to our age that gives us a sense of connection and allows us to place ourselves in the universe—a sense of place that is sorely needed. Swimme and Berry provide a highly readable exposition of what they call "the universe story."[11]

Physics as Process

New theories extend old ones as our knowledge of the physical universe expands. When a superb theory arises, such as the theory of relativity, then this process leads us to a new vision of the cosmos, a transcendence of the old theories. The development of both the special and general theories of relativity demonstrates that science itself is a process. Even such revered concepts as Newton's laws are subject to modification. Science is contingent and ever-evolving. No ideas are fixed for all time.

Space and time become seen in a new way, as space-time. Matter and energy form one entity: mass-energy. Time itself depends upon the presence of matter or the observation of a moving clock. The universe is predicted to expand by the general theory of relativity, and indeed it does, whereas it was considered static in the Newtonian worldview. The theory also predicts gravitational radiation, and evidence for it has recently been established.[12]

Again, quantum mechanics is of more general applicability than classical mechanics. The latter is regarded by physicists as included in quantum mechanics when quantum numbers are large. Classical mechanics is often a useful approximation for our macroworld but, at the atomic and nuclear level, quantum mechanics is essential. It gives us a description that is strange but compatible with process thought—with events as primary, making choices among alternatives, an openness, and a mysterious interconnectedness.

More recently we are beginning to understand complex systems consisting of trillions of atoms with self-organizing nonlinear behavior. Again, new ideas for this new level of complexity are emerging.

In cosmology, old models of the universe are replaced in order to explain new observations. The Copernican system replaced the earth-centered system of the Greeks and of the first millennium A.D. The steady state model of the universe, popular in the 1950s, could not explain microwave background radiation and was eventually superseded by the present big bang model.

The universe is evolving, as well as our knowledge of it. The human enterprise is evolving. Physics, and science generally, is a part of that enterprise, and is an ongoing evolutionary process.

Notes

1. D. Kleppner, "The gem of general relativity." *Physics Today* (April 1993), p. 9.

2. A. Watson, "Quantum spookiness wins, Einstein loses in photon test." *Science* 279: 481 (25 July 1997).

3. As Abner Shimony points out (*Search for a Naturalistic World View, Vol. II* (New York: Cambridge University Press, 1993, p. 291), Whitehead's conceptual framework of quantum theory is that of the Bohr model and its antecedents. No reference is ever made in his works to the new and much more successful quantum theory of Schrödinger, Heisenberg, and others. Shimony argues that this leads to disagreements between quantum mechanics and Whitehead's view of an occasion of experience when one considers a system of identical particles or attempts to define all characteristics of an actual occasion with indefinite precision.

4. James Glantz, "Exploding stars flash new bulletins from distant universe." *Science* 280: 1008 (15 May 1998).

5. P. Sistema and H.Vucetich, "Time variation of the fundamental physical constants: Found from geophysical and astronomical data." *Physical Review D* 41: 1034 (1990), and "Time variation of the fundamental physical constants II: Quark masses as time-dependent variables." *Physical Review D* 44: 3096 (1991).

6. T. Dumour, G.W. Gibbons, and J.H. Taylor, "Limits on the variability of G using binary pulsar data." *Physical Review Letters* 61: 1151 (1988).

7. Charles Hartshorne, *Creative Synthesis and Philosophic Method* (La Salle, IL: Open Court Publishing, 1970), 210.

8. Rosen, Joe, "Response to Hartshorne concerning symmetry and asymmetry in physics." *Process Studies* 26/3–4: 318 (1997).

9. Let us assume that a cubic centimeter of beach sand contains 10,000 grains. Then a cubic meter will contain a million times this, or 10 billion grains (or 10^{10} grains). Assume a beach 1 kilometer long, 100 meters wide, and 1 meter deep. This beach contains 100,000 cubic meters of sand, and therefore 10^{15} grains of sand are on the beach. If we assume there are 100,000 kilometers of beaches in the world and each contains 100,000 cubic meters of sand, then the total number of grains of sand in all the beaches is 10^{20}.

10. Henry P. Stapp, *Mind, Matter, and Quantum Mechanics* (Berlin: Springer Verlag, 1993), 127.

11. Brian Swimme and Thomas Berry, *The Universe Story: From the Primordial Flaring Forth to the Ecozoic Era* (San Francisco: Harper, 1992).

12. Kleppner, "The gem of general relativity."

5

Dialogue for Part I

The dialogue presented below occurred in response to John Jungerman's paper.

Stapp: At the beginning of your talk, you mentioned "increase in order," whereas the normal idea is that entropy increases and that, on the whole, order decreases. In your opinion, is there any evidence against the normal idea that entropy increases, hence order decreases, on the whole?

Jungerman: The laws of thermodynamics still apply so far as I know. On the other hand, we in this room are all evidence of order, a particular order that goes against the second law. We use the low entropy from the sun and the plants and animals that we consume to create order in our bodies. We are indebted to those low entropy sources for our own existence. You can ask where that came from and for that you'll have to go back ultimately to the big bang, which itself must have been a low entropy source, so that we could exist and be in accord with the second law of thermodynamics. I am certainly not proposing to abolish that 'sacred' law of physics. On the other hand, we do see elements of order and we ourselves are examples of that; it's what Gregory Bateson called the 'sacramental' or what Schrödinger called "negative entropy."

Stapp: Do you then see evidence for a process that works against disorder?

Jungerman: It seems to me that self-organizing systems create order against the second law, but that doesn't mean that overall the second law is violated but locally it is in these systems. If you look at it globally, the second law works, but if you look at it locally, there is order being created so it depends on your perspective.

Jungerman: I would like to address to Henry Stapp the same question: Is there anything in your work that illuminates the idea of emergent order against the second law of thermodynamics?

Stapp: Von Neumann points out that a quantum system, represented by a density matrix and evolving according to the Schrödinger equation, doesn't change its entropy: the order doesn't degrade. He is not doing any fine graining or anything like that. You just look at the whole Schrödinger state. Basically the system doesn't change. It is just a unitary evolution and thus the entropy does not change. On the other hand, if you have a collapse of the wave function, the entropy does change: the collapse increases the negentropy. (*Negentropy* denotes negative entropy, which is a measure of order.) This effect gives you the potential of injecting order into the universe: each collapse picks out of a host of possibilities, something that's very special. This puts negentropy into the universe: every time a collapse occurs, there is this apparent increase of the negentropy of the universe. I find this a very attractive idea because then the universe could start in a very uniform state, without this tremendous negentropy (hence great order) that ordinary thermodynamics would require. The order could be put in bit by bit as these collapses occur. So there's a possibility that these collapses, if they really occur, are what is responsible for the fact that negentropy can run downhill all the time: it's because it's being increased all the time by the collapses.

Fagg: I heard a talk a few months ago by Freeman Dyson on the subject of entropy. The second law of thermodynamics applies to a closed system and, as John was talking about, if you make the system global it still applies, even though locally you can get these negative entropy results. Dyson subscribes these days to an open universe. If so, on a really global or cosmic scale, the second law cannot really apply.

Valenza: For a moment, let's shift our measure of order away from the classical physicist's notion of entropy. Another notion of order is provided by an information-theoretic notion. I sit in front of my computer screen, black-and-white, and it has a million pixels on it. A state vector for that system would mean a million bits of information—a string of *on* and *off*. On the other hand, if one of Dr. Jungerman's hexagons is on my screen, I could say that I need a million bits of information to specify it, but I could also say that with just a few bits of information a generative mechanism could be specified from which the pattern evolves and, in that sense, there is a very high level of order on that computer screen. That notion of order is not equivalent to entropy (or negative entropy), and in a completely deterministic system that measure of order becomes an invariant. With that idea of order, is the order of the universe increasing or decreasing?

Chew: I can only venture a guess based on my model [see chapter 8]. If you consider the universe to include all this non-material history, then there is a sense, a quantum-mechanical sense, in which it may be that nothing changes, but I've got this huge non-material component and this tiny material component, and any question you ask will almost surely be focused on this material component. Thus things could look very different from this statement that nothing is changing.

Finkelstein: If we're going into these borderline areas far outside ordinary experience, then we must be careful in carrying our concepts along. I suggest that the concept of entropy makes sense for systems in equilibrium and in no case is it really meaningful to speak of the entropy of the universe. To determine the entropy of a system, you have to be able to carry out a thermodynamic cycle on it or, if you put it in informational terms, there has to be someone outside the system getting the information. The universe, by definition, therefore does not have an entropy. We speak at best metaphorically when we speak of the entropy of the universe.

Noemie Kenna: I was originally in physics but am now working in theology and religion. I think that the Whiteheadian concept of experience is very pertinent to this discussion because in material sciences we know that fluctuations in thermodynamics have a distinct impact on atomic structure at the finest level. Each fluctuation causes permanent change in the material that forever alters the future of the process. This relates to what John said about the universe becoming more ordered and that everything in the universe is an accumulation of all of the past experiences that have had physical impacts on the matter. This means that you have creativity in which suddenly new forms come into being. This idea of experience is where science and philosophy are coming together (see chapter 17).

Part II

Order and Emergence

6

Constraints on the Origin of Coherence in Far-from-Equilibrium Chemical Systems

JOSEPH E. EARLEY SR.

Consideration of how dynamic coherences arise from less organized antecedents ("concrescence") is one of the central concerns of Whitehead's thought.[1] Other chapters in this volume consider how his concepts may apply in various branches of physics, including quantum mechanics. But many of the coherences that are of human interest are macroscopic. Earlier interpreters generally regarded Whitehead's actual entities as exclusively submicroscopic. There is now a growing consensus that Whitehead's conceptual scheme ought also to be applicable to larger items. The origin of dynamic coherence in far-from-equilibrium chemical systems has been widely studied experimentally and theoretically, and is now rather well understood. The chemical level illustrates many features that are also characteristic of more complex aggregations (biological and social, for instance), but chemical systems are more amenable to experimental and theoretical investigation. This chapter considers some aspects of the origin of chemical coherence.

Logicians and analytical philosophers tend to be more comfortable with mathematics and mathematical physics than with chemistry. This seems apparent in current treatment of wholes and parts. The term *mereology* (coined by Leśniewski) is "used generally for any formal theory of part-whole and associated concepts."[2] A "mereological whole" is an individual that is composed of parts that are themselves individuals. Any two (or more) individuals can compose a mereological whole—the star Sirius and your left shoe, for instance. Concerning such aggregates, D. M. Armstrong's "doctrine of the ontological free lunch" seems valid:

> whatever supervenes, or . . . is entailed or necessitated, . . . is not something ontologically additional to the supervenient, or necessitating, entities. What super-

> venes is no addition of being. . . . Mereological wholes are not ontologically additional to all their parts, nor are the parts ontologically additional to the whole that they compose. This has the consequence that mereological wholes are identical with all their parts taken together. Symmetrical supervenience yields identity.[3]

This kind of mereology may well be quite appropriate to deal with some questions of mathematics,[4] but it seems quite inadequate for dealing with situations that chemists encounter. Chemical combination is not well understood in terms of mere addition of properties of components. Chemical combination generates properties and relations that are not simply related to the properties and relations of the components. Entities that are important in other branches of science, and in other parts of culture, have all the complexity of chemical combination—and more. It does not seem that the standard approach to mereological questions has wide applicability, much less *metaphysical* generality.

It may well be that every system of reckoning must have a unit—some thing, or class of things, that may properly be taken as simple (not composite), at least for the sake of the reckoning. For instance, Armstrong observes:

> the existence of *atoms*, whether particulars or universals, is held to be a question for science rather than metaphysics, and one that we should at present remain agnostic about. The world divides, as Wittgenstein wrote, but it may divide *ad infinitum*, and there be no terminus, even at infinity.[5]

If, strictly speaking, there are no "simples" (atoms, in the classical sense), but yet we need to deal with entities that are "relatively atomic,"[6] then we require what Armstrong calls "unit-determining properties"—qualities that give rise to "unithood."[7]

Before getting into what sort of things unit-determining properties might be, we have to decide which level (spatial, temporal, or other) is appropriate to use in this discussion. Whitehead is usually interpreted as holding that there is a fundamental level of description, perhaps at some submicroscopic level of spatial size.[8] But if we hold the alternative position—that there are no classical "atoms," merely entities that are "relatively atomic," then it is not at all obvious that any spatial, temporal, or other level has priority—it seems that there may well be no "fundamental" level. In this alternative view, we ought to be able to consider working at any spatial or temporal level whatsoever. We might even hope that we could identify unit-determining features that were operative at many levels—that might provide warrant for a consistent, adequate, and applicable metaphysics.

In previous centuries, chemists used to say that certain chemicals had high "affinity" for certain materials (those with which they reacted) and low affinity for others (with which they did not react). Zinc was said to have high affinity for acids, copper much less affinity, and gold hardly any affinity for those acids. Tables of chemical affinities were the stock in trade of many pre-twentieth century chemists. In Brussels in the 1920s, Theophile de Donder (1872–1957) put chemical affinity on a firm quantitative footing.[9] Consider a reaction:

$$a\,A + b\,B + \ldots\ldots \rightarrow p\,P + q\,Q + \ldots\ldots,$$

the law of mass action asserts that there exists a constant, *K*, such that:

$$K = [P]_e^{\,p}\,[Q]_e^{\,q} \ldots\ldots / [A]e^{a}\,[B]_e^{\,b} \ldots$$

Where brackets denote concentrations (or activities), exponents are stoichiometric coefficients (from the balanced chemical reaction equation) and, importantly, the subscript *e* indicates that the equation is valid only when the reaction has reached chemical equilibrium. The chemical affinity of a reaction mixture is defined:

$$A = R\,T \ln K / Q.$$

where *Q* is a quotient of the same form as the equilibrium constant (*K*), but involving actual concentrations (activities) of the reaction mixture, whatever they may be, *not* equilibrium concentrations (activities). Chemical affinity (*A*, a "state function") is the measure of the driving force of a chemical reaction. A reaction system with high affinity is said to be far from equilibrium. If *A* is small, the system is considered to be close to equilibrium. At equilibrium, the affinity is zero.

The concept of affinity was not used in Lewis and Randall, the thermodynamics text regarded as definitive in the United States for the middle third of the twentieth century.[10] Many chemists are uncomfortable with the concept. Probably logicians would find it odd as well. Affinity is a property—but a property of what? Not of an individual molecule, nor of a collection of like molecules, but of a macroscopic reaction mixture—a collection of molecules of diverse sorts. This is also true of such relatively unproblematic properties as pressure and volume. But, in contrast to pressure and volume, affinity involves a relation. Affinity is defined with respect to a specific reaction. One and the same reaction mixture might have various values of affinity with regard to the several possible reactions that could occur between the chemicals that constitute the reaction mixture.

It is a matter of discussion among logicians as to whether several distinct mereological wholes can have exactly the same constituents.[11] There is no doubt among chemists that one and the same chemical reaction mixture can produce a wide variety of different results, depending on the presence of diverse catalysts, for instance. A pot full of assorted amino acids can readily generate a myriad of different proteins (or a mixture of them), depending on what RNAs are present.

The single pot would have an array of different values of chemical affinity, one value for each conceivable product.[12]

In a sense, chemical affinity is like "distance." What is the distance from Claremont? Distance to where? Claremont is about 40 miles from Los Angeles, but several thousand miles from Cambridge, England. The quantity "distance from Claremont" is meaningful only if a specific destination is designated. Chemical affinity is a proper "state function" (like volume or pressure) of a chemical sample, but only if a particular reaction is specified. It would not be surprising if some philosophers refused to accept such a notion.[13] Chemical affinity is an odd sort of property—but one that is rigorously defined and useful in both theory and practice.

Ordinarily, chemical reaction mixtures change in such a way as to smoothly approach an equilibrium state—that is, concentrations change (either increase or decrease) while chemical affinity (with respect to the reactions that are occurring) steadily decreases. But certain far-from-equilibrium reaction mixtures behave quite differently. Rather than monotonously changing, concentrations of particular components of these mixtures oscillate regularly and repeatedly around an unstable steady state.[14] Often, these mixtures spontaneously become structured in space, as well as in the time dimension. If the reaction mixture is a closed system (one that does not exchange energy or chemicals with its surroundings), then oscillations of components gradually decrease, and spatial structures fade, as the system gets closer to the condition of equilibrium. Such a gradual decay of oscillation can be avoided by arranging matters so that reactants are continuously fed into the system, and reaction products are removed as they are formed. Such open systems are always far from equilibrium—spatial structures and/or oscillations can be maintained indefinitely while these systems convert high-energy-content materials into low-energy-content materials. These spontaneously organized chemical systems are called "dissipative structures."[15] Such oscillating chemical systems resemble structures of more familiar types, inasmuch as they have the ability to withstand disturbance. A dissipative structure, if caused to deviate from its regular oscillatory pattern, will tend to return to the same pattern.

When Nature involves what seems to be a constant concentration of some chemical, close examination often shows that the concentration is maintained near an average value by some sort of oscillatory network of processes. Many technological devices work in a similar way. A home thermostat does not maintain a preset temperature steadily, but rather achieves the purpose for which it was designed by small oscillations of temperature above and below the target value. We now recognize that many—perhaps all—of the entities formerly thought to be substantial and perduring are, in fact, resultants of networks of processes. The question we need to consider is: What sorts of relationships between components must exist in order that a collection of processes would behave as one unit? Ivor Leclerc discussed this question, but did not come to a specific conclusion of general relevance.[16] We now consider chemical dissipative structures (oscillating reactions) as examples of how unit determination arises. Similar unit-determining properties (better, unit-determining closures of relations) can be seen to occur elsewhere—from the quantum level to the cosmological level, and at many intermediate ones.

In order for a chemical dissipative structure (oscillating reaction) to exist, the system must be far from equilibrium (have high affinity) and there must be no stable equilibrium state accessible to the system. When parameters change, it sometimes happens that a previously stable stationary state (or equilibrium) becomes unstable. For instance, at ordinary air temperatures, heat that is generated by fermentation inside a haystack diffuses out of the pile of dried grass, and a stable, quasi-equilibrium, steady state is eventually reached. However, if the ambient air temperature comes to be much higher than usual, the rate of exit of chemically generated heat would be reduced, and the haystack would heat up. Since fermentation occurs more rapidly at higher temperatures than at lower temperatures, retention of heat in the haystack would cause still further heat to be generated, and generated faster and faster. The nonequilibrium steady state would become unstable. The pile of hay would quickly become quite hot, and might burst into flame.[17]

The circumstance that increase of temperature causes increase of the release of heat, and yet further rise in temperature, is a particular example of *autocatalysis*. A chemical reaction produces a product (heat in this case) that increases the rate of production of that product. Autocatalytic processes abound in nature. Many nuclear reactions produce neutrons that can initiate yet further nuclear reactions. Genes are biological catalysts that make copies of themselves, through quite complicated mechanisms. Human groups socialize their members to engage in behavior that tends to increase the number of members of the group. Resources invested in bull markets produce yet further resources, which enable additional investment. Each such process can destabilize prior steady states—equilibria or quasi-equilibria.

Every dissipative structure must involve at least one autocatalytic reaction. The simplest kinds of autocatalysis (to which others can be reduced) can be represented:

$A + X \rightarrow 2X$	Rate $= k\,[A]\,[X]$	*quadratic* autocatalysis
$A + 2X \rightarrow 3X$	Rate $= k\,[A]\,[X]^2$	*cubic* autocatalysis

Any autocatalytic process can be the basis of a *clock reaction*—a reaction that appears quiescent for a long period of time, and then bursts into activity: a solution changes color rapidly after a time without apparent change, or a haystack suddenly bursts into flame. A virus (an autocatalytic agent) may lurk undetected for years, then replicate explosively. The length of the delay (the "induction period") depends on the initial concentration of autocatalyst. The rate of production of autocatalyst will initially be low if the starting concentration of the autocatalyst is small. But no matter how low that rate is (so long as it is not exactly zero), the concentration of the autocatalyst continually increases, and the autocatalytic reaction steadily gets faster. This gives rise to the long induction period followed by a sudden rapid rise of the autocatalyst concentration. All dissipative structures are based on autocatalysis of one kind or another.

But something else is required to turn an autocatalytic process into a dissipative structure: there has to be some way to "reset the clock"—to return the autocatalyst concentration to a low value. This requires the existence of at least one additional chemical, the *exit species*. The autocatalyst is often called X, and the exit species designated Z. Here are representations of two types of chemical oscillators; both based on cubic autocatalysis, but each using a different strategy to reset the clock. (Both of these are open systems, with chemicals entering and leaving the system.)

$$A + 2X \rightarrow 3X + Z \qquad Z + X \rightarrow P \qquad A \rightarrow X \rightarrow Q$$

In this first set of reactions, Z is a byproduct of the autocatalytic process; Z also has the ability to remove the autocatalyst. As the autocatalytic reaction proceeds, the exit species Z builds up through the reaction shown on the left. In the reaction set shown in the center, Z combines with X to remove the autocatalyst. In the rightmost set of reactions, the autocatalyst is fed into the system and also departs from the system (e.g., decomposes). A is the (constant) supply of reagents; P and Q are products that play no other part in the reactions.

$$Z + 2X \rightarrow 3X \qquad A \rightarrow X \rightarrow P \qquad A \rightarrow Z \rightarrow Q$$

In this second case, Z is not a product, but is a reactant in the autocatalytic reaction. Z is used up as the autocatalytic reaction (shown on the left) proceeds. Once Z is depleted, autocatalysis stops. Both the autocatalyst X and the exit species Z are fed into the system, and both leave the system by decomposition (center and right reactions).

In these two sample cases, autocatalysis is controlled by an exit species. In the first case, the autocatalytic reaction is choked off by buildup of a toxic by-product. In the second case, the autocatalysis is starved by shortage of a necessary reactant. Any chemical oscillation that is based on cubic autocatalysis can be shown to correspond to one or the other of these two reaction types.[18] Oscillations based on quadratic autocatalysis must involve three (or more) variables, usually including a "feedback species," generally called Y. Like reactions based on cubic autocatalysis, oscillations based on quadratic autocatalysis are reducible to only a few basic patterns.

For simplicity, we confine our attention to a version of the second cubic autocatalytic system shown above, with spontaneous linear (not autocatalytic) formation of X from Z and spontaneous decompositions of X to produce an inactive product.

$$A \rightarrow Z \rightarrow X \rightarrow P \qquad Z + 2\,X \rightarrow 3\,X$$

There are only two variables, X and Z. Important parameters in this system include proportionality (rate) constants for each of the three chemical reactions shown, and for the input of Z. We may examine the behavior of the system as two of these parameters are varied—while values of other parameters are held fixed. For parameter values that correspond to one region of this two-dimensional parameter space, there is a stable steady state with high autocatalyst concentration (the thermodynamic branch). Let's say this state is blue. For parameter values corresponding to another region of the parameter-space diagram, there is a stable (self-restoring after disturbance) steady state with low values of the autocatalyst concentration (the flow branch)—let's say that this nonequilibrium steady state is red. For parameter values corresponding to a third region of the two-dimensional parameter space, both of these nonequilibrium steady states are stable. In this region of parameter space, the system is said to be bistable. For parameters corresponding to points in this region of parameter space, the color of the system may be either blue or red, depending on its past history. If this region of parameter space is entered from the blue side, blue will prevail; if it is entered from the red side, red will obtain. (There is a third nonequilibrium steady state in the bistable system, but it is unstable—any fluctuation will cause the system to move to one or the other of the two stable nonequilibrium steady states, the red one or the blue).

A fourth region of the two-dimensional parameter space often exists. If the two variable parameters happen to correspond to a point in this fourth region, sustained oscillations of concentrations occur. In this region of parameter space, both blue and red nonequilibrium steady states are unstable. The system moves repeatedly from states approximating the blue nonequilibrium steady state to something approximating the red nonequilibrium steady state. Concentrations of all chemicals continuously oscillate in time. In such cases, a plot showing the state of the system at various times, in a two-dimensional concentration space (one dimension for each of two concentrations), yields a closed trajectory that goes around a point corresponding to the unstable (third) nonequilibrium steady state. In contrast, parameter values that correspond to points in the other three regions of the two-dimensional parameter space yield a trajectory (in concentration space) that approaches either the red or the blue nonequilibrium steady state.

A proper dissipative structure corresponds to such a limit cycle in some appropriate space. That is, there is a single, closed curve that describes the sequence of states that the system follows over time. This same single, unique trajectory is eventually attained no matter what the starting conditions might be. In the case that we are discussing, the appropriate space is the $[X] - [Z]$ plane. If we wish, we could divide the curve in two-dimensional concentration space that describes this limit-cycle trajectory into four segments, using as dividers the points at which each of the two variables changes direction of motion (reaches a maximum or minimum in concentration).

- In one segment, *X* increases rapidly, while *Z* decreases. This segment is dominated by the autocatalytic reaction.

- In a second segment, the autocatalyst *X* begins to decrease while *Z* continues to decline. In this segment, the autocatalytic reaction competes with decomposition of the autocatalyst.

- In the third segment, *X* decreases to a low concentration and *Z* increases a great deal. The autocatalytic reaction is now largely shut off, but decomposition of the autocatalyst is proceeding, and the feed of Z is significant.

- In the fourth segment, both *Z* and *X* increase somewhat. The concentration of *X* has fallen so low that decomposition of the autocatalyst is no longer important. As the feed increases the *Z* concentration, nonautocatalytic production of *X* from *Z* becomes significant.

- When the *X* concentration passes a certain critical value, the autocatalytic reaction reaches a high rate, and loss of *Z* exceeds the rate at which *Z* is being fed in, so a rapid increase of *X* and concomitant decrease of *Z* begins. This is the phase noticed first above. The cycle is complete. As James Joyce would have it, "*Finagin.*"

As with cycles of other sorts, there is no point at which this cycle may properly be said to begin. Also, division of this cycle into just four phases is quite arbitrary. Closure of the network of processes to yield the set of states of affairs described by the limit cycle is a real and significant feature of this collection of processes, but dividing the cycle exemplifies "the fallacy of misplaced concreteness."

In order for the oscillation to have long-term stability, its trajectory must pass through to the same set of conditions during each oscillation, a stringent condition.[19] Compared to the total range of parameter space, the region in which this condition is met may be quite small. But still, this region is often large enough that considerable tolerance (more or less, depending on the other parameters) exists for variation in parameters. Frequently, setting up such experiments using aqueous solutions requires only "bartender's precision." Variation of parameters, within the oscillatory region, gives rise to changes in the frequency and the amplitude of the oscillations.

Here, then, are the constraints on origin of a dissipative structure in a chemical dynamic system:

- Affinity must be high. (The system must be far from equilibrium.)
- There must be an autocatalytic process.
- A process that reduces the concentration of the autocatalyst must exist.
- The relevant parameters (rate constants, etc.) must lie in a range corresponding to a limit cycle trajectory.[20] That is, there must be a closure of the network of reaction such that a state sufficiently close to the prior condition is achieved at the corresponding part of each oscillation.

If all these constraints continue to be met, the dissipative structure continues to exist, and may serve as a center of agency. Interactions of the system with the rest of the world are quite different in the presence of the dissipative structure than they would be in the absence of that self-organized coherence. In this sense, the closure[21] of a network of relationships that gives rise to a dissipative structure is the unit-determining feature required to secure unithood, in Armstrong's terminology. The effects of the structure as whole are the resultant of the effects of the components, but the concentrations of the components that exist at any instant are the effects of the closure of the limit cycle. This is a definite example of a kind of "downward causation"—an influence on the components arising from the thing those components constitute.

Similar situations, where closure of a network of processes has important effects, occur in many fields. The postulate of de Broglie, central to the development of quantum mechanics, is strikingly similar in some respects to the positions being advocated here. In systems involving electrons and atomic nuclei, there are stringent conditions on the closure of sets of relationships. Once that closure is attained, a system maintains its coherence indefinitely, and can function as a unit in yet higher-level coherences. Networks of interaction also abound in biochemistry, molecular biology, organismic biology, and ecology. In favorable cases, systems in all these areas display unit-making closure of relationships quite similar to those displayed by dissipative structures.

The "toy model" used to illustrate the discussion of chemical coherence given above is basically the same as the model generally used to understand the generation of calcium oscillations in biological cells.[22] These oscillations are known to function in the control of complex biological organs, such as the human brain. Remarkably, it is the frequency, not the amplitude, of the calcium oscillations that is decoded as the controlling signal.[23] The signal is generated by the unit-determining closure of the regulatory network that defines a dissipative structure composed of processes in the brain, and the information that is transmitted results from subtle alteration in the parameters that control that oscillation. There seems to be no way to deal with effects of this sort using standard mereology. This difficulty is of central significance in questions of philosophy of mind and of human action,[24] but its outline is clear even in the simpler chemical cases discussed in this chapter.[25]

If we understand the word *mereology* in its general sense, as a theory of parts and wholes, (rather than as the specific system of Leśniewski), it seems clear that a new, more discriminating, mereology is needed. An adequate logic of wholes and parts must be capable of dealing with coherences of the types considered here, in which closure of a network of relationships gives rise to significant effects (both external and internal) that would not exist, absent that closure.

Notes

1. An earlier version of this paper was presented at the Second Summer Symposium on the Philosophy of Chemistry of the International Society for the Philosophy of Chemistry, Sidney Sussex College, Cambridge University, England, August 3–7, 1998 and (in absentia) at the Silver Anniversary Whitehead Conference, The Center for Process Studies, Claremont, CA, August 4–9, 1998.

2. Peter Simons, *Parts, A Study in Ontology*. Oxford: Clarendon Press, 1987, p. 5.

3. D. M. Armstrong, *A World of States of Affairs*. Cambridge: Cambridge University Press, 1997, p. 12.

4. David K. Lewis, *Parts of Classes*. Oxford: Blackwells, 1990.

5. Armstrong, *A World of States*, p. 263.

6. Ibid.

7. Ibid., pp. 189–191.

8. For an alternative interpretation, see J. E. Earley Sr., "Towards a Reapprehension of Causal Efficacy," *Process Studies* 24: 34–38 (1995); and J. E. Earley Sr., "Naturalism, Theism, and the Origin of Life," *Process Studies* 27: 267–279 (1998).

9. Dilip Kondepudi and Ilya Prigogine, *Modern Thermodynamics: From Heat Engines to Dissipative Structures*. New York: Wiley, 1998.

10. Gilbert Lewis and Merle Randall, *Thermodynamics*. New York: McGraw-Hill (2nd ed.), 1961.

11. Armstrong, *A World of States.*

12. In such a case, the reaction products would be kinetically determined, not thermodynamically determined, and the catalysts, not the affinities, would account for which product(s) were, in fact, produced.

13. It seems worth noting that Peter Van Inwagen asserts that distances do not exist! Seminar at Georgetown University, April 17, 1998.

14. Stephen Scott, *Oscillations, Waves, and Chaos in Chemical Kinetics*. Oxford: Oxford University Press, 1992.

15. Gregoire Nicholis and Ilya Prigogine, *Self-Organization in Far-from Equilibrium Systems*. New York: Wiley, 1977.

16. Ivor Leclerc, *The Nature of Physical Existence*. New York: Humanities Press, 1972; and *The Philosophy of Nature*. Washington: Catholic University of America Press, 1986.

17. Scott, *Oscillations, Waves, and Chaos.*

18. M. Eiswirth, A. Freund, and J. Ross, *Adv. Chem. Phys.*, 190: 127 (1991); *J. Phys. Chem.* 95: 1294 (1991).

19. In certain systems (that are more complex than the simple examples discussed here), this requirement is somewhat relaxed, and the oscillation becomes irregular, even chaotic. In these cases, the closed trajectory in two-dimensional space lies on an object of fractal dimension (a strange attractor) rather than on a single curve.

20. Or to a strange attractor (see preceding note).

21. Joseph E. Earley Sr., "Varieties of *Chemical* Closure" in *Closure; Emergent Organizations and Their Dynamics*, J. L. R. Chandler and G. Van de Vijver, editors, *Annals of the New York Academy of Sciences*, 901: 122–131 (2000); Joseph E. Earley Sr., "How Dynamic Aggregates may Achieve Effective Integration," Advances in Complex Systems, 6: 115–126 (2003).

22. Albert Goldbeter, *Biochemical Oscillations and Cellular Rhythms: the Molecular Bases of Periodic and Chaotic Behavior.* Cambridge: Cambridge University Press, 1996.

23. James W. Putney Jr., "Calcium Signaling: Up, Down, Up, Down . . . What's the Point?" *Science*, 279: 191–192 (9 January 1998); Paul De Koninck and Howard Schulman, "Sensitivity of CaM Kinase II to the Frequency of Ca^{2+} Oscillations," *Science* 279: 227–230 (9 January 1998).

24. Alicia Juarrero, *Dynamics in Action; Intentional Behavior as a Complex System.* Cambridge, MA: MIT Press, 1999.

25. Additional examples relevant to this point are discussed in: *Chemical Explanation: Characteristics, Development, Autonomy*, Joseph E. Earley Sr., editor. *Annals of the New York Academy of Science*, 988 (2003).

7

Whitehead's Philosophy and the Collapse of Quantum States

SHIMON MALIN

Introduction

The January 1963 issue of *Reviews of Modern Physics* contains a paper by J. M. Burgers titled, "The Measuring Process in Quantum Theory."[1] The first four sections of the paper are, indeed, devoted to the quantum measurement problem. However, in the fifth and last section, "Philosophical Excursion," the author points out that Alfred North Whitehead's philosophical system is uniquely suitable for providing quantum physics with metaphysical underpinnings. Burgers' paper turned out to be the beginning of a subdiscipline devoted to the study of Whitehead and contemporary physics.

On March 3, 1964, A. Shimony gave a lecture titled "Quantum Physics and the Philosophy of Whitehead" at the Boston Colloquium for the Philosophy of Science. The lecture contained a detailed, point-by-point comparison between process philosophy and quantum physics, as well as suggested changes in Whitehead's system, designed to accommodate the findings of quantum theory. The text of the lecture, together with Burgers' comments on it, were published in Volume 11 of the *Boston Studies in the Philosophy of Science*.[2]

By the mid-1980s an increasing interest in the subject of Whitehead and contemporary physics led to the participation of leading physicists and philosophers in a conference, the proceedings of which appeared as a book titled *Physics and the Ultimate Significance of Time*.[3] And the lively *Whitehead and Physics* workshops at the Silver Anniversary International Whitehead Conference in Claremont, California, in August 1998, as well as the publication of the present volume, clearly demonstrate that interest in issues raised by comparisons of process philosophy and the discoveries of physics in general, and quantum physics in particular, is still growing.

In the present essay we compare Whitehead's actual entities with "the collapse of quantum states" in light of what Erwin Schrödinger calls "the principle of objectivation." The essay is structured as follows: the next section is a presentation of relevant quantum mechanical concepts and ideas, including "the collapse of quantum states." In the third section we summarize some of the important characteristics of Whitehead's actual entities. Schrödinger's principle of objectivation is the subject of the fourth section. The fifth section is an interlude devoted to the question of whether the indeterminism in quantum mechanics makes room for creativity. In the sixth section we discuss the correspondence between processes of collapse of quantum states and Whitehead's actual entities. The seventh section is devoted to an analysis of the nature and inception of processes of collapse in light of this correspondence. Our results are summarized in the final section.

Quantum States and Their Collapse

How do quantum systems change with time? Quantum systems undergo two kinds of change; one takes place in situations in which the quantum system is isolated from the environment, and the other results when it interacts with its environment. When the system is isolated, the quantum state (the mathematical entity that describes the quantum system) changes in a completely deterministic way, following a mathematical equation, such as Schrödinger's equation. However, when the system interacts with its environment, e.g., when it is being measured, it undergoes a sudden, discontinuous, and unpredictable change, called "the collapse of the quantum state."

The difference between the two kinds of change can be analyzed in terms of superpositions of quantum states. When isolated, a quantum system is subject to "the principle of superposition": If a state A and a state B are possible states for the system, then a superposition of the two states is also a possible state. Consider, for example, a quantum system that consists of a single electron, which may be in located in some region A, and may be located in some other region B; the two quantum states that describe an electron in regions A and B respectively are both possible. In this case the sum, or superposition, of these two quantum states is also a possible quantum state.

Such a sum indicates that if the location of the electron would be measured, there is some probability of finding it in region A, and some probability of finding it in region B. Such a superposition cannot survive a process of measurement (interaction with the environment). If the electron is observed, i.e., its position is measured, the electron will be found *either* in region A *or* in region B, never in both.

A delightful explanation of the principle of superposition is contained in Robert Gilmore's book, *Alice in Quantumland*:

> Alice entered the wood and made her way along a path which wound among the trees, until she came to a place where it forked. There was a signpost at the junction, but it did not appear very helpful. The arm pointing to the right bore the letter "A," that to the left the letter "B," nothing more. "Well, I declare," exclaimed Alice in exasperation. "That is the most unhelpful signpost I have ever seen." She looked around to see if there were any clues as to where the paths might lead, when she was a little startled to see that Schrödinger's Cat was sitting on a bough of a tree a few yards off.
>
> "Oh, Cat," she began rather timidly. "Would you tell me please which way I ought to go from here?"
>
> "That depends a great deal on where you want to get to," said the Cat.
>
> "I am not really sure where . . ." began Alice.
>
> "Then it doesn't matter which way you go," interrupted the Cat.
>
> "But I have to decide between these two paths," said Alice.
>
> "Now that is where you are wrong," mused the Cat. "You do not have to decide, you can take all the paths. Surely you have learned that by now. Speaking for myself, I often do about nine different things at the same time. Cats can prowl around all over the place when they are not observed. Talking of observations," he said hurriedly, "I think I am about to be obs . . ." At that point the Cat vanished abruptly.[4]

A process of measurement leads to a discontinuous and unpredictable choice of one among the terms that make up the superposition. This is the collapse of the quantum state. Pursuing our example of a quantum system that consists of a single electron, let us suppose that the electron is about to impinge on a TV screen. A TV screen can be considered a device for measuring the position of the electron. Before the electron hits the screen, the location of the impingement is not yet determined. Any location may be hit. The electron is in a state of superposition of all the points on the screen. As it interacts with the screen, however, it ends up hitting one particular location. Its quantum state has "collapsed" to this particular spot.

So far we have mentioned a few characteristics of a simple quantum system, such as an electron. Can one come up with a concept of an electron that accommodates these characteristics? The answer to this question is controversial. In the context of the present essay, however, we will adopt the mainstream interpretation of quantum mechanics suggested by Werner Heisenberg. According to Heisenberg, the interplay between the two modes of quantum systems, being isolated and being measured, is an interplay between potentialities and actualities. When an electron is isolated, it does not exist in the actual world; it exists merely as a field of potentialities (e.g., a potentiality for being in region A as well a potentiality for being in region B). When it is being measured, however, the interaction with the measuring apparatus brings it into momentary actual existence. It actually exists at a particular spot on the TV screen. Its momentary existence there is called "an elementary quantum event."

Heisenberg's interpretation implies that the collapse of a quantum state is not a process in time. Space and time represent relationships among *actual* things. The collapse is a process of emergence from potentiality to actuality. Therefore the collapse is an atemporal process that leads to the emergence of an elementary quantum event in space-time.

Whitehead's Actual Entities

We will come back to quantum systems and quantum states in later sections. The present section is devoted to the second ingredient of our analysis: Whitehead's actual entities. These actual entities are, according to Whitehead, "the atoms of reality":

> 'Actual entities'—also termed 'actual occasions'—are the final real things of which the world is made up. There is no going behind actual entities to find something more real. They differ among themselves: God is an actual entity, and so is the most trivial puff of existence in far-off empty space. But, though there are gradations of importance, and diversities of function, yet in the principles which actuality exemplifies all are on the same level. The final facts are, all alike, actual entities; and these actual entities are drops of experience, complex and interdependent.[5]

Listed below are some of the most important characteristics of actual entities.

1. They are neither purely subjects nor purely objects; they have both subjective and objective characteristics, as we will see.
2. They are short-lived, flashing in and out of existence in space-time. The *apparent* existence of enduring objects is due to many collections of actual entities, coming one after the other in quick succession, like frames in a movie.
3. Each actual entity is a nexus of relationships with all the other actual entities.
4. An actual entity is the process of its own self-creation.
5. This self-creation involves accommodating and integrating within itself (comprehending, or *prehending* in Whitehead's terminology) all the previous actual entities as "settled facts" that cannot be changed, and all the future actual entities as potentialities. This process of self-creation involves a sequence of phases, which are delineated and analyzed in detail in Whitehead's magnum opus, *Process and Reality*.[5] An example: Listening to an orchestra playing a symphony involves, at each moment, accommodating the sounds produced by the orchestra. This accommodation depends, in turn, on many collections of past actual entities, such as pre-

vious knowledge and training in music, associations with the symphony, etc. Notice that at each instant there is one experience.

6. The end product of the process is *one* new actual entity, one "throb of experience." The fundamental building blocks of the universe are, then, elementary experiences. We live not in "a universe of objects," but in "a universe of experience."

7. Subjectively, i.e., for itself, an actual entity is a pulse of experience. In particular, the end of the process of self-creation is called "the satisfaction of the actual entity." Although its subjective existence is momentary, objectively, i.e., for other, future actual entities, it is a settled fact: The fact that it did happen cannot be erased. "The end of . . . [its] private life—its 'perishing'—is the beginning of its public career."[6]

8. The point that is crucial for our purposes is this: The process of self-creation of an actual entity is not a process in time; it is, rather, an atemporal process leading to the momentary appearance of the completed actual entity in space-time. Quoting Whitehead: "In the process of self-creation which is an actual entity the genetic passage from phase to phase is not in physical time . . . the genetic process is not the temporal succession. . . . Each phase in the genetic process presupposes the entire quantum."[7]

Schrödinger's Principle of Objectivation

Schrödinger's principle of objectivation is the third ingredient of our analysis. In the latter part of his life Erwin Schrödinger, one of the founding fathers of quantum mechanics, published a number of books on a variety of subjects: the general theory of relativity, cosmology, thermodynamics, philosophy, the relationship between science and humanism, and even biology. His book *What Is Life?* was influential among both physicists and biologists, and played an important part in the process that led to the discovery of DNA. One of the lesser known among these late books appeared in 1958 under the title *Mind and Matter.* It contains the text of the Tarner lectures, which were delivered for Schrödinger at Trinity College in Cambridge, England, in October 1956 (Schrödinger could not go to England to deliver his lectures because of ill health).[8]

The third lecture is titled "The Principle of Objectivation." It is devoted to a presentation and discussion of an unconscious assumption that underlies our perceptions, as well as our science:

> By this [i.e., by the principle of objectivation] I mean what is also frequently called the "hypothesis of the real world" around us. I maintain that it amounts to a certain simplification which we adopt in order to master the infinitely intricate problem of nature. *Without being aware of it and without being rigorously system-*

atic about it, we exclude the Subject of Cognizance from the domain of nature that we endeavor to understand. We step with our own person back into the part of an onlooker who does not belong to the world, which by this very procedure becomes an objective world. [italics added][9]

Schrödinger's point is similar to what Whitehead called, in his book *Science in the Modern World*, "the fallacy of misplaced concreteness."[10] I look at a tree. The statement "there is a tree out there" is an *abstraction.* The *concrete* fact is "*I see* a tree out there." Indeed, the first statement, "there is a tree out there," may not even be true. When I say it, I may be hallucinating. In contrast, the second statement, "I see a tree out there," is valid even if I am dreaming or hallucinating. Schrödinger and Whitehead's point is that we are so used to mistaking the abstract for the concrete that the fundamental fact of experiencing is excluded from scientific analysis. With this exclusion the universe becomes, for science, a lifeless object. *Thus the standard view of the universe as inanimate is a characteristic of the scientific method, not of the universe.*

Does Quantum Indeterminism Make Room for Creativity?

The present section is devoted to the last hurdle we have to surmount in order to be in position to compare the findings of quantum mechanics with process philosophy.

As Whitehead points out, "the Newtonian worldview allows no room for free will or creativity. The Newtonian framework is completely deterministic."[11] The state of the universe at present and in the future is predetermined, down to the smallest detail, by its state in the distant past. By contrast, quantum mechanics is not completely deterministic. Processes of collapse involve unpredictability. Pascal Jordan, who collaborated with Max Born and Werner Heisenberg in the discovery of quantum mechanics, was the first to point out that this element of unpredictability makes room for free will. Erwin Schrödinger, however, relying on the work of the philosopher Ernst Cassirer, rejects Jordan's claim. He writes:

> According to our present view the quantum laws, though they leave the single event undetermined, predict a quite definite *statistics* of events when the same situation occurs again and again. If these statistics are interfered with by any agent, this agent violates the laws of quantum mechanics just as objectionably as if it interfered—in pre-quantum physics—with the strictly causal mechanical law.[12] (italics original)

If Schrödinger is right, then Whitehead's system and quantum mechanics are incompatible. Creativity, the most general characteristic of the universe according to Whitehead, has no place in the quantum mechanical scheme of things. But is Schrödinger right? I believe that Schrödinger is wrong. In principle there is no contradiction between the intentionality of agents and the statistical predictability of the outcome of many events. Thus drivers drive with the intention of avoiding accidents; and yet insurance companies' business is based on the correct predictions of the statistics of accidents.

Let us analyze this issue in a simple context. Consider a large number of people, each of whom is tossing a coin once. Consider the following two scenarios: (1) The people have no control over the results of the tossings; hence these results are random. (2) Each participant has complete control over the result; he or she can determine the result to be heads or tails. And it so happened that roughly half the participants intend the result to be heads, and the other half tails. In both of these scenarios, the results of the tossings would appear to be random.

In practice, these considerations are hardly relevant to the behavior of quantum systems. In all likelihood, the creative intentionality of electrons is very minimal indeed, and the result of their collapses is truly random (subject to an appropriate statistical distribution). In principle, however, it is important to point out that there is no contradiction between quantum physics and Whitehead's philosophy. Pascal Jordan was right. The quantum indeterminacy does make room for creativity and free will.

Correspondence Between the Collapse of Quantum States and Whitehead's Actual Entities

According to Heisenberg's interpretation of quantum mechanics (see Quantum States and Their Collapse), the quantum mechanical universe is an interplay of potentialities and actualities. Quantum states represent potentialities, some of which are actualized in acts of measurement. A measurement brings about, through the process of the collapse of a quantum state, what John Wheeler called "an elementary quantum event." Such an event is "a flash of actual existence"; it appears out of a background of potentialities, and disappears back into it almost immediately.

This concept of collapse is remarkably similar to Whitehead's concept of actual entities. According to both quantum mechanics and Whitehead's paradigm, reality is not made of enduring objects, but of flashes of existence that disappear almost as soon as they appear. Furthermore, in both accounts, the processes that lead to these appearances are atemporal! There is, however, an important difference between actual entities and processes of collapse. While actual entities are throbs of experience, processes of collapse have no subjective aspect—they are completely objective. This is due to the fact that quantum mechanics, like all of our science, is based on what Schrödinger called "the principle of objectivation" (see above). If processes of collapse have, in fact, a subjective side, there is no way for this subjective side to appear in their quantum mechanical description. We can summarize the situation by saying that if we consider processes of collapse as representing the objectivized aspect of actual entities, the agreement between Whitehead and quantum mechanics is perfect.

The Collapse of Quantum States: A Whiteheadian Analysis

The conclusion we have reached in the previous section has consequences regarding the process of collapse. One remarkable consequence is this: For the past 70 years, physicists have been searching for the mechanism of the process of collapse, and no such mechanism was found. From a Whiteheadian perspective this is hardly surprising. If the collapse is an aspect (the objectivized aspect) of the self-creation process of an actual entity, it cannot be the result of a mechanism, since a mechanism precludes creativity. There could possibly be some mechanism(s) involved in a partial determination of the outcome of a collapse. A complete determination of such an outcome is clearly incompatible with process philosophy. This is an important conclusion, but a Whiteheadian analysis of the process of collapse can go much further. Consider the following question: What brings about the beginning of such a process? Physics has provided no answer. Let us consider, however, the beginning of the atemporal process of self-creation that an actual entity is. According to Whitehead, this beginning, the arising of a subjective aim, is rooted in God:

> God is the principle of concretion; namely, he is that actual entity from which every temporal concrescence receives that initial aim from which its self-causation starts. That aim determines the initial gradations of relevance of eternal objects for conceptual feelings; and constitutes the autonomous subject in its primary phase of feeling with its initial conceptual valuations, and with its initial physical purposes.[13]

What can be the reason for the arising of such an initial aim in the case of actual entities that correspond to processes of collapse? To arrive at an answer to this pivotal question, let us contemplate what the universe would have been like had there been no processes of collapse. To describe such a universe, a universe without collapse, we cannot help stretching the language. This imaginary universe is far from the actual one; our language was not designed for its description. Please bear this in mind as we proceed.

To begin, a universe without collapse is a universe of potentialities rather than actual events; the transition from the potential to the actual never takes place. Suppose, however, that potentialities could become actual without disturbing the superpositions we discussed earlier; in other words, suppose that in every measurement the whole superposition, rather than one of the potentialities that make it up, becomes actual. What then?

Consider the example of an electron impinging on a TV screen. The electron has hit the screen, but no atom has been singled out. In a sense, then, the electron has hit all the atoms. In our universe, a universe with collapse, one atom is singled out. In a universe without collapse, however, the superposition remains a superposition. Every atom on the screen is being hit "to a certain extent."

What next? In our universe the hitting of one atom starts a series of processes that end up with a visible flash at the location of this atom. In a world without collapse there will be trillions and trillions of flashes; flashes everywhere, each flash existing, however, only to a certain extent, because each flash is still just an element in a superposition of flashes. Each of these flashes is, to a certain extent, sending up myriad photons that affect the physical world around the screen in immensely complicated ways. The eye of the viewer, for example, would receive to a certain extent myriad photons not from one location on the screen, but from all locations. The same applies to the atoms in the walls and in the furniture—they too will be affected by immensely complex superpositions of photons coming from everywhere on the screen.

In a universe without collapse, then, every interaction leads to a tremendous increase in complexity, and there is no process that works in the reverse direction, i.e., toward simplification. Let us come back now to our own universe. I suggest that the answer to the question, "Why does God initiate processes of collapse?" is this: *The function of the collapse is to simplify.* If there was no collapse, each and every interaction would have made the universe immensely more complicated, and this trend toward increasing complexity would have continued unchecked. In the real world there is a process that reverses this trend—the collapse! Because of the collapse, the level of complexity increases and decreases as quantum systems switch back and forth between the potential and the actual. An electron that is moving toward a screen is in a state of superposition of possibilities. The impingement on one particular atom on the screen means that the superposition has collapsed, the electron has become actual, and its quantum state has become simpler—the electron is interacting with a single atom. As a result of this interaction, and the processes that follow, photons are emitted. Each photon is a new field of potentialities—a superposition of possibilities as to the choice of the atom it (the photon) will hit. The situation is complex again. Now there are new collapses—each photon interacts with one particular atom. Things are simpler again. And so on.

Collapses occur because they are essential ingredients in the interplay of opening up and closing down. As the actual becomes potential, there is an opening up into a superposition of many coexisting possibilities. As the potential becomes actual, there is a closing down: Most potentialities are denied actualization as one of them is singled out. The chosen one becomes an actual event, only to change immediately into a superposition of possibilities for the next collapse. And so the dance goes on.

Conclusion

Born, Heisenberg, and Jordan's discovery of quantum mechanics, as well as Schrödinger's discovery of wave mechanics, was contemporaneous with Whitehead's discovery of process philosophy. During the period in which he developed his system, Whitehead was unaware of the work of the quantum physicists. This makes the correspondence between process philosophy and the findings of quantum mechanics all the more remarkable.

This correspondence is not complete. And it cannot be complete because of the principle of objectivation, which precludes subjective aspects from scientific inquiry. The correspondence between actual entities and processes of collapse is, however, as close as this limitation of the scientific method, i.e., the principle of objectivation, allows it to be.

Identification of processes of collapse as the objectivized aspects of actual entities has far-reaching consequences. It precludes the existence of a mechanism that completely determines the outcome of a collapse, and it suggests a raison d'être for processes of collapse: their existence brings about a balance between complexity and simplicity in the unfolding of the universe.

Notes

1. J. M. Burgers, "The Measuring Process in Quantum Theory," *Reviews of Modern Physics*, 35 (1963): 145.

2. A. Shimony, "Quantum Physics and the Philosophy of Whitehead," *Boston Studies in the Philosophy of Science*, Vol. II (1968).

3. D. R. Griffin, ed., *Physics and the Ultimate Significance of Time* (Albany: State University of New York Press, 1986).

4. Robert Gilmore, *Alice in Quantumland* (New York: Copernicus, 1995), 49.

5. A. N. Whitehead, *Process and Reality* (New York: The Free Press, 1978), 18.

6. V. Lowe, "Introduction to Alfred North Whitehead," in *Classic American Philosophers*, ed. M. H. Fisch (New York: Appleton-Century-Crofts, 1951), 404.

7. Whitehead, *Process and Reality*, 283.

8. E. Schrödinger, *What Is Life? with Mind and Matter and Autobiographical Sketches* (Cambridge: Cambridge University Press, 1992).

9. Ibid., 118.

10. A. N. Whitehead, *Science and the Modern World* (New York: The Free Press, 1967), 51.

11. Ibid., 77.

12. E. Schrödinger, *Nature and the Greeks and Science and Humanism* (Cambridge: Cambridge University Press, 1996), 164–165.

13. Whitehead, *Process and Reality*, 244.

8

A Historical Reality That Includes Big Bang, Free Will, and Elementary Particles

GEOFFREY F. CHEW

Introduction

This chapter forecasts development of a "natural science" based not on material reality but rather on a Whiteheadian 'historical reality.'[1] Such a science would come to grips with the hitherto-inaccessible phenomenon of free will. The term *free will*, as used here, means any influence on universe history not describable as "materialistic"—i.e., not describable in terms of "matter" controlled by local causality. In the foreseen scientifically meaningful historical reality, there is parallel with what David Bohm has called "implicate order."[2] The envisaged broader reality would combine local Whiteheadian process with quantum principles and global evolution of the universe (cosmology). There would be explicit differentiation between the local time of individual occasions and a global time providing meaning for the 'present'—a concept not recognized by classical physics but essential to quantum mechanics.

Material reality, currently epitomized by representation of the universe through the elementary particles of grand unification theory (GUT), will be recognized within global historical reality as a localized causal component of universe history—the only portion of reality that science has so far been able to define. I expect quantum cosmology to elaborate Whitehead's process so as to allow material reality (e.g., causally interrelated elementary particles) to be recognized not as a priori but as an aspect of a universe history that includes the impact of free will. (Importance of global evolution has led to my preferring the word *history* to Whitehead's term, *process*. The term *process* carries a local connotation.)

Whitehead is identified at this conference primarily as a philosopher, not as a scientist. Why do I expect his approach eventually to reshape natural science? One consideration is that Whitehead's 'process' invoked a concept whose precise definition requires mathematical language of a type developed by physicists, not by philosophers. A more important substantial consideration is that, in the face of growing human appreciation of evolution, physics cannot indefinitely sustain a stance that disregards the history of the universe. 'Process' puts history up front while managing to recognize material reality as an important feature of "mature" regions of the universe, even though not the only feature and a feature not present everywhere.

Whitehead's 'process' involves what we shall call here "pre-events"[3] individually localized (in a sense that requires careful attention) in Hilbert space. (The term *Hilbert space* characterizes the mathematical language with which physicists state quantum principles.) At a Whiteheadian pre-event something "happens," although not to matter. The notion of matter is subordinated to the notion of 'happening.' 'Matter' corresponds to certain very special patterns of pre-event patterns that may or may not develop. The vast majority of pre-events do not belong to matter patterns. The foregoing may be restated in familiar physical language by saying that Whitehead subordinates energy to 'impulse.'[4] Localized energy is not a priori; the 'occasions' that build our universe are localized impulses. When a pattern of localized impulses exhibits a certain persistently repetitive regularity, the repetition provides meaning for energy. But all impulse patterns need not be repetitive: Historical reality has both material and nonmaterial content, the latter predominating.

Because free will, whatever the precise meaning of this term, has impact on history, historical reality includes free will. The "commonsense" belief that free will locates outside material reality might be an illusion; perhaps free will is no more than a manifestation of causally behaving elementary particles; perhaps historical reality adds nothing to materialism. I do not believe such to be the case.

One influential consideration stems from the cosmological idea of the big bang. Studies of the big bang notion by many different workers using many different approaches concur in concluding that in the extremely early universe, ordinary physical concepts fail. Historical reality, if it is to embrace early as well as late universe, must include nonmaterial components.

A second set of considerations stems from my personal efforts during the past two years to represent historical reality through a chain of pre-events after big bang and before the present, each pre-event locating an impulse. A promising pattern has been found for representation of elementary particles, but all pre-events in a Whiteheadian chain cannot congregate in such patterns. I shall expand below on this assertion.

I should dearly like to report discovery of a pre-event pattern associable with locally exercised free will. This I cannot do. But there is far more than material content in historical reality. Let me now consider more precisely the meaning I attach to 'material reality,' making a link to the traditional meaning of natural science.

Material Reality

The term *material reality* or *materialism* as used in this chapter refers to those aspects of our universe characterizable as material in the general sense introduced by Einstein. In Einstein's sense, matter is synonymous with localized energy, so light and gravity as well as atoms are forms of matter. Both fields and strings are material in the sense of this chapter.

Ever since the astounding successes of Isaac Newton 300 years ago, physics (originally called "natural philosophy") has been based on energy localized in space and time. Twentieth-century science was forced by experimental observations of matter at small scales to introduce a set of principles called "quantum mechanics"; principles that entail noncausal uncertainty aspects that continue today, after decades of study, to baffle the wisest humans. (The noncausal implications prevented Einstein from ever accepting the completeness of quantum mechanics.) Often these "weird" aspects are called "nonobjective," and I agree that they should be so described. But quantum mechanics heretofore has failed to dislodge localized energy—i.e., matter—as the underpinning of natural philosophy. What this chapter calls "material reality" includes not only gravity, light, electrons, and any other forms of matter (such as strings) but the Hilbert-space representation thereof, as described in the most advanced physics graduate courses. Maintaining Newton's materialistic worldview, quantum physicists have heretofore supposed that all Hilbert-space labels refer to 'matter'. (To what else might these labels conceivably refer? The reader will by now not be surprised to hear that Hilbert space labels may refer to history.)

The achievements of a science based on matter have been so impressive as to raise hopes for an eventual "materialistic theory of everything." How is it conceivable that such phenomena as consciousness and what appears to be free will might be manifestations of localized energy? A persuasive consideration is that decisions affect the behavior of matter: I decide to raise my arm and matter responds. Given such linkage between matter and what seems to be free will, is it not reasonable to seek understanding of all reality through energy? Success for a materialistic theory of everything would mean absence from universe history of nonmaterial content. Free will would be an illusion.

I have by now repeatedly indicated my failure to envisage complete success for materialism. I believe there to be a major nonmaterial component of historical reality.

A Model of Historical Reality

My expectations for the future of science have been influenced by a cosmological model I have been pursuing over the past decade in a search for the quantum mechanical meaning of space and time. Roughly two years ago I began to connect the model with Whitehead's idea of a chain of history—a global uni-

Table 8.1. Sample Time Scales Within Historical Reality

Our present age (here and now, since big bang)	: 10^{17} sec
Human lifetime	: 10^{9} sec
Scale of human consciousness	: 1 sec
Period of atom	: 10^{-15} sec
Minimum duration of elementary particle	: 10^{-25} sec
Age step between pre-events along history chain	: 10^{-43} sec

verse history comprising a huge collection of discrete ordered pre-events. I found in the model explicit realization of a key Whiteheadian supposition: that objects (i.e., matter) correspond to regular localized repeating patterns of large numbers of occasions. The model locates an impulse at each pre-event; spacing between successive pre-events is on the scale of Planck time—roughly 10^{-43} seconds. The identity of an object, such as an electron or the reader of this book, resides in the detailed structure of its repeating pre-event pattern.

To give a sense of the tremendous range in time scales incorporated into our very inclusive history, Table 8.1 provides a sample.

Energy derives from regular and enduring collections of impulses. But all impulses do not lead to energy. I found in the model that most pre-events fail to build rigid patterns exhibiting the causal regularity characteristic of matter, i.e., characteristic of energy. This model of historical reality, in other words, has explicit nonmaterial content. In fact, because a single chain of Whiteheadian pre-events constitutes the history of the universe, with every pre-event in principle influencing and being influenced by every other pre-event, it is at first sight difficult to imagine any meaning for materialism in historical reality. How might there be individually identifiable particles obeying local Einsteinian causality within a global Whiteheadian chain of history?

The mechanism provided by the model depends on screened magnetism in combination with a meandering of the history chain back and forth in local time. Within space-time regions that are large on a Planck scale while tiny on a human scale, a tightly woven fabric of causally related pre-events is possible. Stability of this 'material fabric' is magnetodynamic. Different fabric patterns correspond to different forms of matter. Elementary particles in the GUT sense correspond to tower pre-event patterns,[5] with the width of a tower being roughly a hundred times larger than the Planck length ($\sim 10^{-31}$ meters). This factor relates to the ratio of elementary magnetic and electric charges. A particle scale roughly 10^{18} times larger than Planck scale controls "vertical" tower structure.[6]

The model classifies local patterns of pre-events into two different categories, one suited to material (tower) fabric and the other not. The latter admits irregular patterns associable with a "vacuum" that provides geometry for space-time at suitably large scales. Particle rest mass is conjectured to stem from coupling between Whiteheadian irregular vacuum patterns and regular matter (tower) patterns. Vacuum patterns may constrain material history by damping quantum fluctuations at scales large compared to particle scale—a damping described in Copenhagen quantum mechanics as "wave-function collapse."

Robustness of matter pre-event fabric depends in the model on strongly-confining magnetic impulses.

Interpretation needs to be found for "loose vacuonic strands" of history that lie outside material fabric even though comprising pre-events that, individually, are similar to those building material reality. The vast majority of history consists of such loose strands, which influence material history, although not in the manner, according to the laws of materialistic physics, by which one piece of matter influences another.

Freedom from the causal constraints of tightly woven material fabric suggests for loose event strands a historical role that is unstable and unpredictable by current science. This role cannot be described by a science that recognizes only the material component of history. Might loose strands underlie what I have called "locally exercised free will"? I have been unable to find a persuasive argument to the contrary.

How can scientific study of loose event strands be even imagined? How can questions, to say nothing of answers, be formulated? I believe the response to this query to lie in mathematics and requirements of consistency. In my model, appreciation of the very special pre-event fabric interpretable as matter comes through mathematics. Ordinary language, with all its nouns, is so prejudiced by material reality as to be unsuited to the asking and answering of questions about historical reality, but mathematics provides a language.

One immediately identifiable limited goal, closely connected to material reality, is a mathematical meaning for observation within historical reality. Any observation is a part of history, and human ingenuity may be able to identify the essential mathematical characteristics of a Whiteheadian pre-event pattern that qualifies as an observation on one piece of matter carried out by another.[7] An associated pre-event pattern, less of a challenge to define, would be 'record of observation' or 'memory.' Artificial intelligence theory, already under development, promises here to be of assistance. Historical meaning of observation will interlock with but probably not be contained within material fabric. Record of observation, on the other hand, promises to reside in this fabric. Memory, after all, is a component of an object's identity.

Any individual measurement is a portion of history, and its pre-event pattern links the measurement to surrounding history. Why did the observation occur? Such a question parallels the question, Why did elementary particles emerge from an early universe where particles had no meaning? Contemporary physics-cosmology is attacking the latter question with ideas such as inflation.

Why should science not contemplate the reason behind observation? You may say, "Because the reason involves free will and science is helpless to deal with free will." But according to my model, free will controls the history in which matter emerges from nonmatter.

The acausal aspect of loose pre-events does not mean such pre-events are above the law—following no rules. They are part of a single Whiteheadian chain of history. Furthermore, the meandering of loose pre-event strands of history, even though acausal and locally unpredictable, is subject to global constraints: Free will is cosmologically constrained.

A global character need not put free will investigation beyond human reach. Cosmology has proved to be a fruitful area for human enterprise, even though some of my most respected physicist colleagues assert inability to "think about the entire universe."

Recognition of observation as simply part of Whiteheadian history would remove wave-function collapse as a separate puzzle. I see one mystery, not two. Connection between free will and quantum mechanical collapse has often been conjectured. What about free will's cosmological aspect: its connection to big bang? Like all models, the one I have been studying allows arbitrariness in global initial conditions. Once it is acknowledged that mysterious global conditions influence loose pre-event strands throughout history, one appreciates that none of the speculations in this chapter diminish the role of God in the universe. But agreeing that God's role may be manifested through free will does not mean to me that effort is futile to improve understanding of Whiteheadian pre-events located outside the fabric of material reality.

Is historical reality independent of observer? When, in a criticism of Copenhagen quantum mechanics, Einstein insisted that "the moon is really there," he expressed desire for a universe whose meaning transcends observers and measurement. Historical reality, by recognizing observers and measurements to be merely aspects of history, accords with Einstein's wish.

A curious quantum mechanical feature of the historical reality model, a feature brought to my attention by Henry Stapp, is that it provides a global meaning for 'the present' through a 'center of the universe.' The homogeneous universe principle has constrained cosmology both before and after recognition of universe expansion. No notion of a preferred spatial location, qualifying as 'center,' has proved appealing. Such an idea would conflict with the relativistic philosophy that dominated twentieth-century physics and cosmology, though not with the fact that the observed universe fails to be homogeneous at finite scales. Spontaneous symmetry breaking, a feature of quantum physics, is believed to account for "clumping of matter" in the universe. In my Whiteheadian model, breaking of symmetry also selects a center of our universe which, although difficult to detect, provides meaning for 'the present'—a meaning that applies to the entire universe and facilitates global quantum mechanics.

Conclusion

David Bohm wrote an intriguing little book distinguishing what he called "implicate order" from "explicate order." Although Bohm ignored evolution of the universe, I like to associate the material classical reality of localized energy with Bohm's explicate order. Explicate order would comprise that very special component of classical historical reality that is bound up in tightly woven material fabric. Implicate order would comprise complete quantum historical reality—superpositions of meandering chains of pre-events, only a tiny portion of which lie within material fabric. The remainder is inaccessible to objective science and embraces free will.

It may be useful to appreciate that in our history, the total number of Whiteheadian pre-events is huge but not infinite, depending (unsurprisingly) on the global age of the universe. Because the total number of pre-events increases with global age, it is plausible that qualitatively new pre-event patterns will develop indefinitely. Remember that the tightly woven fabric of materialism could not exist when the universe was very young. Conversely, when the universe is much older than it is today, pre-event patterns may develop with a significance that goes beyond that of matter and presently appreciated manifestations of free will.

Momentarily, my efforts are focusing not on free will but on details of the material fabric interpretable as light. My long-standing identity as a materialistic physicist compels me to give higher priority to light than to free will. After light, the mystery of rest mass must be attacked. Then comes gravity. But exertions in my physicist identity will not cause me to forget free will. Understanding anything about a pre-event pattern associable with free will would for me eclipse in satisfaction any imaginable physical illumination generated by contemplation of historical reality.

Note Added in Proof

The spirit of the text survives in my present thinking, although, alas, model development over the past three and a half years would cause me now to use very different language. The major development has been to associate Whiteheadian pre-events with Feynman paths, rather than with Hilbert space. I now understand Hilbert space to be essentially materialistic—based on enduring process—whereas Feynman paths embrace the nonenduring process on which my essay focuses. In my model, path space is hugely larger than Hilbert space.

Notes

I am indebted to Philip Clayton for a critical reading of the original version of this paper, leading to sharpened terminology in the present version.

1. A.N. Whitehead, *Process and Reality*. New York: MacMillan, 1929.

2. D. Bohm, *Wholeness and the Implicate Order*. London: Routledge and Kegan Paul,1980.

3. Pre-events are quantum occasions each labeled by a (finite) set of Hilbert-space parameters.

4. The psychological meaning for the word *impulse*—a precursor to exercise of free will—is curiously suggestive of free will's inclusion within historical reality. In physics the word 'impulse' describes a sudden transfer of momentum—a 'blow.'

5. Particle birth or death at tower ends (bottom or top) corresponds to the more usual notion of 'event'—not the Whiteheadian occasion here called "pre-event." The time scale characterizing particle birth and death lies far above the Planck time scale—even though far below human scale. Grand unified theory refers to any of the theories attempting to find a common basis for quarks and leptons and their associated force fields.

6. G.F. Chew, *Coherent-State Dyonic History*, *Pre-Event Magneto-Electrodynamics* and *Elementary-Matter Propagation Via 3-Scale Square-Tower History Lattice*, Berkeley Lab Preprints (1998).

7. The characteristic that identity of observables not be destroyed by the observation promises to depend on smallness of the elementary electric-charge unit. Further essential to meaning for 'observation' is diluteness of material history, a feature absent in the early universe.

8. Bohm, *Wholeness and the Implicate Order*.

9

Whiteheadian Process and Quantum Theory

HENRY P. STAPP

Introduction

Quantum theory has been formulated in several different ways. The original version was Copenhagen quantum theory, which was formulated as a practical set of rules for making predictions about what we human observers would observe under certain well-defined sets of conditions. However, the human observers themselves were excluded from the system, in much the same way that Descartes excluded human beings from the part of the world governed by natural physical laws.

This exclusion of human beings from the world governed by physical laws is an awkward feature of Copenhagen quantum theory that is fixed by "orthodox" quantum theory, which is the form devised by von Neumann and Wigner. This orthodox form treats the entire world as a quantum system, including the brains and bodies of human beings. Some more recent formulations of quantum theory seek to exclude from the theory all reference to the experiences of human observers, but I do not consider them, both because of their technical deficiencies and because they are constitutionally unequipped to deal adequately with the causal efficacy of our conscious thoughts.[1]

The observer plays a central role in both Copenhagen and orthodox quantum theory. In this connection, Bohr, describing the 1927 Solvay conference, noted that:

> an interesting discussion arose about how to speak of the appearance of phenomena for which only statistical predictions can be made. The question was whether, as to the occurrence of such individual events, we should adopt the terminology proposed by Dirac, that we were concerned with a choice on the part of "nature," or as suggested by Heisenberg, we should say we have to do with a choice on the

> part of the "observer" constructing the measuring instruments and reading their recording.[2]

The point here is that two very different kinds of choices enter into the determination of what happens.

1. First some particular question must be posed.
2. Then nature gives an answer to that particular question.

The second kind of choice is described by Dirac as a choice on the part of "nature" as to what the outcome of a given observation will be. For this kind of choice, quantum theory gives a statistical prediction: it specifies, for each possible outcome of the observation, the probability for that outcome to appear. This is the famous statistical element in quantum theory.

But that choice of outcome is out of human hands, and it is not the focus of this study.

The first kind of choice is also essential to the quantum process. It is the choice by the experimenter of which aspect of nature to probe. In the context of an experiment being performed by a scientist on some external physical system, this choice by this experimenter of which experiment to perform is decided by some process going on in the experimenter's mind/brain. In the Copenhagen interpretation, that mind/brain process is placed definitely outside the system being investigated. But if, following von Neumann, we take the view that quantum theory ought to cover all physical systems, including human brains, then the system that is determining which question will be put to nature becomes part of the system being studied.

Posing the Question

The starting point of this study is the fact that contemporary quantum theory is ontologically incomplete. Two fundamental questions remain unanswered. The theory requires that a sequence of specific questions with "yes" or "no" answers be put to nature, whereupon nature promptly delivers an answer. The relative statistical weights of the two possible answers, "yes" or "no," are then specified by quantum theory. But what is not specified by contemporary quantum theory is:

1. what determines which questions are put to nature, and
2. what determines whether the individual answer to a posed question is "yes" or "no"?

The objective here is to begin to answer these questions, adhering to the naturalistic principle that the actually occurring experiences "supervene on" the entire history of the physical universe, which is the full history of the evolving quantum state of the universe. However, that condition of supervenience—although it means that given the full physical history, the full experiential history is fixed—does not determine whether our experiences enter as causes or effects; or as both or neither.

To make the following discussion of these issues clear to physicists, I shall use the language and symbols of quantum theory. However, I shall try to explain things in a way that others can understand if they merely regard the symbols I use as pictorial abbreviations of the ideas that I describe.

The (physical) state (of the universe) is represented by the (density operator) S. A possible experience is labeled by the letter e. The connection of this experience to the mathematical formalism is via the correspondence;

$$e \rightarrow P_e$$

where P_e is the projection operator

$$P_e = SUM_e \mid i > < i \mid.$$

Here SUM_e is over the maximal set of basis states i that are compatible with experience e.

The basic dynamical connection is this: If S is the state of a system before experience e occurs, then the state of the system after this experience occurs is:

$$S \rightarrow P_e \, S \, P_e$$

This change is called the "reduction of the wave packet": the reduction is to the unique new form that incorporates the restriction imposed by the new knowledge supplied by the experience e, and that introduces no other information into the state.

There is a basic difference in philosophy at this point between the Copenhagen view espoused by Bohr et al. and the view proposed by von Neumann. Bohr assumed that the state involved in the quantum description is the state of some relatively small system that has been prepared in some specified way by experimenters, and that the projection operator P_e acts in the space associated with that small system. The surrounding world was not represented in the theoretical description except by way of the scientist's specifications, via ordinary language, of the experimental setup. Thus the whole quantum procedure was considered to be merely a procedure that allows scientists to make statistical predictions about what would appear to themselves under those well-defined observationally specified conditions. However, the great bulk of the physical world was not represented in the quantum description, except via our descriptions of our experiences. This radical restriction on the scope of science, and of its description of nature, was rejected by Einstein and many others.

Von Neumann adopted the view that one ought to assume that, because measuring devices and human bodies are made up of atoms, the laws of quantum theory, if universal, ought to work also for these physical systems, and hence in principle for the entire physical universe, and that any partial description was artificial. By following through the mathematics of that more global perspective, von Neumann showed that one could indeed suppose that the laws of quantum theory applied to the whole physical universe (at least in the nonrelativistic approximation and for the somewhat primitive idea of the makeup of the world that prevailed in the 1930s), and that the projection operator P_e could then be supposed to act on those degrees of freedom of the universe that correspond to the brain of the observer. This transformation $S \rightarrow P_e\, S\, P_e$ selects out from S and retains only those states of the brain that are compatible with the knowledge that constitutes experience e.

This procedure allows one to recover, in principle, all the predictions of the technically simpler, but ontologically unsatisfactory, Copenhagen theory. It provides a conception of the universe that is in accord with all the predictions of quantum theory, and is in general accord with the classical idea that there is a causal chain in the physical universe that links the observed event in the external world to the brain of the observer of that event, and that this connection leads—under appropriate conditions of alertness and attention, etc.—to a corresponding mind-brain event of the kind we know.

The only reductions of wave packets that are needed in the von Neumann picture, in order to reproduce the predictions of the pragmatic Copenhagen interpretation, are reductions associated with human experiences: these give the increments in "our knowledge." Of course, it is unacceptably anthropocentric to single out our particular species in a general ontological approach. So I assume that this process in human brains is just a special case of a general natural process. However, I focus here on that special case, because we have direct access to the subjective aspects in that case.

Von Neumann builds into his formulation the demand that a specific question must be posed by invoking his famous Process 1. To understand this, note that for any P, the following identity follows from simple algebra:

$$S = PSP + (1 - P)\, S\, (1 - P) + PS\, (1 - P) + (1 - P)\, SP.$$

"Posing of the question" is represented by the von Neumann reduction (i.e., by the von Neumann Process 1). For some possible experience e,

$$S \rightarrow P_e\, S\, P_e + (1 - P_e)\, S\, (1 - P_e).$$

The two "interference" terms, which involve both P and $(1 - P)$, are dropped.

The first term (after the arrow) is the part of the state S that corresponds to the definite outcome "Yes, experience e occurs now!" The second term corresponds to the definite outcome "No, experience e does not occur now." The other two terms are stripped away by the von Neumann Process 1.

This action on S defines which question is put to nature. Nature will then give the answer "yes" with probability

$$Tr\ P_e\ S\ /\ Tr\ S = SUM_e < i \mid S \mid i > /\ SUM < i \mid S \mid i >.$$

Here Tr denotes the trace operation, the sum SUM_e is defined as before, and SUM is the sum over all members of the basis set.

Quantum theory makes this definite statistical prediction about which outcome will appear, after the definite question is posed. But it does not specify what the question will be, beyond the requirement that answer "yes" must correspond to some identifiable experience. Which question is posed is in the hands of the observer. This freedom places in the hands of the observer great power to control the course of physical events in his or her brain, without in any way conflicting with the constraints imposed by the known laws of nature. The argument for this follows.

Light as Foundation of Being

There are many theoretical reasons for believing that our experiences are correlated mainly to the electromagnetic properties of our brains. Our experiences have a classical character, and the closest connection of quantum mechanics to classical mechanics is probably via the so-called 'coherent states' of the electromagnetic (EM) field.[3] These coherent states integrate a vast amount of information about the motions of individual atomic nuclei and electrons; motions that cannot be expected to affect our thoughts except via their integrated activity.

These coherent states are probably the most robust feature of brain dynamics, with respect to perturbations caused by thermal and other noise.[4] I shall not go into more detail here, except to say that the coherent low-frequency part of an EM field in the brain can be decomposed approximately into mesoscopic modes, each of which behaves like a simple harmonic oscillator. The coherent state description is in terms of this collection of mesoscopic harmonic oscillators. For each such oscillator, the ground state is a certain gaussian state in both of its internal variables p and q: exp $\{-qq/2\}$ or exp $\{-pp/2\}$. This gaussian "cloud of possibilities" is centered at the origin $q = 0$ and $p = 0$ in both q and p. If one shifts this state so that it is centered at some other point (Q, P), then this center point will move around a circle of fixed radius with constant velocity, which is just the motion in these variables that a classical particle would follow for the simple harmonic oscillator case.

I shall assume that the mind-brain connection is via these coherent states of the EM field, and will examine the effects on the brain of mental action by considering the effects of mental action on these low-frequency mesoscopic coherent states of the EM field in the brain.

Effects of Mental Action on Brain Behavior

I first show that, within the framework of quantum theory, the mere choice of which question is asked can influence the behavior of a system, even when an average is made over the possible answers to the question. This demonstration is intended for physicists and is quite short. Other readers can perhaps get the gist.

The issue is this:

Can $X = Tr\ [QPSP + Q(1 - P)S(1 - P)]$ depend on P?
Take $Q = sz$, $S = (1 + sz)$ [With sx, sy, and sz, the Pauli sigma matrices]
If $P = S/2$ then $X = 2$.
If $P = (1 + sy)/2$ then $X = 0$.

This just confirms, as a matter of principle, that it matters which question is posed—and answered—even if one averages over the possible answers. Thus the gross behavior of a system can depend in principle upon which questions the system is asking, internally, where the gross behavior is obtained by averaging over the answers that nature gives to these questions.

I give two examples of how one's behavior could be influenced in this way, simply by controlling, via one's attention, which question is posed.

The first example is an application of the quantum Zeno effect. This effect is well understood, theoretically, and has, at least in a certain sense, been confirmed experimentally.[5] The point is that, according to quantum theory, a very rapid sequence of posings of the same question "freezes" the answer: if the answer to the first question is "yes," then the answer "yes" will, according to the quantum principles, keep on occurring. Thus the mere fact that the question is asked repeatedly in rapid succession keeps the system in the subspace where the answer is "yes," even in the face of strong mechanical forces that would quickly take it out of that subspace if the questions were not being asked.

This effect might be connected to the psychological experience that intense concentration on an idea tends to hold that idea in place. For example, if one is holding up some heavy object, then intense mental focus of attention on an experience *e* of willful effort could produce a very rapid sequence of such experiences *e*, each resulting in a collapse of the wave function associated with the brain to a state compatible with this experience *e*. The effect would be to guide the evolution of the brain state, holding this idea in place, in spite of strong purely physical forces that would to tend move the brain away from this state.

A second example is this. Suppose we are representing the brain, insofar as its interface with consciousness is concerned, by coherent states of the EM field. This state is a Gaussian state represented by $N \exp\{-[(q - Q)(q - Q)/2]\}$, where N is a normalization constant.

Suppose I ask the question: Will I find the state to be $N \exp\{-[(q - Q')(q - Q')/2]\}$? The probability that the answer is "yes" is the square of:

$$N^2 \int dq \exp\{-[(q - Q)(q - Q)/2]\} \exp\{-[(q - Q')(q - Q')/2]\} \\ = \exp\{-[(Q - Q')(Q - Q')/2]\}.$$

For small Q the probability is $(1 - (Q - Q')(Q - Q'))$.

Suppose one has a large distance L in Q space, but breaks the distance into n small intervals, for which the above approximation is adequate, and asks the succession of questions: Is the state the Gaussian centered at the end of each of the succession of intervals?

Then the probability, at the end of this process, of finding the state to be the Gaussian centered at L is $(1 - (L/n)(L/n)n)$. In the limit of large n this is unity: the mental effort of focusing attention in this way will, with high probability, according to the statistical rules of quantum theory, have changed the state of the brain to this other state in spite of the absence of any tendency for this to happen via action of the Schrödinger equation.

These effects may seem strange. But the point is that there is a loose connection in quantum theory: the physical principles themselves do not specify which question is posed. This opens up the logical possibility that, strictly within the bounds of orthodox quantum theory, our conscious thoughts per se could be entering into the mind-brain dynamics in a way reducible neither to purely mechanical effects governed by the Schrödinger equation of motion nor to the random effects of nature's choices of outcomes, nor to any combination of these two effects. There is a rigorous need for some third process, which I call the 'Heisenberg process', and which selects which question is put to nature. This process is not reducible to the 'Schrödinger process' of evolution between jumps via the Schrödinger equation, or to the 'Dirac process' that selects an answer once the question is posed. Thus in orthodox (*vN/W*) quantum theory there must be these three processes entering into mind-brain dynamics: mind-brain dynamics has a "tripartite causal structure," one component of which is naturally mental or experiential, since it must choose a question that has an experientially recognizable answer.

What Determines Which Question is Posed?

What sort of process might one imagine to be filling this logical gap in contemporary quantum theory?

Once one becomes open to the notion that maybe our conscious thoughts have a reality in their own right, it becomes apparent that there is a natural causally efficacious place for them in quantum mind-brain dynamics. The point is that, according to the basic quantum precepts, the occurrence of a conscious thought associated with a quantum system is supposed to cause a reduction of the state of that system to the reduced state that is compatible with the increment in knowledge that constitutes that conscious knowing. In *vN/W* quantum theory, this reduction will be a reduction in the brain state of the person who has the thought. This newly actualized brain state must tend to actualize the functional properties implicit in the conscious thought: it must initiate the brain activities that the thought feels are being initiated. Thus the evolution of this brain state must generate messages going out to various motor centers, if the thought is about generating actions. But, in any case, the Schrödinger evolution must also be generating instructions for the creation of a brain state corresponding to a succeeding thought. However, the natural diffusion caused by the Heisenberg uncertainty will entail that the quantum state actually generated by the brain process will be somewhat fuzzy: a host of possibilities will be created. But this diffusion can be counteracted in part, and the process kept on a focused and intended track, by asking—i.e., attending to—the right questions.

We know that often, when we have a thought that initiates an action, we also initiate a monitoring that will test to see whether the action is proceeding as intended. That command to monitor is an instruction to "attend" to some question at some later time. I propose that in general our thoughts issue, as part of their intentional aspect, commands to "attend" in the future to certain questions, and that these directives supply the missing component of the quantum dynamics: *they* pose the particular questions that are put to nature. Then the necessary posings of the questions become an aspect of quantum mind-brain dynamics.

Since the question to be posed is supposed to be of the form "Is an experience of such-and-such a kind occurring?" it would appear that the question really ought to be part of the mental, rather than physical, side of the mind-brain dynamics. The aspect that makes the mental side essentially different from the physical aspect governed by the Schrödinger equation is that the latter process is mechanical: it is governed by *local* causal connections. But conscious thoughts correspond to global properties of brains. They are associated with the action of the operators P, and these operators must be nonlocal because any local projection operator P would introduce an infinite amount of energy into the system.

The essential point here is that quantum theory has a lacuna that can very naturally be filled in such a way as to allow our thoughts to exercise real, though not absolute, control over the mechanical aspects of mind-brain dynamics. This bringing of the experiential aspect of nature into the causal structure is very much in line with the ideas of Alfred North Whitehead.

Whitehead and the Quantum Mind/Brain

It may be useful to expand upon this final point concerning Whitehead. To bring the issue into focus, I shall first briefly summarize the main points described above about the connection of mind to brain within quantum physics, and then consider the possible relevance of Whiteheadian thought.

The main point of the above discussion is that the original Copenhagen interpretation of quantum theory placed the human experimenter outside the system that was described in the mathematical language of quantum theory. In that early approach, the human experimenter has three roles. The first is as a receptacle for the experiences that are the database of science. The second is to pose to nature a sequence of particular "yes–no" questions that nature immediately answers by either returning the experience associated with the answer "yes," or by returning no experience. The third role is as a scientist who communicates to his colleagues "what he has done and what he has learned," and thus contributes to the ongoing scientific enterprise. Our interest here is in the second role as "the chooser of what question to ask." This choice is not fixed by any laws or processes that are contained in or entailed by Copenhagen quantum theory. This is, first, because Copenhagen quantum theory places the human observer outside the system governed by the quantum laws, which are the only known precise laws, and second because in Copenhagen quantum theory the human observer is treated in a practical way, and in practice the choice of which experiment to perform is effectively free: it depends on the observer's intentions and a host of inscrutable and empirically uncontrolled factors.

Von Neumann considered a sequence of different ways of dividing nature into two parts: the first part is the observed system, which is described in terms of quantum mathematics, and the second part is the observing system, which is described in terms of human experiences. Von Neumann showed that it made no practical difference which of the various placements of the dividing line between the two parts of nature he used. In each case, the choice of which question to put to nature was represented by his Process 1, which was outside the control of known physical laws, but was ascribed to the mental processes of the participant/observer. Von Neumann's final placement of the dividing line put all systems composed of atomic constituents and physical fields—such as the electromagnetic and gravitational fields—on the side described in terms of quantum mathematics, and thus placed only the consciousness of the participant/observer outside the mathematically described physical world. However, there must be a dynamic connection between the mind and the brain: the mind of the observer is obviously connected to what is going on in his brain, and his choice of which questions to put to nature influences his brain in ways controlled in principle by the quantum laws, once the neural correlates of our conscious experiences are known. This connection is via the quantum Zeno effect, which shows explicitly how the choice of questions, and their timings, can influence the course of events in the probed system, which, in this case, is the brain of the participant/observer.

This description summarizes the quantum mechanics of the mind/brain system within the von Neumann formulation of quantum mechanics. It opens the door to the natural incorporation of causally efficacious conscious thoughts into contemporary basic science. But it only opens the door: it does not specify how the brain and the conscious thoughts act together to formulate the necessary questions, and their timings. This is a problem to whose solution Whiteheadian process theory might contribute.

A scientist with physicalist leanings would be prone to try to create some sort of "purely physical process" to produce the needed questions. This would make our conscious thoughts into either redundant side effects, or redescriptions of processes that can be completely described in purely physical terms. But either of these options renders mysterious the fact that experiences exist as the psychological realities that populate our streams of consciousness. The fact that, at least within the Copenhagen and von Neumann approaches, the questions posed are supposed to have experienceable answers means that experiential realities must play some special sort of role in the dynamical processing. But in that case, quite independently of the ultimate truth about the basic nature of these realities, it makes good theoretical sense to take these realities as they are actually are known to us, namely as constituents of human streams of conscious experiences, and weave them into the physical world described by quantum theory in just the way that such experiences fit into the quantum framework, namely as efficacious agents. But they must also conform to philosophical requirements of rational coherence. That establishes the link to Whitehead, whose central endeavor was to create a rationally coherent scheme that weaves experiential-type realities into physical reality by making them integral parts of eventlike actual occasions, which are naturally identifiable with the psychophysical events that are the basic realities of von Neumann's formulation of quantum theory.

The brain is of course the carrier of potentialities generated by past events. It carries the physical correlates of appetites for various particular resolutions of the conflicting potentialities. While the physicalist might seek to explain the process in purely physical terms, the Whiteheadian approach recognizes the descriptions in terms of potential experiences as valid and useful descriptions of real aspects of nature. It makes use of the huge partial understanding of mind/brain process provided by exploring the structure of human experience itself. Quantum theory does not encourage the idea of taking the physically described aspect of brain process as dynamically complete. It calls for a psychophysical process of the creation of events that actualize experienceable potentialities.

A central Whiteheadian-type notion is that each actual event must have both an aspect that fixes, experientially and physically, what is taken from the past, and also an aspect that creates experiential and physically manifested potentialities for the future. Each event has an intentionality that does not just take randomly from the past, but rather takes selectively from the past particular potentialities whose actualizations create potentialities for future events that serve some purpose. Quantum theory seems to allow mind/brain events of this kind, and Whiteheadian thought can be viewed as an effort to provide the beginning of an understanding of the structure of such events.

Whiteheadian thought also provides an approach to the question of how one extends quantum dynamics to times and regions where consciousness as we know it is absent. All of nature is composed of actual occasions of the same genus, and thus even primordial events must have aspects that have some kinship to our conscious thoughts.

Notes

1. H.P. Stapp, "Flagstaff Talks," (1999); available from http://www-physics.lbl.gov/~stapp/stappfiles.html ; Internet.

2. N. Bohr, "Discussions with Einstein," in *Albert Einstein: Philosopher-Scientist*, ed. P.A. Schilpp (New York: Tudor, 1951), 223.

3. J.R. Klauder and E.C.G. Sudarshan, *Quantum Optics* (New York: Benjamin, 1968); R.J. Glauber, "Coherence and Quantum Detection," in *Quantum Optics*, eds. S.M. Kay and A. Maitland (New York: Academic Press, 1970); H.P. Stapp, "Exact Solution of the Infra-Red Problem," *Physical Review* 28D (1983): 1386–1418; T. Kawai and H. Stapp, "Quantum Electrodynamics at Large Distances," *Physical Review* 52D (1995): 2484–2532.

4. O. Kuebler and H.D. Zeh, "Dynamics of Quantum Correlations," *Annals of Physics* 76 (1973): 405–418; H.P. Stapp, *Mind, Matter and Quantum Mechanics* (New York: Springer-Verlag, 1993), 130; W.L. Zurek, S. Habib, and J.P. Paz, "Coherent States via Decoherence," *Physical Review Letters* 70 (1993): 1187–1190.

5. W. Itano, D. Heinzen, J. Bollinger, and D. Wineland, "Quantum Zeno Effect," *Physical Review* 41A (1990): 2295–2300.

10

Dialogue for Part II

The dialogue presented below occurred in response to presentations by Shimon Malin, Geoffrey Chew, and Henry Stapp.

Stapp: Let me ask about your vacuum. In Whiteheadian thought you have the impression that experiential or objective events are at work here, so is there any possibility that experience as we know it would be understandable and explainable as a property of your vacuum? Ordinarily, physicists think of the vacuum as particles without any experiential quality. Would you say that your vacuum could be something different?

Chew: Yes. I'm coming more and more to believe that it could be.

Nobo: I'm very excited about Professor Chew's paper because it brings physics closer to features of Whitehead's thought that are generally neglected among Whitehead scholars. These are features having to do with a theory of extension which is metaphysical—that is something prior to the becoming of actualities, prior in a supersessional sense, and which gets structured as a result of the becoming of actualities and those structures, in some cases, can be construed as spatio-temporal. The internal structure of the occasion or the pre-event mirrors certain extensive-genetic relationships in the becoming of *eventities* [see Bibliography: Nobo, 1986]. But the immediate issue concerns entropy or disorder increasing as a function of particles. We're talking about a different kind of information if I read Chew right. It's an information regarding the history and relationships of pre-events. You could have order increasing there that is not manifested in the decreasing order of particles.

One of the things I'm interested in is that what you're calling the historical element seems to correspond to what Whitehead called supersession, and a supersessional order is much more primitive, more basic, than temporal order. It gives rise to it and is connected to the theory of extension in that the supersessional order is encoded into the very structure of the extensive standpoint of the eventities, or the actual occasions. The whole history of the universe up to the becoming of that eventity is projected into its internal structure so that the information about that history is in that event—the complete history of the universe up to that point.

That temporal dimensions or durations arise presupposes not only patterns of eventities, but their interaction. We can correlate those patterns that we can measure as so much physical time, but the supersessional relation is much more basic. I'll formalize what I am talking about so physicists can make use of it [see Nobo, ch. 17].

Stanley Klein: It's very clear to many people that religion has a powerful role in shaping human behavior and influences the way the culture is going to go. It's my feeling that with a stronger coupling between knowledge of science and its images, and expressions of spirituality, the world would become a better place. And it seems to me that Whitehead's process thinking can play an important role in that.

Thirty years ago at Lawrence Berkeley Laboratory, Geoffrey was doing very adventuresome thinking. Those were the bootstrap days, and one of the things that came out of the S-matrix approach that he pioneered was that it got coupled into the popular domain. Fritjof Capra's books and a number of others linked bootstrap theory with Buddhism and Eastern ways of thinking, and that actually had some influence on the big world, on theology, and how people thought about the universe. Physics has some amazingly beautiful understandings, incredibly beautiful symmetries that haven't made it to the outside world. It might be that the Whitehead intermediary, process thinking, is enough—it might be other things. What is it that we physicists can do to better connect with philosophy and theology?

I don't think that we need to go to Chew's complex 4×4 matrices and pre-events to get a sense of how physics has a strong non-material aspect. In fact, the present standard quantum mechanics that we know and love has a dual structure and a non-material element. Henry points out this duality in his eloquent writings. The public needs to see better physics images and metaphors than we've provided. I came to this conference because of my belief that a coupling of Whitehead with quantum mechanics is the avenue to link physics better with philosophy and the "man in the street."

Murray Code: Whitehead in *Science and the Modern World* points out that there are two abstractions which we must use if we are to explain both matter and spirit, and that puts me in mind of the kind of duality here. Materiality and spirituality are a typical duality. The problem of language keeps coming up here whether or not our understanding keeps improving as we become more knowledgeable about smaller and smaller things. To be more specific, I was intrigued by Chew's comment about what it is that causes patterns to have duration, and what is the observation of one pattern by another. This implies to me that communication is in some sense a fundamental part of the world. Have you ever thought about Peirce's semiotics in terms of trying to understand the vacuum and what is going on in the non-material side? In this non-material world, communication appears to be occurring, but the language for this is missing.

Chew: I do not have any language yet. I agree with you completely—it's an extremely important issue.

Robert Valenza: In your pre-material, pre-event world, is there some identifiable aspect of your theory, some parameter lurking in the 16 slots of that matrix that says in the non-material world why pre-coherence, if that is the right word, even emerges? This is why a larger time-scale reality evolves in such a way that human biological systems find it very effective to use particle and substance ontology in conducting our hunting and fishing, and shaping each other.

Chew: There is in the model, as presently formulated, need for a huge dimensionless parameter, of the order 10^{30}, and I do not know where that parameter comes from. But it is essential for the universe to manifest the qualitative features that it does, that there be some huge parameter that sets the big ratio between scales that is so essential. You would not get anything like the universe we know, if you did not have that huge parameter built in, and I don't know where it comes from. I'm hoping that some beautiful discrete mathematics will say that a number like that is picked up.

Lawrence Fagg: Is that at all related to the cosmological constant?

Chew: No, it's not—it's related basically to what sometimes people say is the smallness of the gravitational constant expressed on particle scale: the ratio of the particle scale to the Planck scale.

Nobo: In Whitehead the universe, from the metaphysical point of view, is an open universe. Regarding Klein's issue about theology, it is certainly possible within the Whiteheadian metaphysical scheme to have God be the cause of what is interpreted as the big bang. It is not strictly Whiteheadian, since I take an aspect of the universe to be eternal, supersessionally antecedent to any becoming, and out of that there is spontaneous becoming, which is the primordial actuality.

But it's not God until the decision that constitutes itself into a creator of the first set of what we might call worldly occasions. It's partly a self-constituting decision and partly a transcendent decision, which uses the same conceptuality which afterward we would use to see how one actual occasion can be, in an important sense, the determining cause of why another occasion comes into being or is begotten by the universe. The interesting thing is that you don't need a concept of an infinite density. You just need eventities, a finite set such that each member of that set gives rise to an indefinite number of other eventities so you have a quick expansion of the members of this growing history of eventities which not too much later would give rise to particles and enduring entities.

Fagg: When you're talking about infinite densities you're talking about the initial event?

Nobo: Physics has to read an infinite density into the big bang because it has to have those particles come out of that infinite density, but I'm saying that in Whitehead that is not necessary. You have a finite set alpha of eventities, but they very quickly, in a few generations, can expand at an indefinite rate and account for the expansion of the universe.

Chew: A feature of my model, which physicists typically find very puzzling, was only emphasized in my own consciousness fairly recently as a result of discussions with Henry Stapp: there is an important meaning for the present. That the present has a width—not precisely defined here as a width. In the model, this width is more or less what I call the observer scale. So if you want to associate a number with it, it's something like 10^{-15} seconds, that's a huge interval in terms of the basic step size, which you remember is 10^{-43} seconds, so you can get all sorts of patterns within the width which is the present but it's still very, very tiny on the scale of human consciousness. So it's perfectly consistent with the normal idea that the present is well defined but at the same time there being lots and lots of room there for complicated patterns to develop.

In the model all the new history develops in the present—there's a definite region. You've got all this past history and there's a little strip there—that's where new history develops. However, if you push the age of the universe back too far, you will cause the distinction between past and present to disappear. If you try to talk about an age of the universe that is less than 10^{-15} seconds (cosmologists often do this) there isn't any distinction between past and present in that region—normal psychological time at least is blurred.

Eastman: The time width being discussed seems very analogous to a notion that David Finkelstein introduces in his book *Quantum Relativity* and that he refers to as a "chronon." Does this relate directly to what Geoff has brought out here?

Finkelstein: The chronon is a quantum of time [and an elementary process]. There would be a technical problem if there were such a thing as "the present." Of course, relativity makes the idea of the present very observer-dependent, particularly for systems in motion. I'm sure that's taken into account in all these discussions. But there is also a large difference in the sizes we give our time units.

Since the philosophy of organism is a cell theory of actuality, it makes sense to ask if the cells have any sizes associated with them, and, if so, what is this typical size? Many physicists think that the fundamental cell has a size that is roughly the Planck length or the Planck time. If you look at any operational way of actually measuring fields at points determining locations of events and so on, you never get down to the Planck length. There are many, many obstructing effects. Even black hole formation cuts off some 10 or 12 orders of magnitude above the Planck length [when the field is measured throughout the experimental volume rather than just at one point]. If you consider the Compton effect and limits on the cell size, the only way that anyone could possibly think that the Planck length was the fundamental cell size is to imagine a universe as nothing but gravity—defining the Planck length and time or the numbers just made from Planck's constant, gravity, and the speed of light—and that would be a totally inconsistent picture of nature, because then there would be nothing to measure gravity with. Gravity is one theory that can't be of everything.

So, if there are cells, I think that they're much bigger than the Planck length. And then the question of where the Planck length comes from might have been answered by Weisskopf in his very first paper on renormalization, where he points out that the Planck length is e^{-137} (note: $e = 2.71828\ldots$) times typical particle sizes (137 comes from the coupling constant of electromagnetism). This looks as if gravity might be a statistical phenomenon, as many other physicists have suggested.

Eastman: What would be more fundamental?

Finkelstein: I'm not sure that the idea of 'fundamental' is fundamental. What is fundamental is an axiom, and physics is experimental, not axiomatic; there is no one outside physics to formulate its axioms. But I am still trying to model all processes as patterns of chronons.

Chew: Jorge Nobo's comments remind me of two features coming out of my model. One is that relativity is not a feature of the model. The universe has a center in the model, and relativity is an illusion associated with these huge scales. To put it crudely, it's like saying that the portion of the universe to which we are accessible is such a small portion that even though the universe has a center there's no way of us discovering that because we're confined to looking at a relatively small part of it. But it's quite essential in this model that relativity not be an exact idea; it's only an approximate idea which is based on the huge ratios.

The second feature is that gravity is not a priori either. The notion of gravity depends on the notion of energy, and the notion of energy doesn't arise until you get patterns that are big enough to show this persistence. On what scale is gravity first situated? My guess is that it is between what I call particle scale and observer scale. It's somewhere in that region that gravity begins to have a meaning, but certainly not at the lower scales.

Klein: Do your pre-events have dynamic properties? How does one maintain causality, and how do pre-events become real events?

Chew: Well, those are constraints on what I call the history lattice. There are electrodynamic constraints which have the usual ideas of causality built into them.

Klein: Is there a Lagrangian for pre-events?

Chew: Lagrangian is a notion related to energy, so there is no Lagrangian.

Klein: So how do you maintain causality?

Chew: As a constraint on the lattice. There has to be a certain relationship between the pre-events and the lattice that is causally consistent. There is 'time.' If you look at my preprint on magneto-electrodynamics, you will see a detailed description of this.

Stapp: Two questions: If gravity only comes in later, why is the Planck scale defined earlier?

Chew: In the model, I would say that there is a unit that is the spacing of the pre-events along the chain and that is a basic constant of the model. I'm hoping that it will turn out later that this unit is related to gravity. But it's wishful thinking that the model will describe gravity.

Stapp: My second question is in regard to the remark by Murray Code. He says that physicists, on the whole, deal with the physical and that's certainly very true. On the other hand, the very point that Stanley Klein was making is that quantum mechanics, in the way it's really formulated, brings in an observer. There has been a lot of effort to get the observer out of quantum mechanics. The founders of quantum theory found it necessary to bring the observer in. It is true that Bohr dismissed them in a certain sense by saying that we can't deal with biological systems and therefore diverted the attention of physicists away from the observer and experiential aspects of reality. It seems somehow to be in the works since the beginning, as Stanley was stressing, and at least a few physicists are concerned particularly with this question of how the observer plays a role in the whole thing. Quantum mechanics certainly opens the door for this in a very natural way.

Klein: Here's the problem. The equations of physics are reductionistic. If you go outside this room and ask a typical scientist about what is causing me to raise my hand, they will give you a chain of arguments that is mechanistic based on how neurons are processing. It is very deterministic because quantum effects are quite negligible. We have a science story with no place for free will. The big question facing this science-theology connection, the subject-object connection, is how do you get a totally reductionist ontology to also allow total free will? That's the problem. Did Whitehead offer a solution? If so, I would love to know where to read that. To have total reductionism and still have total free will is difficult. I think that quantum mechanics does provide a framework for tackling that problem. I believe that some of the things that Henry Stapp and Abner Shimony have written about are in that direction—I have to advertise Shimony. In Penrose's latest book, Shimony has an important essay on an augmented Whitehead—one combining Whitehead and quantum mechanics to address this very tricky business of how you get a reductionist theory to be not reductionist.

Fagg: So you're saying that quantum mechanics qualifies the idea of a total reductionism?

Klein: Yes, the rules for quantum mechanics are very precise, except that you have this funny duality. There is the realm of the observed and the realm of the observer. You can't reduce the observer to the observed. Von Neumann showed that the quantum split was movable. You can move the split up high so that the equations of quantum mechanics and biology apply to my moving my hand. So that does look pretty reductionist. But there will always be the collapse that isn't reductionist.

Finkelstein: I would like to qualify the remark that quantum theory is totally reductionist. Indeed, it's the least reductionist of our physical theories in that for the first time, for example, one could know everything possible about a system of particles and know nothing whatever about the position of any particle in the system. This is totally unthinkable in classical physics. In classical theory, every property and symbol is simply a collection of 'and' and 'or' combinations of properties of individuals. This expresses the reductionism of classical physics. And that's just not the way it is in quantum theory. If you have a benzene ring, you don't know where the bonds are.

Klein: Let me first say that I agree with Finkelstein that the intrinsic duality of quantum mechanics has loosened the reduction. The act of observation in quantum mechanics (collapse of the wave function in some interpretations) is not reducible. I think the presence of a movable split makes that nonreduction more palatable. However, other than the collapse event, the chain from quantum electrodynamics to human behavior seems reducible. Just because you don't know where the bonds are in benzene doesn't mean it isn't reductionist. The big problem here is carefully defining the word *reductionist.* Philosophers have adopted the word supervenience to have a more precise, but similar, meaning. I suspect that if we have a disagreement on whether human behavior is reducible to atoms, it is because we are defining reduction differently.

Finkelstein: But it isn't the split that solves it—it's no longer reductionist.

Fagg: When you say *split,* do you mean the split between the observer and the experiment?

Klein: Yes.

Nobo: Quite often the indeterministic aspect of quantum mechanics is dismissed as irrelevant to the free will issue in philosophy because statistically it just about cancels out. But that assumes already that our stream of experience is nothing but neurons and ultimately particles interacting. The advantage of this approach is that our stream of experience is analyzable into the same basic types of entities that Professor Chew is talking about. Here I identify eventities with pre-events. That argument doesn't hold if at the quantum level we can speak of individual events that are in principle unpredictable and if our experiences of events are, in some sense, exactly at that same level, then they are, in some respects, self-determining. We have to attribute self-determination to ourselves (what Griffin calls hard-core common sense), but we can also attribute some measure of self-determination to more elementary, less complex occasions. Reduction is also somewhat of a sham—there are a lot of promissory notes in the reductionism of science.

Klein: The subjective aspect, of consciousness or feeling, is where I think quantum mechanics, by working between consciousness and events, has a place for the observer, the subjective. But what I'm talking about is the objective aspect. Is there anything that happens that can be measured and that isn't reductionist?

Nobo: You already threw the baby out with the bath water—you said anything that is measurable. Why does everything have to be measurable in reality? Obviously there is reduction in the sense of reducing higher-level laws or properties to lower-level laws or properties, but I don't think that you can do this across the board, even for measurable things.

Klein: Are events measurable for Whitehead?

Nobo: It depends on whether you are talking about actual occasions, societies of actual occasions, or interactions of these. I think that they have measurable features, but not all the features are measurable.

Stapp: This is in regard to this question of reductionism. As we have seen, it can be defined in different ways and depending on the definition you might get different conclusions. Perhaps my talk ought to have been entitled "Is Mind Slave to Body?" Is there some possibility that it is not slave to body? That there is something, given a complete description of the system's universe, that is not determined. In quantum mechanics there are two simple answers. If you have statistical laws, perhaps mind could bias the statistics. But that would be outside quantum mechanics. If you want to stay with quantum mechanics, it says that there is no bias in these laws, you should stick with these laws. There is another way that you might evade this question by saying, "Well, if you have only statistical predictions as to what is going to happen, you have some possibility for an intrusion of some choice from outside quantum mechanics which is nonetheless controlling the way things should go." But I would also regard that as outside the bounds of quantum mechanics. That's essentially another way of saying that this freedom you want to give is not a true freedom. As shown in my paper, there is a place for something outside the physical universe that can come in and change the physical system.

Open dialogue (panel plus audience):

Clayton: Geoffrey, from your comments this morning and from previous discussions, I understand that you question the future of dualism—you do not want to have a place for an observer or something outside the model. That would mean that whatever you finally say about quantum mechanics, it wouldn't be Copenhagen or any interpretation with an observer outside. Whitehead didn't have such dualism either. That makes you two allies against a neo-Whiteheadian rationalism that seeks to incorporate dualism.

Chew: I think that's fair. Let me be more explicit with this phenomenon called Everett branching, which is the bugaboo of quantum mechanics. If you try to say that there is nothing beyond quantum mechanics then you get stuck with the crazy "many worlds" picture. So how do I imagine this might be dealt with in the model? I hope it will turn out that many worlds means many *physical* worlds. If you insist on writing your wave function just in terms of Hilbert space parameters that have to do with matter or substance, then you get stuck with these puzzles. Imagine that the complete wave function includes all these things which I vaguely call "vacuum" and which play a part in the whole dynamic, and imagine that these are built up of Whiteheadian actual occasions which have the same set of 16 parameters as actual occasions that form matter. However, they just form different patterns, and these patterns don't fall into the categories that science has so far been able to deal with, characterized by this persistent matter. Maybe the wave function will finally exhibit the feature that the influence of all this non-material history picks out one of the usual Everett branches and by some interference effect gets rid of the other ones. Thus you end up with just one material branch, which cannot be understood if you insist on working with a Hilbert space that only has material labels on it. You have got to use these non-material aspects in the Hilbert space or you'll never have a chance.

Clayton: That's an important clarification, because the vacuum comes out now as non-material and the word *vacuum* might be misleading.

Chew: Yes, physicists do talk about vacuum and to that extent it could be misleading, but the meaning of *vacuum* here is quite different than usual.

Klein: Geoffrey has done something very much like David Bohm. He has a different language in which a non-material component has the flavor of a

Bohmian quantum potential that knows where the two slits are to guide the pre-events.

Marcus Ford: My question is, if Whitehead were here, how would he respond? Might he say that Professor Chew is allowing the Eliatic camel to get his pugnacious nose under the Heraclitean tent and Dr. Chew has missed the main point of *Process and Reality* wherein I, Whitehead, have tried to say there that the least bits of the universe are not bits at all and the building blocks are not blocks? We're talking about events, and the glue that holds them together is known as "prehensions," and when prehensions occur there is an interpenetration so that there are no distinct lines between self and other, between this and that. We start with an event that remembers a past, anticipates a future, and is part of a society of other events that interpenetrate. Every individual has a temporal as well as a spatial dimension. Would he say that you've missed the point in *Process and Reality*, or would you say that I've missed the point?

Chew: I could not detect anything in your list of Whiteheadian principles that my model doesn't satisfy. The model starts without any meaningful endurance or substance and has to build meaning for both out of patterns and change. My understanding is that that's okay with Whitehead.

Ian Barbour: It seems to me that Whitehead and Hartshorne put considerable emphasis on differences in the organizational levels of the world as organizational duality or, better, organizational pluralities, not just two kinds of organization. I'm not sure that Whitehead would expect the kind of action at the sub-quantum level or pre-event that Geoffrey was talking about. I'm very sympathetic with emphasis on the event character of objects, but what I think is the way that Whitehead protects against reductionism is not by positing something going on at the very, very bottom or one that is *mental* or *spiritual* in itself, but by positing rather radical differences in the way that patterns of events at higher organizational levels work to exert causative influence, you might say, from the top down. In other words, to provide a lure. I get the impression from Geoffrey's presentation that some of these non-physical elements can be explained from the bottom up.

Chew: That's not clear. In the model, there is a big bang boundary condition. Even at the big bang, there already are a huge number of pre-events; the history chain is already very long. It is simple in a sense that has to do with loop structure, but it's also complex. There has to be built in this huge factor that I talked about, so the boundary condition at the big bang has to know about these huge gaps in scale that are going to develop. Thus, it is not entirely building up from the bottom. You must have built in already considerable organizational complexity and anticipation in the boundary conditions for all this complexity.

Isabelle Stengers: When does the Hilbert space enter into your model—at the very beginning?

Chew: Yes, so far I have not been able to exclude a Hilbert space at the beginning. But you are not committed to material labels on the Hilbert space. And at this stage there is no commitment to stationary states, which is a completely separate question. What I have is basically the direct product of 16 simple Fock spaces.

Finkelstein: Probably your coupling to Hilbert space is more general than that.

Tanaka: There is a sharp distinction between events and actual occasions in Whitehead's philosophy. 'Actual occasion' is a metaphysical concept whereas 'event' has a physical, experiential basis. How should one think about temporal possibility?

Chew: One needs to distinguish at least two concepts of time in this scheme. I talked about the age of the pre-events. I mentioned that they don't have to keep advancing along the chain. There is a sequence so that the chain has a sense of always advancing, but as you move along the chain the age doesn't have to keep moving forward. In fact, the characteristic that distinguishes matter as opposed to vacuum is that to describe matter you need the notion of event, already a cluster or pattern of pre-events bouncing back and forth. Otherwise, you cannot represent matter. All of that takes place at particle scale that is way, way below observer scale, which is where measurement first arises. You can't have a notion of measurement at particle scale; you must have huge patterns that are much, much bigger, and then when you get to those very big patterns then there is a second time which controls this "present" idea. Remember I said that there is a width to the present and the whole band keeps moving forward, and that's the time in which the age of the universe, if you like, keeps moving forward. But within that band you still have all that bouncing back and forth, which is building the meaning of individual events.

When the age of the universe moves forward one step, it opens up some room for new history to develop. Question: what controls that new history, what limits it? And there is an influence from the immediate past reflected in the constraints, the lattice constraints—causal electromagnetic constraints. You are just not allowed to have histories which violate these, but there are still plenty of possibilities which obey these. There the model goes along perhaps naively with the standard quantum mechanical idea that there is a unitary operator which acts on the wave function that is there before the step is made and that tells you how the wave function will be in the new step. This involves the entire history, including all this vacuum stuff, and it is not controlled by the purely material component and certainly not by the immediate past.

Clayton: What about a modern theory of causality which goes beyond Whitehead? That's an insightful critique about Whitehead trying to take into account self-organizing systems, and it is a response to your precise question, "What about a broader theory of causality?"

Jungerman: I think that different levels of nature can have their own laws, and all is not reducible to elementary particles.

Chew: I want to make some remarks about this notion of measurement or observation that has taken on such importance in quantum mechanics. My comment is that, from the historical point of view that I am currently favoring, there's nothing special about an observation. It's just simply a particular pattern of history, a localized pattern of history, a collection of events that have certain characteristics. Among them is the characteristic that is the subpattern which corresponds roughly to the observer and the subpattern which corresponds roughly to that which is observed. And although they get interlocked, they don't mess each other up so much that the identities of the two get spoiled. You can qualitatively say this, you can imagine it, but to quantify that, to say exactly how much messing up you will allow in an observation is pretty obscure, and I'm certainly not in a position to do that. But, in order for such patterns to exist at all, it is nontrivial.

One requirement is that the material component of the universe has to be rather dilute. You happen to be over there and I happen to be over here in order for me to be able to look at you and not completely get entwined with you and mess you up. And that depends on the fact that this is an extremely dilute phase of the universe that we're functioning in. That diluteness by the way is specifically related to what I said was the relatively small component of history which is material, most of it being non-material. Most of the rest of the universe that isn't matter dominates the total of whatever we have available to put history into. So diluteness is one essential characteristic and that isn't present in the very early universe. It has to expand a great deal before you can even begin to talk about observation.

The second requirement is the smallness of the elementary electric charge. Henry Stapp emphasized that brain activity is electromagnetic, but certainly any pattern of history that you could call a measurement is electromagnetic. It is mediated by photon arcs, and at the ends of the photon arcs you have these very tiny electric charges that are causing those photon arcs to either be born or die without disturbing terribly whatever it is they're running into. So observation and measurement, which is the cornerstone of a large part of Copenhagen ideas, is not obvious. It's a very special feature of our portion of the universe. And of course if it weren't there, we couldn't accumulate knowledge that depends on such local conditions.

Measurement is electromagnetic. There could be no such thing as measurement without the very special properties of electromagnetism such as the extremely small mass of the photon and the small charge of the electron. The matter that is built out of these ingredients has to have certain characteristics or you couldn't imagine measurement. But all these discussions about collapse of the wave function and all the puzzles of quantum mechanics that go along with notions of measurement almost always are carried out without any recognition or attention to the fact that this is an electromagnetic process. That seems to me to be a deficiency. I've always believed that if you could really understand how these special characteristics of electromagnetism are essential to measurement, you would make more progress in the puzzles and paradoxes of quantum mechanics.

In the particular model that I've been pursuing, there are two distinct meanings for time. There is the global time that applies to the whole universe, and then there is the local time that applies to what I call pre-events or histories within the universe. Again, if there really are two meanings for time, you better pay attention to that when you start struggling with these tricky issues.

In my model there is a double cone in which the history of the universe is located for a given age of the universe, and the size of the double cone is parameterized by the age of the universe. But the history is the interior of the double cone so as the universe keeps expanding the size increases and the present keeps moving forward and it's in the present where measurements are being made. Remember, the present has a width, so you still have to specify those local times inside the width of the present to have a complete specification of what's going on.

Klein: What seems to be needed is to add the subjective element of experience to quantum mechanics. Something *panexperiential* has to be added. I have a feeling that electromagnetism is essential to this. Photons may be the extra ingredient, not only for measurement but also for the extra elements that arise in process thinking.

Stapp: My question is whether Geoffrey's vacuum is physical or mental, and it seems that it could very well be mental or have mental aspects. The endeavor that I am involved in is actually using quantum mechanics as it exists today, and looking at the most simple ontology (von Neumann, I believe), where the mind is left out of the part of nature that is normally regarded as being described by a wave function and the evolution of the wave function. It may be that in your other picture it's there and that's quite possible. The idea that something could come out of nothing at all, it seems to me, is an absurdity. The idea that this chance decision comes out of nothing at all I certainly reject. So I tried to carefully pose the question not in terms of total freedom of the universe but freedom of the part that we associate with experience, from the part that is well defined in quantum mechanics, which is this evolution of the wave function of the physical part of the universe. So I could formulate the question in a particular way by identifying the physical with the evolving wave function as it's understood today in quantum mechanics, and ask about freedom relative to that.

Klein: Since Geoffrey picked up on this photon business, I would like to go along with that in the context of the quantum Zeno effect that was brought up. The thing that I would urge everyone to read in connection with the quantum Zeno effect is in a book review by Joe Rosen on Henry Stapp's *Mind, Matter and Quantum Mechanics* [*Process Studies*, 26/3–4, pp. 328–330], and the last paragraph is very interesting. By the way, it's a very laudatory, positive, and well-written review. Stapp goes through a process that I also hope that David Finkelstein would do, which is a translation of Whitehead's language into the language of physics. In particular, he examines three aspects of Whitehead's thinking and their parallels in quantum mechanics. But then in Rosen's last review paragraph he says, "Well, let's look at how the brain works." He gives the identical structure to Whitehead and to quantum mechanics, including the global, long-range, coherence, etc. And so let me do that with Zeno and maybe hook in photons.

The idea of the Zeno effect is that if you attend repetitively to a single idea, you can freeze it and stop it from evolving. Well, there are simple brain mechanisms that could freeze an idea. Similarly, it's very easy to make a neural network with attention that could shift as well as freeze an idea. And so I would like people first of all to look at Rosen's very well-written paragraph and I would like Henry, David, or others to comment on this translation of Whitehead having parallels with quantum, but then it might also have parallels in classical neural dynamics with feedback.

Stapp: Yes, I would like to answer, because that's just exactly the point of freedom that I was emphasizing in my last talk. For the chaos mechanism or any of those neural networks, if it's a classical system, then that mechanism causing it to freeze in place or to move is itself controlled by the system just prior to it. The physical in and of itself is doing the whole job. There's no freedom of anything outside, whereas I claim that in quantum mechanics there's a difference. If you define the physical system in terms of the wave function or state vector then there is something outside that is not determined by that physical system: namely, what question is going to be posed, and that freedom is able to change and control the behavior. So I claim there's an essential difference between the chaos mechanism and the one that I'm talking about.

Chew: I want to introduce a word to the discussion that I have not yet heard but which I feel is an extremely useful word that natural science has come up with which philosophers don't seem to like. I've always felt that this was a defect in most philosophy, but I understand that Whitehead's philosophy is an exception. The word is *approximation*. If you start thinking from a physical science point of view, you realize that no statements about the universe can be absolutely exact. They must all be approximate because, for example, as I mentioned earlier, measurement is inherently an approximate idea. There is no such thing as exact measurement, and since all of our ideas are based on observation, observational knowledge, and so on, it's difficult to believe that there are any statements that humans can make about the universe that could be exact statements. And this connects to these ideas that have been discussed; for example, the smallness of the elementary electric charge. It appears to be very, very important but because you depend on it, the fact that it isn't zero means that all these lovely statements you make based on its smallness are not quite exact. They are all approximate because it's not zero, and the same thing can be said about the largeness of the age of the universe. We have all these scientific concepts that are based on the notion that tomorrow is pretty much the same as today. Anything that happens today could be reproduced tomorrow, and that's what natural science is based on. But that is based on the idea that the age of the universe is infinite. That the universe is not expanding. But it's changing. Tomorrow is not the same as today.

There is no exactly reproducible experiment. There is no statement about the universe that you can make that is absolutely exact, given the fact the age of the universe is finite. You can go down the list with many other parameters. Science has come up with these huge ratios which we call dimensionless numbers which give you some understanding of why certain approximations can be extremely accurate. When you have a ratio as big as 10^{17}, it makes certain approximations so accurate that humans don't bother to keep remembering that this is only an approximation. The human lifetime isn't long enough to make it worthwhile to worry about the fact that the age of the universe is finite. In any case, I believe that it's important to keep this word *approximation* in the vocabulary, and I'm very pleased to hear that Whitehead does this. I hope he consistently agrees that no word that he defines so carefully can have an absolutely precise meaning. It has to have an approximate meaning.

Clayton: Well, he does, roughly.

Finkelstein: I'm delighted by the lovely things that Geoffrey just said.

Malin: My question to Henry is the following. You spoke about freedom in what I think were two phases; one was the freedom to ask the question. You seem to imply that the collapse of the wave function comes from the consciousness of the observer. If you're implying that, you must be aware of the well-known objections to that position.

Stapp: No, I'm not implying that. I tried to draw a very sharp distinction between two kinds of choices. First, there's a choice that seems to be assignable to a small system, like a human being or a brain or something like this. Quantum mechanics seems to allow that choice—the choice of what question to ask is assigned to this small system. Since it's a choice of basically what experience

I'm going to look for, it's very much like directing your attention. Basically, when I direct my attention to some extent, it's asking, "Is it this?" So I'm making a close alliance between the choice of asking the question and the choice of a small system to attend to, but the answer to the question for large systems is apparently far from local. It's not connected with the observer, and that's the quote that I had. It's connected with nature, so nature answers the question. But maybe it's the small system that has the option to ask which question nature will answer. So there is a division of labor.

Malin: Let me make one more comment. If we say with Dirac that nature makes a choice, then of course the question that people have been working on for 70 years or so is how does nature make the choice, and is there a mechanism? And it seems to me that, if we believe in Whitehead, we must say there cannot be a mechanism that completely determines it because, if there is, there is no place for creativity, which is one of his most fundamental concepts.

Stapp: The importance of what I have said today is that you can answer the one question without answering the other. This question of how nature chooses is really a big question, but maybe we can tie the question of how we human beings pose the question to something that is much more accessible to study, in fact scientific study.

Clayton: Do you agree with Malin that one can't have a fully deterministic mechanism and still have creativity?

Stapp: I believe in the law of sufficient reason. There has got to be some reason behind every choice. Nature in its totality must have some mechanism for answering these, but it doesn't have to be local. According to quantum ideas, the whole universe can be involved in that choice. So that's a big, big question. But if we can focus on another question, the observer's choice, and make that associated with a local human system or some other local system, then we have something we can attack.

Nobo: I was struck by the fact that Whitehead uses the phrase "organic mechanism" in *Science and the Modern World*. It may be an unhappy phrase, since mechanism has the connotation of complete determinism, but I think you can interpret it to mean a partial determinism. That the earlier partly determines the later but leaves "elbow room" for self-determination, and so there is a mechanism and this is the wholistic aspect. It's the universe as a whole determining earlier phases of an event but leaving later phases open.

Malin: In other words, it's not complete determinism?

Nobo: No, it is a mechanism of sorts, but not a complete determinism. It strikes me that there's a parallel to the business of asking a question. It may be that one event puts a question to the next event. It's a question and an imperative. Do this if you can. In Whitehead, an "experience" really cuts across what you were calling "subjectivity" and what Whitehead calls "physical experience." The physical experience is the more mechanical, deterministic, earlier-to-later; the subjective experience is where the freedom is, but that freedom puts a question to the next event, so to speak. I'm playing with an idea here. If we think of every event being partly informed by the past and partly self-informing or self-forming, then that self-formation together with the earlier information can put a question to the future. In Whitehead, every subjective aim has a transcendent component. The aim is at intensity of experience in the present and in the relevant future, so that this transcendent element is the question being put to the future.

Stapp: Let me first say that the events that I'm talking about, that you can call occasions or actual occasions, are more in line with Whitehead's earlier thinking where he was talking about human experience that is associated with a large organism, and not these tiny little things that are even smaller than particles. Once you start with the idea of a human-type experience, of course, you realize that that's not enough to build a whole world out of, because we weren't always around, and so you have to have some generalization of it. It's natural to go the route that Whitehead did down to these tiny little things. I'm not at all sure that's really the way you need to generalize. In fact I see a lot of problems with it, which I won't go into, but I don't like that way at all. I think events are a much higher-level thing even if they're not human. The way it works at that level, as both Whitehead and Heisenberg were saying, is that as this event or occasion occurs, it creates potentialities for the future.

Nobo: That's Whitehead's intent.

Stapp: I'm thinking of things at this higher level and not at the mechanical level where it has these powers which are almost rigidly formed. That's a way to go too, but I'm certainly not focusing on that level. Although it's potentiality, and although it's a good way to talk about it the way Heisenberg talks about it, I find it difficult to think that there's nothing in the whole universe, or the whole totality of existence, that is controlling it. So I think that you can draw a distinction between physical and non-physical on the basis of quantum mechanics. In other words, the physical is the part that is represented by the quantum mechanical wave function, and that only represents potentialities. But there's got to somehow be a bigger, deeper reality that finally decides it, and it's the whole universe somehow that's coming into play. But that's a big problem.

Nobo: For Peirce, matter was mind hidebound in habit. In Peirce and in Whitehead, there's a possibility of thinking that one level of reality becomes habit ridden. You have constants, then you have a level that builds on that, which for a while has freedom, indeterminacy, and novelty, but it too becomes hidebound. Then you go to a higher level and, if we can make that hierarchy of levels meaningful in theory, then you could have your constants at one level, and your tendencies, trends, and so forth at another level. It's possible that once something becomes a constant, it's always a constant. But there's always a level where you wouldn't have the determinacy, mechanism, and so forth. My question to Professor Stapp is how would you answer someone who says that the question that you're posing is itself determined by earlier states of the brain?

Stapp: If, by earlier states of the brain, you mean the quantum mechanical description of the brain in terms of the wave function and the state, then it's not. There has to be something outside the quantum mechanical system that poses the question to the quantum mechanical system. That's how it works. Quantum mechanics by itself doesn't determine what the question is going to be—there's something outside. That doesn't mean it's outside of nature. It means outside this narrow definition of what is physical, namely what the physicists call physical, and the generalization of classical physics to these wave functions. It's outside that 'physical.' The physical is local, and evolves according to local deterministic equations. It's outside that narrow definition of physics, although the definition is broad in the sense that it covers all of quantum mechanics, except for the decision and experiences of the observer. If you call those things as being outside the physical, then these choices are not made by the physical brain.

Fagg: Dr. Stapp, in the early part of your lecture, you mentioned a whole list of subjective things like thoughts and feelings and so forth, as arising and then dying. Let me note that there are some feelings that go on for a whole lifetime—traumatic experiences, for example. The other thing is that I simply did not understand this business of repeated observations as developing and freezing. Is this a freezing in the brain, or in the system that the brain is occupied with observing?

Stapp: Yes, I was suggesting that our experiences are associated particularly with certain aspects of the brain which identify the electromagnetic properties, electromagnetic field, of the brain, and I list a number of advantages of it. An individual particle is so tiny it's hard to think that it's affecting thoughts very much, so the states of the electromagnetic field somehow average over lots of things. When you look at it in this way, you're getting a nice averaging that seems to be more closely connected to our experiences. There's this particular feature of the brain that I'm focusing on, and saying that's the particular feature that's connected with experience. Then you're able to talk about these Gaussian wave packets that I mentioned.

Focusing your attention, asking a particular question repeatedly, translates into "Is this particular Gaussian wave packet here?" In other words, you're asking some particular question about that wave packet. Now the rest of the brain can be doing all sorts of things while you're asking this particular question. If this is a particularly important feature of the brain, which is controlling lots of other things that are going on, your freezing of that particular aspect the brain remains in place. For example, if you're holding up a heavy weight, the feeling you have is that by focusing your attention on some particular thought, you are working against some sort of force that is wanting to do something else, and you're holding the weight in place. I'm suggesting that you're holding in place some particular aspect of the brain that's in charge of lots of other things, and it's not the whole brain that's being frozen, it's just a particular aspect of the brain that's being frozen.

Audience: Professor Stapp, are you saying that at some level consciousness plays a major role in the unfolding of events? Perhaps it's more like the maps that we create of events. It's basically just a map, like Whitehead would say, "the map is not the territory." He was not that invested in the way that we explain natural laws, but rather saying the territory itself is much larger than our map of it. The map did not create any event in the territory. What happens when we as human beings do something that is "unprecedented and impossible"? I'll give you an example. When John Kennedy announced that we were going to the moon in 10 years, this was something we as human beings had never done before. We had no map. The territory was unknown to us. Yet what we did in that process is that we imagined ourselves being successful doing it, and entered into an experiential acceptance that we were successful. Essentially, we went from an acceptance to an accomplished fact, an event that was pulling us toward it. We allowed that event to organize not just our thinking, but our action that led to that event, to the consequence we had desired to achieve. How do you explain that? Where does that fit in the logic of science, and the logic of physics?

Stapp: It seems to me that it fits very naturally and easily into this point of view. This point of view really is saying that our experiences are creating tendencies for future things to happen, and if you have a certain structure of your experience, if you're talking about a human experience, it has a lot of the future in it. Most experiences have intentionality in them. I'm about to raise my hand, or I'm intending to raise my hand. There's always this "looking forwardness" of your experiences, and of course they can look a long way ahead to being on the moon, and you put into action that whole sequence. That sort of thing fits beautifully into this view of the universe.

Audience: So you don't think that there's an element that comes in there that is not anticipated by your experience? I think that our experiences are not a very good future, because all our experiences can tell us is what we have already done. They can only give us a limited outcome, and if we had never done something before, our experience would not lead us into doing things that we have never done before.

…uman brain has the capacity to pull together ideas from …mbine them in different ways. You do get new ideas no …efore, and you could make it a mystical thing if you …nat's really so necessary. It seems to me that the human …these potentialities that might account quite naturally for the way …ideas come up. This idea that I just told you today, I didn't have several months ago, but a lot of tendencies were there, and I talked to some people, and I think it ultimately emerged out of a confluence of influences acting on me.

Malin: Schrödinger, in a wonderful little essay called the "Principle of Objectivation," makes the point that all of science, including quantum mechanics, works under the assumption that the perceiver, or what he calls the "subject of cognizance," is kept out of the picture. That's how science starts, and I think the best fit of Whitehead into physics would be to consider physics as a projection of Whitehead's idea into the objectivized domain. In other words, if you do that, if you remove the subject of cognizance from the picture, then the objectivized picture you are left with gets to be very close to quantum mechanics.

Mark Germain: Neurologists and psychologists want to point out that, in terms of talking about the brain, there is an empirical datum that we can look at. Two things seem to support Professor Stapp's ideas as they were presented today. One of them is that the intentional matrix is actually the place in the brain where consciousness can most uniquely be located. The intentional matrix is primarily in the brain stem, and slightly higher, and if you have stroke or damage to any part of the brain, you will recover complete consciousness, you will be completely awake. You might have less to be conscious of because you've lost a portion of the brain, but you still would be completely conscious. But if the intentional matrix is examined at one of its more fundamental levels, then the loss shows. You can literally see this in people floating in and out of consciousness, on the basis of the way some things press on this part of the brain.

There is a lot of theory and evidence that supports this idea of attention being the basic control parameter in conscious process. The other thing relevant to Professor Stapp's model is the fact of actually discrete brain states. The brain goes through chaotic process for the most part, and there's a period of change in the state of the brain that is followed by a relative period of stability, and then change from stability and so on. So you can break down these discrete states. As to what extent they might correlate to states that Professor Stapp introduces would have to be determined experimentally. I think that a lot of things that he's saying can be flushed out experimentally because they do have correlates with what is actually happening.

Clayton: This changing focus of attention which is not identified with any particular brain state is not a way of speaking that a neuroscientist would be comfortable with, is it?

Germain: Well, the intentional matrix basically raises the brain to a certain energy state, it's called "activation," and the activation in terms of energy can be thought of in a lot of different ways. Neuronal matrices form basins of attraction, and the state then is drawn into a basin of attraction based on an energy minimum. For instance, when a rabbit sniffs, they have this process of activation that forms attractors that represent different odors of what things are. Along the lines that Professor Stapp has said, then you have the experience and you have to match the experience with the brain state, and that's the process by which the states actually go into one of these specific attractors which match the odor.

Audience: What do we do about individuals with multiple personalities or persons under hypnosis?

Germain: There's a process that leads to consciousness and states that are the end product of that process, but that process can vary. For example, in someone who has multiple personalities, each state is actually the becoming of an identity. Reality is discontinuous in a Whiteheadian sense, so the identity is actually matched with the brain states in the same way that experiences match with the brain state. If you act in a certain way with your multiple personality and if that's personality A, then personality A will be the one that comes into play. There's a lot more complex process theory involved with that, and how identity works and how it relates to brain states. As far as hypnosis goes, hypnosis is basically bringing down the activation energy of the brain state. You're more or less in a state of quietness. The direction of your questions is coming from the therapist who is leading you along in the process. So in that case, the executive if you will, the one asking the question, is actually outside of you.

Keeton: Geoff Chew was indicating that in his current model for quantum mechanics and relativistic explanations of the world of experience, there is a very small component related to matter and a very large component that is non-material. So the model includes elements which, from a classical standpoint, had explanatory meaning or content relative to the material aspects of the model, but for the non-material it's a little more difficult to talk about, for instance, their causal relationship to experiences. What I'm hearing Germain say in explaining the rise of consciousness, and locating consciousness, Whitehead would describe as committing the "fallacy of misplaced concreteness." How would you describe the non-material, non-physical aspect of consciousness?

Germain: Actually, I think I tend a little more toward a subjective model that reality is basically units of experience, and experiences are all interrelationships which occur in a hierarchy. In other words, electrons have very little experience, and yet it builds up in larger and larger systems, which then can express a higher level of experience. There is no concrete collapse of the wave function. The actual event is subjective, and we are subjective entities. In a sense, this is almost like the many-worlds idea, except that there is only one mind.

Clayton [to Stapp]: When you've heard these comments do you say, "Yes, that's what I'm trying to say," or are you resisting these points?

Stapp: Well, I thought I was in agreement with him until this last remark. I am certainly thinking that we are not living in a many worlds situation, and that there are these actual occasions that collapse the wave function to some particular form. That's certainly the basis of the way I've been talking, and I think it's quite compatible with Whitehead. That is, that an actual thing occurs and it somehow uses up the potentialities in one particular sort of way, and the other incompatible possibilities are no longer possibilities. Of course, they would be potentialities for other events. There are definite decisions that are made. That it happens *this way* and not *that way*, so it's not a many world interpretation. It's not that all the things that quantum theory says are possible are all existing simultaneously, and there's no choice anywhere.

Germain: Those are two ways of looking at it and I don't know how to describe what's between them, but if you look at the view that there is this sum of all possible worlds, then it really wouldn't be possible for part of the wave function to perish. It has a primordial existence in some sense.

Eastman: [to Finkelstein] You mentioned the many-worlds interpretation. It seems to me that for Everett and others who take that interpretation as literally about actualization are implicitly making a philosophical claim. This claim is basically a denial of real potentialities. You have the wave function propagating and creating all of the multiple worlds and actualizing every one of them. In other words, they treat 'potentiality' as itself multiply instantiated and multiply actualized. In this way, they are implicitly making the philosophical claim that there is no 'real' potentiality. Whitehead wants to say that potentiality is real although not actual—there are many 'real potential' worlds and not many 'actualized' worlds. Taking a position regarding the reality of multiple worlds necessarily involves making a philosophical claim. Physicists who buy into a particular interpretation of the many-worlds concept are implicitly making such a philosophical claim, perhaps without knowing it.

Finkelstein: For me the whole business about multiple universes is the result of a graduate student who never really met quantum mechanics. He met this collapsing theory and wants to do something about it. He didn't think of giving up the idea of objects all together. It's a very Whiteheadian, Heisenbergian, Einsteinian idea. So he took this other road. I can't make any sense of it. I can't follow the road. I don't see the point. It seems to me ignoring the real point of quantum theory. I agree completely with your analysis. In fact, I was very informed by what you said.

Eastman: There is effectively a denial of 'real' potentiality in most many-worlds interpretations?

Finkelstein: Yes, thank you very much for pointing that out.

Joseph Bracken: I would like to ask Henry Stapp about the relationship between electromagnetic events going on in the brain area and what is usually called the mind. You were saying that the interaction goes on at the electromagnetic level in there and you didn't say much about brain cells. Is it your feeling that brain cells don't produce their own results to influence us, and the only feedback is at the electromagnetic level?

Stapp: My idea here was that with these neurons with calcium and potassium ions, and all of these charged particles moving in and out at a very rapid rate, and gates opening and closing, there's a tremendous amount of activity in every individual neuron. With all these billions of neurons, there's a tremendous amount of electrical activity. This electrical activity produces an electromagnetic field and this field is kind of a way of integrating and getting a picture of what all these things are doing. It's a consequence of what they're doing. It's not that you're leaving them out, you're saying that there's a higher-level description that is a consequence of what all those neurons are doing. The suggestion was that consciousness really interfaces with that higher-level description rather than going directly down to individual ions that are going through some microchannel in some neuron; that's too low a level. Consciousness doesn't directly interface with that; only indirectly, by virtue of the electromagnetic field that it has created.

Bracken: I think I follow that, but I'm just wondering whether you felt that there isn't another level in which cells as a whole produce a result, a field or a concrescence, in addition to the general electromagnetic field produced by their parts?

Stapp: That's a good question. The point is that in quantum mechanics, as well as in classical mechanics, there is a question of the choice of variables and, for example, if you have two particles you can choose to talk about this or that particle, or you can choose to talk about the center-of-mass of these two particles. These are two ways that you can talk about the same system, and sometimes it's useful to do one thing or the other. When you are talking about the collapse, there is a question about the variables in which the collapse is going to occur and, if you say that these collapses are at this very high level of integration, it's kind of like saying that the center-of-mass of something is at some place. There's still a tremendous amount of other process that could be going on. It's like dealing with a kind of relative coordinates. There are so many ways to slice the process up into what variables you're going to use. So, yes, it's quite conceivable that there could be lower-level events going on that are not incompatible with the high-level events. They occur basically in higher-level variables like the ones I'm talking about, namely, the electromagnetic field, which is more overall.

Clayton: And indeed these lower levels could serve the same function that consciousness serves?

Stapp: I think you have to say that, if you want to follow this idea that experiences are real things, and you want to say it has this role in the world that quantum mechanics in this simple point of view gives to it. If you say that it basically makes a choice that this happens, then you also need something further down the line—other things that are possible that won't have this experiential quality. And if it can happen in simpler systems, you can certainly ask the question, "Well, can't these simpler sort of collapses occur in our brain also?" I can't see a reason why it couldn't happen and not be incompatible with the higher-level kind of events that we are conscious of.

Here I am and I suddenly say, "I'm going to raise my hand or lower my hand." Within this scheme I'm proposing, it's quite possible that an experiential event says "raise," and it raises because an experiential event actualized a certain state of the brain, and that has consequences extending through other mechanisms of the brain, which cause these neural impulses to go out. The conscious thought is organizing the brain in such a way to make this thing happen, and that's at the very simplest level.

Part III

Fundamental Processes

11

The Primacy of Asymmetry over Symmetry in Physics

JOE ROSEN

Introduction

Charles Hartshorne includes in one of his books (1970, pp. 205–226) a chapter, "The Prejudice in Favor of Symmetry," in which he discusses the symmetry and asymmetry of logical relations. Here are some extracts from the chapter, not in their original order. I think they give the essence of what Hartshorne is aiming at.

> Symmetry is in a sense a lack of order. (221)
>
> From symmetrical relations asymmetries cannot be derived! (218)
>
> Look to the asymmetrical relations and the rest will tend to take care of themselves. (223)
>
> Here, as always, symmetry is a partial or abstract aspect of what, in its concrete wholeness, is an asymmetry. Yet in its partial role symmetry is as ultimate as asymmetry. (221)
>
> This pattern, *symmetry within an overall asymmetry*, we meet again and again. I see in it a paradigm for metaphysics. (210)

Especially in the chapter summary:

> In this chapter I have argued that non-symmetrical concepts are logically primary, and symmetrical concepts derivative. Yet both are needed to make an intelligible philosophy. The two things to avoid are taking symmetry as primary, and failing to do justice to symmetry in its proper subordinate role. Metaphysicians have

> tended to commit both of these mistakes in different aspects of their systems. (226)

And finally:

> Another example: . . . In other words, the appeal to symmetry is not a sufficient argument. (I am aware that in physics it is often taken to be so. I have not, in my ignorance, been able, so far, to interpret this in terms of the views set forth in the present essay.) Asymmetry is basic in formal logic; with what right does one assume the reverse everywhere else? (222–223)

Hartshorne actually has more to say in the chapter, such as comments on process taken as creation of novelties and, following Čapek (1961), on the primacy of time (asymmetric) over space (symmetric). But the above sampling will suffice for our purpose.

In the present chapter I respond, as a physicist, to Hartshorne's complaint about appeals to symmetry in physics, showing how and when such appeals are justified. Then I show that, just as in Hartshorne's "paradigm for metaphysics," even in physics asymmetry is primary.

Symmetry in Physics

For those readers who feel the need to get up to speed in symmetry, and especially symmetry in science, I dare suggest two books I have written for that very purpose. One is a semipopular introduction (Rosen, 1975), while the other is considerably more sophisticated (Rosen, 1995).

The appeal to symmetry as a sufficient argument in physics, with which Hartshorne expresses his dissatisfaction, is usually based on either of two principles. One principle is simplicity, also called Occam's razor, which is:

> Unless there is evidence to the contrary, we assume the situation is as simple, i.e., as symmetric, as possible.

For example, are the laws of nature, such as the speed of light in vacuum, the same in all directions? With no evidence to the contrary, we immediately assume the simplest, that the laws of nature are indeed the same in all directions. That also goes by the name "isotropy of space." The symmetry here is symmetry of the laws of nature under all rotations about a point.

How is it that symmetry and simplicity are practically synonymous? Consider the previous example. Clearly, having the same laws of nature in all directions is a simpler situation than having different laws of nature in different directions. Instead of a different physics for each direction, we more simply have the same physics valid for every direction. That is the essence of the simplicity-symmetry connection. Symmetry involves some kind of regularity, so the same physics repeats itself. That is simpler than a different physics for each instance.

For a more general consideration, start with Hartshorne's statement (1970, p. 221), "Symmetry is in a sense a lack of order." Order is practically synonymous with distinguishability, discriminability, irregularity, and heterogeneity. Indeed, symmetry is inversely related to order, distinguishability, discriminability, irregularity, and heterogeneity. On the other hand, distinguishability, discriminability, irregularity, and heterogeneity are inversely related to indistinguishability, indiscriminability, regularity, and homogeneity. So symmetry is directly related to regularity and homogeneity (to keep things brief). Thus, we have "the more symmetry, the more regularity, and the closer to homogeneity," and vice versa.

Now, if we take simplicity to mean requiring as little physics as necessary, then regularity and homogeneity do imply simplicity; the same physics is repeated, rather than additional physics being required. The more regular and the more nearly homogeneous, the less physics needed and the simpler the situation. Thus simplicity and symmetry are directly related.

Such considerations are tied to the logical principle of insufficient reason:

> Absent reason for a difference, rather assume no difference.

For the example of the speed of light in vacuum, with no supporting or contradicting experimental evidence, let us assume the speed of light in direction A is different from that in direction B. Then the speed of light will be greater in one direction, say A, than in the other. But what reason do we have to assume this rather than the opposite, that the speed of light is greater in direction B? With insufficient reason, we are guided to give up the assumption and instead take the speed of light to be the same in both directions.

Now, it can happen, and indeed it has happened and will surely happen again and again as we delve deeper into nature's workings, that the acquisition of new experimental evidence obviates the principle of insufficient reason. An excellent example has to do with symmetry of the laws of nature under mirror reflection (or more precisely, spatial inversion). For ages it had been taken for granted that the same laws of nature govern all physical systems and their mirror image systems. Stated in other words, if any physical process is allowed by nature, then the mirror image process is also allowed (or both forbidden). After all, with no evidence to the contrary and plenty in support, why assume otherwise? Well, in the 1950s nature was shown to be laughing at our simplifying assumption, as reasonable as it had been, when experiments revealed that in certain cases nature does indeed discriminate between systems and their mirror image counterparts. (Those cases involve the weak nuclear interaction. See, for instance, Ne'eman and Kirsh, 1996.)

The other principle upon which symmetry arguments in physics are often based is the symmetry principle (Rosen, 1995, 104; 1975, 108):

> The symmetry group of the cause is a subgroup of the symmetry group of the effect.

Or less rigorously:

> The effect is at least as symmetric as the cause.

The symmetry principle is derived from the existence of causal relations in science (or in any other field, for that matter). The idea is that, knowing the symmetry of a cause, one has excellent reason to assume the same symmetry (and possibly even more) for any effect of the cause. (I call that the minimalistic use of the symmetry principle [Rosen, 1975, 106]. The symmetry principle can also be used, knowing the symmetry of an effect, to place a bound on the symmetry of its cause, which I call the maximalistic use [Rosen, 1975, 121].)

As a physics example of a symmetry argument based on the symmetry principle, we might take any electrical circuit consisting, for simplicity, solely of interconnected batteries and resistors. For this example the batteries, the resistors, and their connections uniquely determine the currents in all the wires. So the connected circuit is a cause and the currents are its effect. Thus one can take any symmetry the circuit might possess and validly use it as sufficient argument for the currents to possess the same symmetry. If, for instance, the circuit has reflection symmetry and is thus equivalent to its mirror image, then the currents in the circuit will possess reflection symmetry too. (Fortunately, the weak nuclear interaction is not involved here.)

That works as well outside physics, wherever there are causal relations and the symmetry principle is valid. In mathematics, one can consider an equation as a cause and the set of its solutions (but not any individual solution!) as an effect. If, for example, a polynomial equation in one unknown x contains only even or only odd powers of x, it is symmetric under change of sign of x, $x \to -x$. By the symmetry principle, the set of roots of the equation must also possess that symmetry. So we are assured that the nonzero roots of such an equation consist solely of positive-negative pairs.

The symmetry principle can be formulated equivalently in terms of asymmetry:

> Any asymmetry of the effect is an asymmetry of the cause.

Or also:

> The cause is at least as asymmetric as the effect.

The Primacy of Asymmetry in Physics

Here is a summary of an argument presented in full in Rosen (1995, pp. 157–161). In the final analysis, what symmetry boils down to is that a situation possesses the possibility of a change that leaves some aspect of the situation unchanged. Expressed most concisely:

> Symmetry is immunity to a possible change.

Thus the two essential components of symmetry are:

1. *Possibility of a change*: It must be possible to perform a change (although the change does not actually have to be performed).
2. *Immunity*: Some aspect of the situation would remain unchanged, were the change performed.

If a change is possible and some aspect of the situation is not immune to it, and the system rather would change concomitantly were the change carried out, we have asymmetry. Note that the same situation might be both symmetric and asymmetric under the same change, depending on which of its aspects are being considered. For instance, the statement "Beauty is truth, truth beauty" preserves its meaning but changes its graphical appearance under beauty ↔ truth interchange.

Change is the production of something different. For a difference to exist, in the sense of having physical meaning, a physical gauge for the difference, a standard, a frame of reference, is needed. So the existence of a standard is necessary for the existence of the difference and the possibility of change. And the nonexistence of an appropriate standard makes a putative change impossible.

To be able to gauge a difference, a standard cannot be immune to the change that brings about the difference, to the change for which it is intended to act as reference. Otherwise it could not serve its purpose. As an example, consider proton-neutron interchange. That change can be gauged by referring to a standard proton, say, preserved in the vaults of the National Institute of Standards and Technology (NIST). If your system consists of a single neutron and you perform proton-neutron interchange, you have indeed made a change; before, your particle was different from the NIST standard, while afterward it became the same. In this example, the standard proton is indeed affected by proton-neutron interchange; it changes into a neutron, which is something else.

On the other hand, a standard proton-neutron pair cannot serve as such a gauge. If you attempt proton-neutron interchange on your system consisting of a single neutron, you find no change at all. Before your attempt your particle was the same as one of the standard pair, and after your attempt it was still the same as one of the standard pair, so you made no change at all. And the useless standard pair is indeed immune to proton-neutron interchange, under which it remains a proton-neutron pair (or neutron-proton pair, if you will, but still the same).

Thus for a change to be possible for a situation, there must be some aspect of the situation that is not immune to the proposed change and can serve as a standard for the change. So for a situation to possess symmetry, it must have both an aspect that can change (serving as a standard for the change and giving the required possibility of a change) and an aspect that does not change concomitantly (giving the required immunity to the possible change). In other words, the possibility of a change, which is a necessary component of symmetry, is contingent on the existence of an asymmetry of the situation under the change. Hence the result:

Symmetry implies asymmetry.

Asymmetry is a necessary condition for symmetry. For every symmetry there is an asymmetry tucked away somewhere in the world.

As an example, consider a uniform equilateral triangle. We normally ascribe it the symmetry that its appearance is immune to 120° rotation (about the axis through its center and perpendicular to its plane). We normally do that, because we have plenty of standards for 120° rotation, such as the walls of the rectangular room (asymmetric under 120° rotation), so 120° rotation is a possible change. But if the equilateral triangle were a universe unto itself, there would be no standard for 120° rotation, so it would not be a possible change, and the triangle would then not possess the symmetry we normally ascribe it. The symmetry exists because the *total* situation, that of the equilateral triangle together with its surroundings, does possess aspects that are not immune to 120° rotation and can serve as standards for 120° rotation. The equilateral triangle is symmetric only in the context of its surroundings.

We can summarize our results concerning symmetry, change, immunity, standard, and asymmetry in the following diagram, where arrows denote implication:

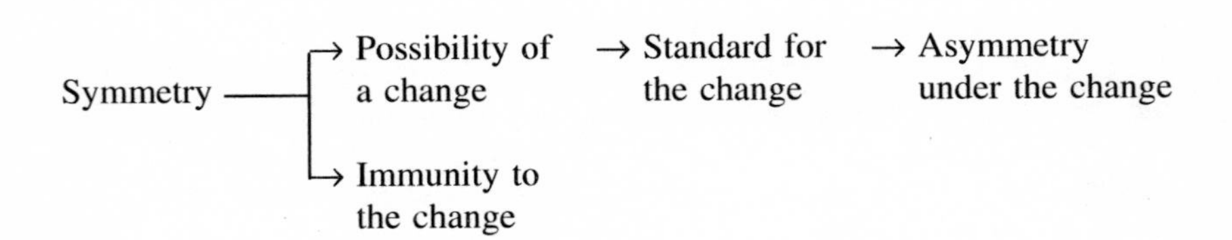

Thus, for there to be symmetry, there must concomitantly be asymmetry under the same change that is involved in the symmetry. The existence of the symmetry depends on the existence, somewhere in the world, of a corresponding asymmetry.

Such considerations can be extended. Any finite physical system might, at least theoretically, possess perfect symmetry, i.e., immunity of all aspects of the system to some change. The standard for the change would, as usual, lie in the system's surroundings. The universe, by definition, has no surroundings, so any standard for change of the universe must be contained within the universe itself. Thus a putative perfect symmetry of the universe would be a self-contradiction. As all aspects of the universe would be immune to whatever change the universe is supposed to be symmetric under, there would be no aspect that could serve as a standard for the change. So the change would not be possible, and there would be no symmetry. Hence the cosmic conclusion:

> The universe cannot possess perfect symmetry.

As an example, consider spatial displacement symmetry, immunity to displacement from here to there. As far as we can tell at present, the laws of physics are displacement symmetric; the same laws seem to be valid everywhere. Could the universe then possess perfect spatial-displacement symmetry? If it did, all aspects of the universe, and not only the laws of physics, would be immune to spatial displacement. But then there would be absolutely no difference between here and there, there would be no standard for spatial displacement, and spatial displacement would be impossible. As a matter of fact, it is the inhomogeneous distribution of matter in the universe that allows differentiation among locations and serves as a standard for spatial displacement.

References

I would like to express my thanks to Tim Eastman for bringing Hartshorne's chapter to my attention.

Čapek, Milič. 1961. *The Philosophical Impact of Contemporary Physics*. New York: Van Nostrand Reinhold Co.

Hartshorne, Charles. 1970. *Creative Synthesis and Philosophic Method*. La Salle, IL: The Open Court Publishing Co.

Ne'eman, Yuval and Yoram Kirsh. 1996. *The Particle Hunters*, 2nd ed. Cambridge: Cambridge University Press.

Rosen, Joe. 1975. *Symmetry Discovered: Concepts and Applications in Nature and Science*. Cambridge: Cambridge University Press. Expanded and reprinted. New York: Dover Publications, 1998.

Rosen, Joe. 1995. *Symmetry in Science: An Introduction to the General Theory*. New York: Springer-Verlag.

12

Spacetime and Becoming: Overcoming the Contradiction Between Special Relativity and the Passage of Time

NIELS VIGGO HANSEN

For us believing physicists, the distinction between past, present and future is only an illusion, even if a stubborn one.[1]
—Albert Einstein

Passage versus Physical Extension: A Classical Problem

Modern science has taught us that the passage of time doesn't really fit into physical reality. In the world of our experience, there is an obvious difference between the facts of the past, the acuteness of the present, and the open possibilities of the future. But in the universe disclosed by modern physics, the notion of a "now" seems to be inconsistent, let alone the "passage" of this now through the continuum of time.

Since the time of experience seems to be at odds with the scientific concepts of space and time, some have drawn the conclusion that everyday notions of change and becoming are illusory. Others have taken this inconsistency to show that scientific abstraction blocks an understanding of the depth of fundamental questions of existence and temporality. Others claim that a coherent understanding of time is a metaphysical chimera.

This chapter is an attempt to outline a fourth response that is more adequate. It points to a way of overcoming the contradiction by realizing that it depends on certain tacit assumptions in the interpretation of physical continua of space and time, and of the temporal aspects of experience. Without these assumptions, even strong notions of dynamism and becoming can be compatible with the special theory of relativity. The suggested solution is a radically processual and relationist interpretation based on Whitehead's process metaphysics. It involves a reading of special relativity as a source of new and deeper (radicalized rather than weakened) understanding of temporality, in technical as well as existential applications.

The idea that there is some kind of conflict between a systematic understanding of time and the intuitive or experienced sense of change and becoming is not new. In fact, it has very ancient roots: it can be traced back at least to the origins of Western philosophy, e.g., in Zeno and Parmenides. But in the context of twentieth-century physics there is a specific version of the classical problem that seems more immune to classical solution models. The modern Whitehead-inspired solution suggested in this chapter involves a reconstruction of some ideas central to Western thinking about time. I suggest the reason why the process interpretation has not been considered seriously yet has to do with the power still exerted in our secularized culture by a certain theological framework for our ideas of time. However, the suggested solution to the problem of relativity and becoming does not require the assumption of any theological framework.

The Contradiction and Two Traditional Responses—A Synopsis

I will briefly review the structure of the metaphysical conflict, and two of the three traditional kinds of response. The remaining third major group of classical responses (which I will classify as "antiscientistic temporalism," a view that is sometimes erroneously ascribed to Whitehead) will then be treated in some detail, before I unfold Whitehead's processual reconstruction of time and temporality and its significance for the problem at hand.

The Conflict

The conflict revolves around the issue of the existence of a frame-independent simultaneity relation. The special theory of relativity (SR) explicitly abandons the assumption of the context-independent existence of a unique temporal sequence of the events populating the universe. According to SR, the sequence of events, and with it the definition of time, depends on the choice of inertial system, the working definition of some real or ideal unaccelerated entity to count as unmoving for the sake of measurements and calculations. If two events are situated in such a way that a light ray or any other kind of causal influence from A can reach B (that is, if the speed of light is enough to reach B from A, so that B is within A's "light cone"), then their sequence is always A before B although the measure of time elapsed may be different (cf. the famous "twin paradox"). If B is outside A's light cone, the situation is much more ambiguous: you can define frames of reference within which it is before, simultaneous with, or after A—whatever you prefer. Furthermore, according to the principle of special relativity, there is no feature of the physical universe that could make the choice of inertial frame anything but arbitrary. The frame used for measurement and calculation is selected for convenience in the context. In fact, physicists and engineers routinely operate with several frames convenient in different respects, using the mathematical hub of SR, the Lorentz transformations, as the translation procedure.

The reason why this relativistic ambiguity of simultaneity and sequence is in conflict with standard notions of 'dynamic time,' 'temporal becoming,' or the 'passage of time' is that the passage of time has traditionally been strongly associated with the image of a universal ontological state of affairs. If time really passes, and if there is a real difference between past, present, and future, we tend to assume that our present is part of a universal present: an instant or very tiny timespan which is now everywhere. Consequently, for every event in the entire cosmos there would be an absolute temporal fact about its belonging to one of three regions: it is now in the future, in the present, or in the past. But for this to be the case in every event that can be here-and-now, there must be a unique series of events for the temporal regions to sweep through. Equivalently, there must be an ontologically unique relation of simultaneity defining B as simultaneous with A if it is present when A is present. Clearly this global temporal fact contradicts the principle of the equal status of all possible inertial frames, each defining a different sequence of events, and a different relation of simultaneity.

How could a real "now" pass through time, or through events in time, if the definition of time and the sequence of events are only conventionally defined according to local preferences? Given SR, time and event sequences cannot really possess dynamic and modal properties: past, present, and future must ultimately be illusory, as Einstein puts it in the initial quote. (For a classical explicit version of the argument that SR and dynamic passage are incompatible, see Putnam's "Time and Physical Geometry."[2])

I think Einstein, Putnam, and many others are mistaken: temporal becoming fits in simply and beautifully with SR. But in order to see this, we will need to construct an alternative to a metaphysical prejudice common to almost all Western thinking about time.

The Antimetaphysical Response

The antimetaphysical response to the contradiction avoids such speculation, claiming that metaphysical speculation gets us into trouble in the first place. On this view, everyday "lived" time and technical calculated time are part of different language games and practices, and it is a mistake to require them to cohere exactly. In particular, there is no single entity which is referred to in all the different contexts where the word *time* is used. Rather than trying to ground things in abstract entities, we should look at concrete practices.

This response expresses the important insight that the unchecked hypostatization of local, specialized bits of language creates metaphysical problems (cf.

Whitehead's analysis of "the fallacy of misplaced concreteness"). However, this insight is not enough to sort things out. The attempt to create coherence between language and practices, and the creative metaphorical use of language to do things beyond its "native" field of practice, is not something we could simply learn to renounce. If we could, language would become dull and impotent. It is interesting to notice here that even the most brilliant contributions to the reorientation of philosophy toward concrete sociolinguistic contexts, such as Wittgenstein's, are thoroughly speculative in the sense that they are mediated by creative interpretative application of certain terminologies and metaphors—"language game" and "family likeness"—far beyond their native use. Furthermore, the adoption of an antimetaphysical attitude by less brilliant minds often seems to lead only to the suppression of explicit metaphysical discussion, not to overcoming implicit metaphysical assumptions. Therefore, the antimetaphysical response to the conflict is very often combined with one of the other classical responses: either assuming the metaphysics of passage or the metaphysics of temporal extension as unquestionable, and subjecting only the other side of the conflict to antimetaphysical scrutiny. In fact, it appears that in questions of time, change, and process it is particularly difficult to avoid metaphysical assumptions. Whitehead's suggestion (echoing Hegel's) is that we can make the critical insight useful by turning it explicitly constructive, making our inescapable metaphysical activity the philosophical project rather than the problem. We must learn to understand abstraction as a concrete process.

The Scientistic Atemporalist Response

The scientistic atemporalist response accepts the expressions of fundamental scientific theory, here SR, as revealing the structure of reality. Consequently this response implies that the "passage" of time is some kind of illusion, a "myth," a derivative phenomenon to be explained on the basis of physical theory. *Temporalism*, here and in the following, denotes that time is "something more" than a kind of extension or a series, whether this something is expressed in terms of passage, ontological or modal difference between past and future, emergence, or becoming. Conversely, *atemporalism* accepts the existence of a temporal continuum or series, but denies this "something more."

In order to offer a coherent solution to our problem, scientistic atemporalism must show it is conceivable that the phenomena, experiences, and practices of the language of temporality (terms such as *change* and *tomorrow* as well as more speculative ones: *emergence*, *nowness*) can be reconstructed or explained away on the grounds of basic physical theory. Indeed, some atemporalists claim to have accomplished this task, but only by taking temporality in a very restricted sense.

Some confusion has arisen because the temporalist notion of becoming is intuitive, and more or less vague. Some atemporalists as well as some temporalists have confined the discussion to one isolated aspect of temporality, the "now" allegedly passing through a continuum of extended time or a sequence of events. Grünbaum, Williams, and other atemporalists have argued that the function of "now"—and with it the other pure temporal modalities, "future" and "past"—can be fully accounted for in terms of indexicality, that is, the place and perspective of an event within a sequence of events of ontologically equal status. If temporality had no more to add to such indexicals than the reference to a completely disembodied inner feeling that "now is now" and "now passes," then the atemporalist reduction would have a strong case. And it would not be using much of the furniture of fundamental physics, just the continuum of physical time.

However, it is obvious that concrete temporality is not just characterized by *position*, the sense of presence of a particular event and absence and distance of events, but also by *orientation*, the sense of difference between past events (remembered and/or traceable to some extent) and future events (never remembered, not traceable, and considered open possibilities). This modal asymmetry is very evident in the experienced difference between our access to past events (memories and traces, at least partial) and to future events (no memories and traces, but an experience of openness to planning and unplanned surprises). A scientistic atemporalist account of this must combine the indexicality with some appeal to the physical and biological accounts of cognition and volition. What is of interest in our discussion is not the details of such accounts of complex interactions, but only the way that the physicalist reduction must ultimately anchor them in the fundamental laws of physics. This must happen via some invocation of ordinary physical causality. There is a general stream of causes from the world into the cognizing system, forming the one-way stream of traces and memories, and another one-way stream of causes from the mental/neural to the world. For now, I will not challenge the idea that some kind of reductionist account of that type could in principle be correct and complete. But this puts the metaphysical burden on the scientistic atemporalist to account for the asymmetric structure of causality.

There are many attempts at such accounts of asymmetric causality and asymmetric streams and losses of information (and hence of the apparent, lived world's overwhelming asymmetry) based on the assumption of a fundamental microphysical reality of symmetric nature. Invariably, they depend on one of two anchor points in "fundamental" modern physics: the second law of thermodynamics, or the problem of measurement in quantum mechanics (the famous "collapse of the wave function"). I have argued elsewhere that neither of these anchor points serves the purpose of explaining the asymmetric temporal features of the concrete, lived world.[3] Here I must leave out a detailed discussion of various strategies and versions of such explanations, and just state my conclusion: The invocation of thermodynamics or quantum mechanics as resources for a scientistic atemporalist account of temporality either breaks down by simply resting on implicit assumptions of temporalism, or depends on some version of an anthropic or transcendental argument. This latter argument not only construes temporal nowness, passage, and orientation as "mind-dependent" and as having "no physical correlate," but also delegates the larger part of physics' explanatory power into the abyss of "mind-dependence."

In conclusion, it does not seem plausible that a workable model for coherent scientistic atemporalist reconstruction of the structure of apparent temporality has been found. This by no means proves classical temporalism, which is connected with equally serious problems. But it weakens the idea that a coherent understanding can be achieved purely on the basis of physical theory. It pushes us back either into the antimetaphysical idea renouncing our requirements of coherence beyond local language games, or compels us forward to constructing new models of temporalist accounts.

Temporalist Antiscientistic Responses

A third main group of traditional responses to the controversy is based on *temporalism.* Explicit temporalism is very often formulated in reaction to scientistic atemporalist understandings of time.

On temporalist views, immediate experience and participation in life gives evidence of a fundamental immediately evident fact of temporality. 'Change,' 'becoming,' and 'passage' express something that is not only immediately evident, but also necessarily involved and presupposed in the experience and understanding of everything else. So if physical concepts of time are in conflict with this fact, this demonstrates an inadequacy or limitation in the physical concepts. Thus we repeat the structure of the atemporalist response: the idea of a fundamental or true time whose nature is adequately expressed in concepts belonging to one side of our clash between spacetime and becoming—and whose nature is distorted in concepts on the other side. But the roles are now reversed, so that it is time as a continuum, as in scientific and technical use, which is diagnosed as the derivative and perspective-distorted view of time. In this sense, temporalism tends to link with antiscientism.[4]

Characterizing 'Temporality'

Temporalism requires a dynamic concept of temporality to express immediate fact. But what belongs to this fact and its expression? In our discussion, we have used *temporality* loosely to imply a realist interpretation of experienced time, and to include some more or less overlapping commonsense and common language features such as nowness, change, passage of time, modal difference between past and future, and unidirectionality. But is this list complete, and does it point to a singular phenomenon or concept at all? We have already seen a certain ambiguity regarding which aspects of 'passage' the atemporalists should be required to reconstruct as a perspective effect. As we saw, the answer to this question strongly affects the plausibility of the atemporalist argument: if the atemporalist is given the benefit of the doubt, and is only required to account for 'nowness,' the argument is much stronger than if the unidirectionality of memory and causation is included.

So it is part of the philosophical project of a temporalist response to provide a systematic expression of temporality in the first place. Temporalist philosophers may point out that it is exactly because change and becoming have been bracketed out by theoretical thought that we have very explicit concepts of temporal extension but very vague concepts of temporality. But a temporalist response to the clash that does not simply fall back on an antimetaphysical claim of incompatible discourses must assume a twofold task: first, temporalism must be defined and then the construction of physical time, time as extension, must be accounted for.

Classical Temporalism

Some proponents of temporalism in the classical form of 'passage' may not agree to this project at all. They may underestand temporality as explicitly and coherently captured by an unambiguous referent of common speech of past, present, and future, identified as the passage of the now occupying each point through the continuum of extended time. They would not criticize concepts of "technical" time as derived or distorted; they would merely claim the need for adding the concept of passage (analogous to McTaggart's A-series of passage simply being superimposed on the B-series of extended time). In fact, classical temporalism expresses temporality in a way completely *dependent* on an underlying concept of extended time. Hence classical temporalism is a compromise response, attempting to give a realist construal to extension, and to passage-through-extension at the same time. However, our initial question was the problem of incompatibility in such a classical compromise.

Modern Radical Temporalism: Bergson and Heidegger

A tradition of much more radical formulations of temporality took shape in the beginning of the twentieth century, explicitly developing alternative accounts of temporality, not framed by notions of extended time, but based exclusively on the concrete experienced/lived world. Bergson and Heidegger developed particularly explicit and influential suggestions of such a radical temporalism. For radical temporalism, the continuum of time is not an underlying structure that simply needs to be supplemented with temporality in the shape of the "now" point. Rather, temporality itself is basic, and has a character completely different from extension, so that extended time is an abstraction or construction out of it, useful for particular purposes.

Bergson's Natural Temporality

In Bergson's account,[5] temporality is a fundamental common nature of all kinds of existence—mental and biological aspects of human as well as nonhuman nature, organic as well as inorganic, subjective as well as objective. The common ground is continuous change. It is essential to Bergson's point that continuity and change are two aspects of the same phenomenon, that they are ultimately joined just as they are in concrete experience. Bergson's term for this original phenomenon of temporality is *temps duree*. One might be tempted to translate this into "time of duration," but "time of ongoing" is probably more precise. In any case, it is essential to grasp the sense in which Bergson uses "duration" here: it is definitely not meant to invoke the idea of an interval in a mathematical continuum of time. Rather, he would have us look at the actual concrete present as experienced and lived. The suggestion is that if we let awareness "be with" this immediate fact, rather than the abstract understanding of it via acquired concepts of extension, it will clearly exhibit Bergsonian 'duration,' or ongoing continuous change.

This Bergsonian original fact of temporality is described as rooted in immediate experience in a very concrete bodily way. Insisting that it is not a phenomenon of disembodied consciousness, Bergson refuses to reserve it for conscious or even living beings. It is essentially a feature common to everything deserving the word *concrete*. Whenever temporality is absent in descriptions or understandings of some aspect of reality, time has been filtered out: the product of abstraction. Continuing a romantic tradition, Bergson considers abstraction to be foreign and hostile to the concrete fullness of life, which is readily available in intuition and aesthetics. Indeed, he proposes an aesthetic cultivation of sensibilities for concrete fullness, in order to counter restricting intuition through technical fixation and abstraction.

Yet Bergson is not simply supporting art and poetry against science. Instead, he tries to propose an understanding and ideal of real science as something beyond mere abstraction—continuing a romantic line of thought strongly reminiscent of Schelling's. This is essential for understanding Bergson's outspoken interest in scientific developments, particularly those which concern the understanding of time and development, such as biological evolutionary theory, and the theories of relativity.

As a consequence of this insistence on the presence of radical becoming in concrete reality, and of the insistence on the ontological primacy of concrete experience, Bergson arrives at conclusions with direct bearing on Einstein's SR and its interpretation.[6]

First, Bergson claims that SR does not have to be understood to contradict the central intuition of temporality, if temporality is construed along the lines of Bergson's *temps durée*, and if one drops certain restrictions on the full and radical appropriation of relativity. They are restrictions stemming from the implicit continuation of classical assumptions, restrictions Bergson claims Einstein fails to overcome. Therefore, according to Bergson, Einstein was unable to grasp the full radicality of his own theory. As you will see, I suggest we follow Bergson in this line of reasoning.

However, Bergson also suggests that the intuition of radical temporality, allied with the core insight of relativity, reestablishes the idea of a cosmic, ontological simultaneity, the common state of affairs of modality. Here, unfortunately, Bergson has made his case for temporalism inconsistent, easy prey for refutation. I will take a closer look at Bergson's most central and technical argument for this. It contains a rather simple error mysteriously overlooked by Bergson. Perhaps even more mysteriously, it forces Bergson to embrace a McTaggartian picture of the relation between extension and passage.

Bergson versus Einstein on Duration and Simultaneity

Bergson uses one of Einstein's thought experiments to illustrate the notion of inertial systems and the derivation of the famous relativistic effects, including time dilation and relativity of simultaneity relations. In Einstein's example, a train and an embankment correspond to two inertial frames of reference, i.e., they are both unaccelerated, but there is relative movement at a speed assumed high enough for relativistic effects to be notable. The only other elements in the Einsteinian example are two strokes of lightning, singly striking each end of the train. Einstein's project is to deduce the consequences of assuming that the laws of nature, including the speed of light, are invariant to change of inertial frame of reference. Since this results in conflict with the classical assumption of an absolute relation of simultaneity, Einstein avoids that assumption in the first place, and ends up disproving it. Retaining the speed of light as an invariant, he can introduce standards based on it for measurement of length and for simultaneity, unambiguous but *relative* to either train or embankment, i.e., to either frame of reference. Basically this is all Einstein needs to derive the famous transformation rules for converting measurements of time, length, mass, etc. in one inertial system to those in another, given their relative velocity.

Einstein thereby derived the Lorentz transformations, which were already well known. However, Einstein gave them a radically new interpretation as transformation rules between equally valid frames, rather than deformations in relation to a fundamental frame. Bergson was eager to embrace this equal status of frames of reference—which was, for Bergson, the core of relativity and a support for the claim of a true immediate time inherent in every concrete phenomenon.

Bergson complains that Einstein's interpretation of his own procedure takes the computed abstractions for real time and space, and forgets the immediate fact of temporality, which is, for Bergson, what relativity should be about. There is a true or proper time for any frame of reference—this is in accordance with Einstein's assumptions, except that Einstein would object to any ontological significance of "true." For Einstein, it simply means the reference system is fixed to a clock. For Bergson, "true time" is identified with the immediate time for some concrete dead or living thing, or a group of such things, and this is the only time that could possibly be experienced or observed. The modified times and spaces described in the Lorentz transformations are not real because they are not observable. As Bergson claims, they are abstract calculations and constructions for coordination. Thus, the "dilated" time in the train, as calculated by an observer sitting on the embankment using the t' Lorentz transformation, is not real; it is what would be the case for an impossible *observateur phantasmagorique* experiencing things from a perspective (the train's) moving in respect to its own body at astronomical speeds. The real and observable is what is available to a real observer, who is always a concrete living being in a concrete physical environment, and in his or her own real time; the only time real and observable in his or her body's frame of reference.

This leads Bergson to a frontal attack on Einstein's use of the train example (and the generalized mode of derivation it represents). The argument is extended and full of brilliant phenomenological arguments concerning the need for understanding abstractions of time in their connection to concrete temporality. As far as relativity and simultaneity are concerned, the hub of the argument against Einstein can be stated as two rather simple points. First, Bergson insists that the equal status of all inertial frames of reference means that all relativistic effects, such as "time dilation," are perfectly mutual and symmetrical. No matter whether we are in the train or on the embankment, the time of the other system always *appears in abstract calculations* to be dilated, whereas concrete time is not. Since it cannot possibly be the case that time measured or experienced on the embankment is shorter than time measured in the train *and* that it is also shorter in the train than on the embankment, and since by "complete relativity" the relativistic relation between the two real times is taken to be fully mutual and symmetrical, they cannot really be different. Second, Bergson claims it is not legitimate to analyze events happening within some real rigid object (such as the train) using a frame of reference different from that fixed to the object itself (by using the frame fixed to the embankment). Thus Einstein's analysis of the train-in-the-thunderstorm example, assumed to demonstrate that two lightnings simultaneous in the embankment system are nonsimultaneous in the train system, is illegitimate because it tacitly assumes the embankment system to be the true and fundamental framework for analyzing train happenings, thus overriding the constraints of 'complete relativity'. Bergson considers Einstein's conclusions to be false and conterintuitive.

These two main points in the technical aspects of Bergson's argumentation are not just in disagreement with Einstein's theory—they clearly violate and misrepresent the theoretical structures and procedures they purport to discredit. Therefore, it is very difficult to construe this part of Bergson's argument as anything but erroneous. Milič Čapek has written an insightful exegesis of Bergson's critique of SR, convincingly showing many aspects of it to be "still alive," at the very least in the sense that Bergson's claims are worthy of further discussion.[7] What Čapek shows to be alive are, first, the claim of a concrete temporality underlying the production of abstract time and, second, the claim that SR and the Lorentz transformations can be seen as respecting, rather than violating, the constraints imposed by concrete temporality, including the constraints imposed by causality (unique seriality within any chain of causal relations). However, even Čapek has difficulties making the most essential distinction between what is living and what is dead in Bergson's critique. In fact, he blurs that distinction by trying to partly defend Bergson's failed attempt at showing frame-independent simultaneity to follow from, and be necessary for, concrete dynamic temporality. By doing so, Čapek unfortunately invites the continuation of the idea that the issue of dynamism under SR is the same as the issue of simultaneity.

It is interesting in this connection to notice that Čapek points to Whitehead's treatment of the problematic as a restatement of the Bergsonian project of a temporalist reinterpretation of SR, avoiding some errors of Bergson's argument. Whitehead's solution to this problem was not fully expressed until *Process and Reality* in 1929.

It is obvious that Bergson took the scientific discovery of the immense history of nature, and especially the enfoldment of the evolution of humanity into that process, as an insight of great philosophical potential for human self-understanding and self-expression. The fact that such a development of metaphysical understanding in concert with the development of scientific constructions has been a major project for Bergson only makes it so much more striking that he and his great followers were unable to see the implications of SR as a potential for developing and refining the notion of temporality, claiming instead that the intuition of temporality is absolutely and unquestionably contingent on frame-independent simultaneity. What we are questioning in this project must be an extremely deep-rooted tacit assumption.

Heidegger's Temporality Without Nature

Martin Heidegger was very much aware of the fundamental difference between his radical notion of existential temporality and the extended time of physics, measurement, and chronology. In sharp contrast to Bergson, he made no attempt at building compromises, but rather emphasized the need for keeping them apart; not "falling" prey to the tendency of understanding temporality and existence in light of nature. For Heidegger, this tendency is a constant pull toward inauthentic understandings of temporality, not just inauthentic in the sense of a misrepresentation, but also in the more radical sense that it leads us to *live* inauthentically; the impossible attempt at escaping the openness and responsibility inherent in true temporality, which is basically *situated agency*. Furthermore, this tendency always provides us with the most immediate, available, and well-known grasp of time—the time as available within our projects rather than the temporality unfolding them—so that a determined, authentic attitude is required in order to stay clear of it. Thus, in Heidegger's version of radical temporalism, the contrast to notions of extended time is seen as a virtue, something to be insisted upon rather than overcome.

Also, in contrast to Bergson, common sense is taken as mostly an expression of the inauthentic and "fallen" understanding of time, rather than a major resource supporting the expression of more adequate concepts of time. However, this immediate fallen character of commonsense understandings is not their exhaustive characterization according to Heidegger—if we were in such a state of complete illusion, Heidegger's hermeneutic-phenomenological project could not even get started. A more precise rendering of Heidegger's view is that common sense is ambiguous, that it is some kind of mixture of authentic and inauthentic understanding of temporality; a mixture that generally leads us to insert a few patches or echoes of genuine temporality as subordinate features in a picture dominated by inauthentic, "vulgar" time. In this way, a first attempt at a Heideggerian account of the commonsense notion of passage might find in it an echo of authentic temporality, vaguely represented into quantitative time through reduction to a point, the "now," placed in the continuum. In the adequate and authentic understanding Heidegger wants to advance by philosophical means, the roles are reversed so that it is extended time that is enfolded in temporality as a particular "derivative mode." He does not accept such a point-like commonsense notion of 'present' as an adequate representation of true temporality—as most thinkers in the temporalist camp have done. In fact there is an important sense in which Heidegger does just the reverse, focusing on agency in combined terms of future (possibility) and past (situatedness) rather than presence (facts, states, and things). This is significant. It means that Heidegger's account can be taken to imply that presence, and hence co-presence, is not a fundamental and universal feature of strong temporality, but a derivative and local one. As we shall see, this implied insight is really the core of the Whiteheadian reconciliation between relativity and temporality. Heidegger did not explicitly work out the implication that co-presence is a derivative, local, and mediated pattern of relational time, and furthermore he emphatically avoided the connection between temporality and nature, particularly nature as channeled through science and technology.

Indeed, it seems that Heidegger was aware that SR posed particular problems for notions of temporality. In a footnote in *Sein und Zeit*,[8] Heidegger insists that the implications of SR for the understanding of time can only be discussed and understood once we have found and explicated the existential sense of time and take it as the basis. Heidegger may very well have written this with Bergson's hopeless struggle against Einstein in mind, wishing to stay clear of any such commitment toward a particular notion of physical time over another. In Heidegger's view, any construction of physical time is bound to be so much of an externalization and hypostatization of a certain aspect of the temporality of practice that contradiction should be expected rather than avoided. Hence, a genuine coherent understanding of the relation would only be possible by reducing any physical notion of time to the role of a practical bookkeeping device, so derivative as to have no ontological significance beyond that revealed by a hermeneutics of its underlying practice entirely independent of its technical details. On the face of it, this denial is all that Heidegger has to offer us as an answer to the contradiction.

The denial is closely related to a fundamental split in Heidegger's account of the way extended ("vulgar") time gets constructed. The basis is the project character of human existence, which interprets itself inauthentically into the three temporal modalities ("extatic modes" as Heidegger calls them, i.e., temporality projected "out of itself"). This accounts for the structure of the notions of past, future, and present, and Heidegger points out, in close accordance with Prior's idea of a temporal logic, that all other temporal notions can be constructed out of these. But of course this cannot possibly account for the actual content, the events placed in these "regions," or the regularity in these events allowing for a sequentiality of an appropriately rhythmic character to allow for the properties of real chronologies, calendars, and clocks. Therefore, in order to account for these, Heidegger makes a reference to *something* completely different confronting existential temporality as a background for its projects. This something is the rhythmical sequence of day and night, the movement of the sun making the frame for the workday and the "antique peasants' clock" allowing project-involved existing beings to meet at a previously agreed time. This reference to nature constitutes a peculiar break in *Sein und Zeit*'s developments, which have been expressed programatically in existential terms excluding nature until the very end of the book. Just because existence and its temporality is considered so perfectly transcendental, the metrical and topological properties out of which vulgar "time" is made must issue from a completely different origin, the realm of pure unmediated nature. So, while Heidegger takes important steps beyond the traditional Western metaphysics of time in his characterization of existential temporality, he reduces nature to the role of that which is encountered in the fallen/vulgar mode of this temporality.

Heidegger had the profound insight that "presence" is a particularly derivative mode of temporality, rather than the paradigmatic mode of being. But he avoids making the kind of connection between natural and human temporality that could allow this insight to solve the puzzle of simultaneity and becoming. Instead, he amplifies the contradiction by claiming that nature is at the same time the unmediated background of practice, and irrelevant for the temporality of practice.

Bergson's explicit attempt to handle the tension between SR and becoming failed because his notion of dynamic time consistent with "complete relativity" was, in the end, loaded with a classical metaphysics of presence undermining the relativity. Heidegger, although he went further in overcoming this metaphysics, claimed his radicalized temporality to be so much devoid of nature, and particularly so much in contrast with the constructions of natural science, that nobody, and least of all Heidegger, may have noticed the constructive solution to the paradox it makes possible. Generally, radical temporalist responses to the controversy have tended to fail in this respect because of anti-scientistic reservations, and moderate temporalism failed because of its reliance on the classical notion of absolute simultaneity.

Whitehead's "New" Solution: Radical but Local Temporality

None of the accounts discussed so far have dissolved the contradiction between special relativity and temporality by taking both seriously. They have taken either SR, becoming, or both to be without ontological importance, providing no specifications for what is actually or could possibly be the case in the world.

Of course it may well be too naive to imagine that the two kinds of language (the everyday talk about "today," "tomorrow," "a minute ago," etc., and physical calculations involving space/time systems such as the Lorentz transformations) are fully explicated as representations of context-independent facts. There are many good reasons why both of these language complexes are something richer and more interesting than just representations of the same pool of context-independent facts, with differences consisting of selection of different subsets of this pool, as well as procedures of projection and representation. They are something richer and more interesting because they are also aspects of collectives and practices involved in the making of facts and perhaps even the making of conditions for something to be a fact in the first place. I am not going to discuss in this context the various modes and models of constructedness and historicity of facts; I have done so in a previous article on the historicity of scientific objects[9]—for further reference to the very extensive current debates on modes of scientific realism and constructivism, see the most recent works of Andy Pickering or Bruno Latour.[10]

Still, the conflict we have been discussing about temporality is really so basic that it is very hard to see how it can be dissolved through reference to involvement, contextuality, historicity, and constructedness, unless these are taken in the sense of a very extreme kind of skepticism.

If there is any such thing as a fact, and not just a collection of incoherent states of mind and discourse, then either there are temporal facts (e.g., that you have *already* read the previous sentence) or there are atemporal facts (e.g., that your reading of it is *later than* my writing of it). If there are temporal facts, then either they have the universal character generally assumed in Western metaphysics before Einstein, or they do not.

In other words, Bergson was right that no matter how much contextuality is involved, we cannot seriously hold at the same time both that there are concrete facts involving distant simultaneity, and also that such facts cannot exist in the physical universe. Surely one could claim that such immediate facts are eliminated in the production of physical descriptions (if "description" is taken at a sufficiently theoretical level, this is undoubtedly true) but if concrete facts of co-presence are there before clocks and measuring rods are used, they will still be there in the background when these things are employed. If that is the case, then there will be an ontologically privileged system of reference, a set of events concretely simultaneous. What SR reveals is that this will be a very strange sense of concreteness, since it will not be discoverable or observable in any way. It will be one of a multitude of inertial frames of reference that will be, in any observable sense, equally valid. Unless Bergson or others will claim to have intuitive, non-physical access to knowledge about what happens on Mars or in Andromeda right now, the distant part of the "concrete fact" would be entirely devoid of anything really happening in the world or really observed. Indeed, distant simultaneity has difficulties with being concrete in another and much stronger sense than the one in which atemporalists claim passage and presence to be unreal. The sense of what is "present" is concrete, detectable, and real if anything is, just as long as we consider it apart from the question of what may be the case at the same time at very distant places.

This local concreteness is the turning point of Whitehead's radical suggestion of a solution of the paradox by unfolding an explicit account of a more general, concrete, and flexible way of conceiving the form of a temporal fact.[11] The idea is simply that *a concrete temporal fact is not global but local.*

Such a notion of local temporalism would be a very satisfactory solution to the paradox of SR and becoming, if the notion were thinkable at all. As we have seen, what creates the contradiction we have been discussing is that SR replaces the absolute relation of simultaneity and the equivalent unique sequence of events by a weaker constraint on the spatiotemporal order of events. But if real and concrete temporality can be conceived without dependence on the stronger constraint, then clearly the contradiction vanishes. In fact, Whitehead showed that the weaker constraint contained in SR is exactly what is needed.

This idea is so surprisingly simple and solves the paradox so beautifully and completely that it is a very strange phenomenon in the history of ideas that it has hardly been considered by any other contributor to the rather extensive discussion of the contrast between relativity and becoming—not even after Whitehead proposed it.

Why not? I can think of two kinds of reason for this.

The first kind of reason is the way Whitehead presented it. He developed it as an integral moment in his general metaphysics of concrete processuality, in his notoriously difficult magnum opus *Process and Reality*. This major work of twentieth-century philosophy is sadly neglected, but it must be admitted that Whitehead did not succeed in making it a user-friendly introduction to his philosophy. Rather, it is as dense and rich as, e.g., the works of Heidegger, Hegel, or Kant. Furthermore, the hermeneutic work of gaining access to a philosophical work of that kind of density is rarely undertaken unless one has the impression that it will be "deep," usually because of its reputation—which Hegel and Heidegger have much more of than Whitehead. A further complication is that Whitehead wrote another book, *The Principle of Relativity*, presenting an alternative formulation of the *general* theory of relativity (GR) but not the point about SR which is our focus here. This book is not as dense as *Process and Reality*, and the result is that the rare occasions when Whitehead's contributions to the interpretation of the theories of relativity are discussed, the main contribution is almost always overlooked.

The second kind of reason is more philosophically interesting. Why does the appropriation of this simple philosophical point depend on the readability of Whitehead's *Process and Reality*? Why did nobody else discover and present it? It is very difficult to grasp the idea because of an implicit metaphysics, probably one that projects a particular theological content into a tradition which is no longer aware of being theological. Grasping the Whiteheadian alternative was almost like a religious conversion for me—unfathomable before but simple and almost self-evident afterward.

Therefore, in order to explicate the Whiteheadian alternative, it will be expedient to first ask how it is that the traditional notion seems so obviously true—the traditional metaphysical notion that a temporal fact must be global. In other words, why is the question of becoming always posed, by its friends as well as its enemies, as if either there are global temporal facts or there are no temporal facts? It seems to be so deeply enfolded into our notion of time that it is usually assumed without saying. Very few have given any kind of argument for it. But I did find, in two of the most respected twentieth-century proponents of temporalism, some vague or partial arguments for continuing this traditional metaphysics into their formulation of temporalism.

The first proponent is Bergson, who seems to seek support for the notion by an appeal to his very sympathetic idea of solidarity between human, organic, and inorganic existence. They all share in immediate temporality, he says, and he draws the conclusion that therefore they all share one temporal fact. But as soon as we ask if this really follows, it becomes clear that it does not. The stars in the sky can have their own local temporal facts, even if they are just as temporal as ours.

The other proponent is Prior.[12] He explicitly asks if it is possible, since SR seems to imply it, to entertain the notion that there are no global temporal facts. He rejects it with an appeal to our way of conceiving and expressing concrete temporal facts. We sometimes say, e.g., "Thank God, that's all over now," and clearly refer to a very real and significant state of affairs. What we refer to,

Prior claims emphatically, is definitely something different from "Thank God, the end of that event is [atemporally] earlier than my utterance of this sentence"—in other words, we refer to what is in McTaggartian terms A-type facts rather than B-type facts. Prior takes this to demonstrate that there are facts of temporal modality—some things are now, some things are past, and then still other things have not happened yet but *may* co-occur in the future. I would like to point out that even if this kind of phenomenological evidence of temporality is basically valid, it demonstrates only the existence of *local* temporal facts, and no further. Prior's argument makes sense precisely when the past events in question are in the speaker's local past, meaning the concrete past which is able to affect him, that which constitutes his situation, that which is potentially knowable. The abstract idea of a nonlocal past provides no support for his argument. There is no reasonable "Thank God, that's all over now" about events that have happened on Mars during the last minute or in Andromeda during the last million years—events which are not yet part of the speaker's causal past. Of course Prior may say that it is possible to think or even know some scheduled event—say, a friend's exam—to be over at a distant place—say, on Mars. But then we will have to remind Prior that the problem posed by SR is just that there is no physically privileged time system to synchronize Prior's clock with the one in the Martian University. Hence, even if Prior will claim the existence of a metaphysically privileged time system, he will have no way of determining which one it is. He will either have to make an arbitrary guess of when to initiate his relief, or else phase it in gradually over the approximately 10 minutes duration (depending on the variable distance of the planets) of the relativistic ambiguity of simultaneity—or finally, he could postpone his relief until the event is definitely and unquestionably over, that is, when it is part of Prior's causal past. I trust you will join me in recommending the third option to him, the only one that has anything to do with the way feeling works in real life. And of course, this demonstrates only a local temporal fact and no more than that. Why did we inject Mars into the story, when Prior was simply making a point about ordinary temporality with an ordinary earthly example? I am sorry to depart from the solidity of Earth and practice, but it is exactly what we need to do in order to discover what Prior's earthly example *does not* imply: a global temporal fact. Prior's example functions locally and practically, and the kind of temporal fact it demonstrates is precisely local and practical.

Whitehead's reformulation of a temporal fact should not be read as something far-fetched, but rather as a (re-)naturalization: it is taken back into the concrete relations between real processes as they actually happen, and saved from the burden of carrying an instantaneous cosmic state of affairs—an ideal of global simultaneous presence corresponding to the idea of the "God's-eye perspective." But in this naturalization it is important to see that Whitehead preserves, even in a very radical form, the central insight of temporality: the ontological difference between past, present, and future, experienced locally.

Whitehead's approach is to systematize and explicate an existing, vaguer notion of processurality to the extent that traditional metaphysical notions of substance and time do not reinstall themselves as implicit background, as they might do via an idea of a time and space that processes *happen in*. Instead, Whitehead builds a relational account of time and space as a system of certain kinds of relations between processes. The relational interpretation of time is of course a proposal well known at least since Leibniz argued it against Newton's absolute view of time. It is also well known that Einstein, too, was influenced by this relationist tradition.[13] But because Einstein and others have combined a relational view of time with an atemporal view—that is, a view of time as constructed out of before-after relations, with no place for the "now" or "present" type of fact—it is important to stress that the relational account of time and space is not just able to accommodate truly temporal facts, but that also it allows the expression of a more radical notion of temporality than the traditional one based on absolute time. This radicalized notion is that of change happening in an active, immanent sense rather than events situated in a continuum traversed by a now-pointer ultimately external to the events.

So, one central point in Whitehead's system is that processes are themselves active—the temporal modalities are a function of the happening of processes themselves—the basic realities in the world pass from potentiality through actuality into pastness. Each process is a unit of becoming, and according to Whitehead it "becomes in solido"; it is not some temporally extended entity placed on an axis of time along which it could be played, like an audiotape, one stage or movement after the other. The idea is not that it is of zero extension like a point. It is the more radical idea that it is something different from extension—something out of which extension is *manifested.* Within this relational system, according to its properties, it may become meaningful to assign the process a finite extension. I have represented this idea of integral becoming by taking the arrowhead signifying dynamism inside each process unit rather than putting it on the traditional axis of "underlying" time. However, this account of modality should not be taken as only monadic. The other equally important element in Whitehead's processual and relational account of time are the relations to other processes. They are deeply involved in the very sense of the modalities. One may think of the relations represented in Figure 12.1 as a kind of family tree. They are relations between "parent" and "child" processes, causal relations in which the outcome of a parent process adds to the beginning conditions of each of its child processes. Each child process can have many parent processes, and vice versa.

Whitehead's picture implies that such branching and rejoining is generally the case. What it means for something to have already happened, Whitehead says, is that it has produced a determinate result—to terminate is to be determinate—but this result should not be understood substantially. "Result" means it is available to be part of the initial conditions for new processes. Or perhaps

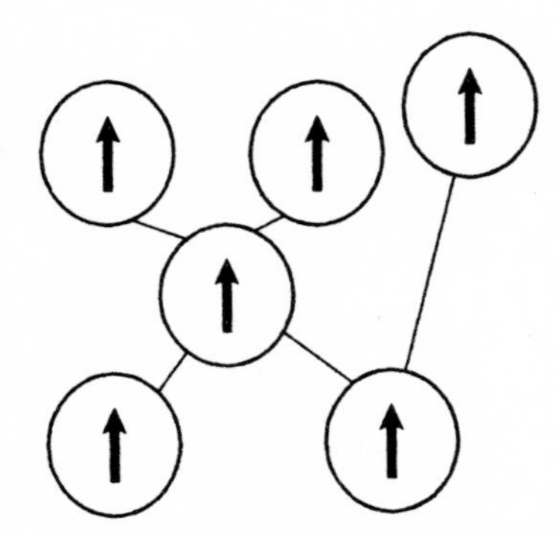

Figure 12.1. Family tree model of causal process.

better, it *is taken in* by these processes—the terminated parent processes are indeed said to be there, in some sense and to some degree "repeated" within the actively new process. It is really all that the new process can "be," apart from its own core of creativity.

Thus, a process is something happening in a particular determinate *situation* in the sense that it is causally influenced by a certain set of terminated parent processes. In turn, the process is something which produces some determinate *result* in the sense that it produces, more or less deterministically and more or less creatively, some coherent and determinate (and in this sense "unified") expression out of the multiplicity of its situation. This unified, determinate expression can be ingressed by some other, new processes.

This "categoreal" account of the simple concept of process is mostly a description of well-known aspects of what it means for something to *happen* in the concrete world. What may be controversial is the linking of it with a relationist account of time and space. Whitehead is explicit about the idea that the concrete dynamism of processes can be understood as the ground of extension rather than the reverse. This is the first element of the Whiteheadian solution to the tension between extension and becoming: the modalities are not really situated in space and time at all, but in the concrete processes whose web of relations gives rise to space and time.

If we attempt to "save" temporality through appeal to such relational concepts and internal processual temporality alone, and then identify the relational web with the time of physics, we would have a strangely dualistic picture—reminiscent of the one I argued that Heidegger creates with the dynamic units reserved for a particular kind of being: human existence. Modality would be an inner state completely out of joint with the world. Temporal facts would not only be local, they would be completely private. And what is more, the content of these private facts would not really have three modal options; it would always be "present" at the only relevant time, the time of the utterance.

Hence it would be tautological. It would be true after all, as Grünbaum claims, that becoming is an irrelevant subjectivistic addition to a world whose content is completely indifferent to it.

The other essential element in the Whiteheadian account is the explication of a strongly dynamic character of the relations in question. That is to say, dynamism in Whitehead's account is not just "inside" the processes, it is also "outside," in their relations. The ways of being related to other processes are an integral part of being a process. Therefore, many of the specifications of processuality in Whitehead's categoreal scheme have the form of explanations of, and constraints on, the ways processes relate. Whitehead describes the features of processural relatedness particularly important to our purpose as: causal, asymmetrically internal character of the relations, and the notion of a causal universe. Causal relations are the primary kind of temporal relations, and they are all that concern us in the present discussion of temporality.

Causality is asymmetric in a way that is deeply involved in the nature of a process. The process is a transition from the possibilities opened by a specific situation into a determinate outcome, and the determinacy of an outcome is nothing different from its availability as part of the situation for new processes. Causality is this one-way transfer from the terminated to the actual process, and it means that the parent process not only gives something to the child process, but in an important sense it gives to itself: it is *in* the new process, repeated in it. In this sense the relation is so strongly asymmetric that it can be characterized as internal from one side, and external from the other.[14]

Closely related to this idea of dynamic causality is the notion that each process adds something, leaves its mark on all of its entire branching tree of child processes. As a corollary, there is no completely identical repetition of any process, or, in Whiteheadian terms, any process "leaves its mark" on the subsequent branching chains of processes. Other cyclic configurations such as Figure 12.2 are not possible. This condition expresses the absence of backward causality and a universal condition of irreversibility. It also means that no character of an enduring object can be absolutely changeless, and that no change can be completely undone. Of course trust may be regained and damages repaired, etc., but not without some other aspects of the situation changing.

Another Whiteheadian categoreal constraint on causal relations essential to our discussion here is that "no two processes can have the same universe" or the same past. This "universe" of any process is its family tree of parent, grandparent, etc. processes: everything that can affect it and be available to it. It typically consists of a few terminated processes that affect the actual process strongly and a larger number of processes with a more marginal influence. The configuration that is ruled out is the kind represented in Figure 12.3: two separate processes with exactly the same universe—for Whitehead they would not be two, but only one process. Two separate processes may share all kinds of subsets of universes, and typically do, but not their entire universe.

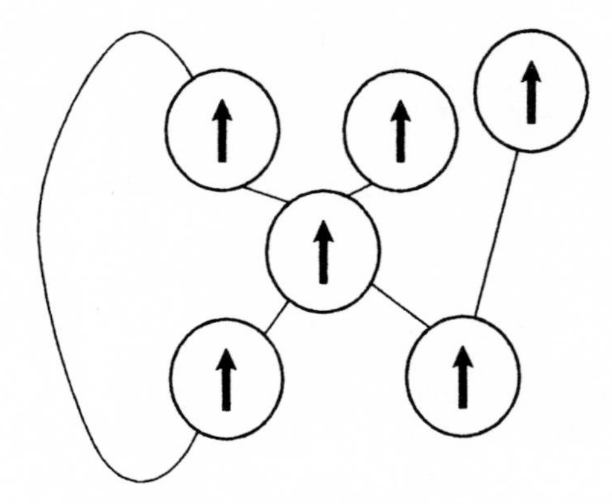

Figure 12.2. Cyclic configurations and backward causality.

It is in this notion of the unique universe of each process that Whitehead's categoreal scheme explicitly states the nonexistence of global temporal facts. Whitehead even uses the term *the principle of relativity* for the principle that the "being" of terminated entities is expressed in their potentiality for the becoming of new process entities—to be part of their specific universes. The terminology seems to imply that Whitehead was aware that this shift in the notion of what it is for the past to "be" solves the puzzle of relativity and becoming, by including relativity in a new and more coherent grammar of becoming. On this account, a fact is truly something that is "done," as the etymology suggests; it is a certain set of terminated processes available to a new actual process. Such a fact is strongly temporal; it involves the pastness of the terminated processes in this universe, and nothing in the future of this actual process. Also, such a fact is not "private" at all, in the sense of just being on the inside of the process unit. On the contrary, the pastness of everything in the universe is "public" in the sense that each of its elements is available to "everybody" in some appropriate causal neighborhood. Each becomes objective by being available as an object to new processes. And finally, such a fact is *local* in the sense that nothing secures its instantaneous transmission into objectivity for the entire cosmos. Thus, the process ontology suggested by Whitehead enables us to formulate a coherent concept of local temporal facts, and hence an idea of becoming with no dependence on simultaneity. This solves our problem in principle; contradiction is resolved.

Some Further Qualifications and Ramifications of the Whiteheadian Solution

First, you may want to raise an objection to the previous move. I went directly from Whitehead's categoreal principle of relativity to the nonexistence of global

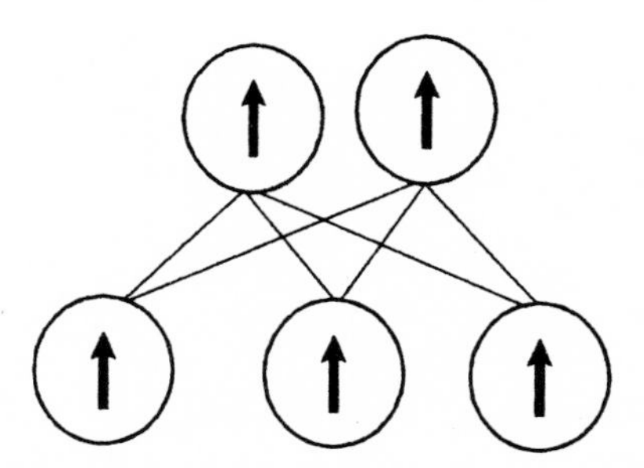

Figure 12.3. Two separate processes with exact same universe.

temporal facts, which does not follow in a strict sense. What does follow is that there are local temporal facts and not necessarily global ones. As I said, this particular principle in Whitehead's categoreal scheme only states that no two processes can have the same universe, that is, the same family tree of predecessors. However, this of course does not rule out the special "degenerate" case of a singular thread of events (Fig. 12.4). In such a case, there would still be a different temporal fact for each of the processes—namely everything that is past when it is actual—and this fact can be said to be "global" in the sense that this fact is then valid for the entire cosmos. However, it is obvious that such a single causal chain does not represent something physically interesting or otherwise useful—except in the construction of very minimalized models (e.g., Markov chains). The constraint involved in this minimalization, that there be no splitting causal trees, corresponds to an absolutely isolated system of zero spatial extension. However, in the general case involving branchings and joinings of causal chains, the Whiteheadian category implies precisely that there are only local temporal facts: the different temporal facts of different processes, analogous to the content of "backward light cones" of different events according to SR. A's universe can be included in B's or vice versa, in the special case that one is a child process to the other, but in the general case (Fig. 12.5), their universes overlap partially. In this general case there is no ultimate determination which one is first, or that they are simultaneous. Whitehead clearly advocates the same metaphysical insight in a beautifully simple form he repeats several times in *Process and Reality*: "there is no unique seriality of events."[15]

I should say a bit more about the sense of the diagrams I have used. It may seem problematic to represent something temporal in a diagram. It amounts to a "spatialization of time"—or, in this case, a spatialization of temporality that is supposed to underlie time. The problem is that different processes and different temporal facts look as if they are together, one beside the other, equally actual, available in the same sense—whereas the point of temporality is just that they are not, or at least that any two processes can only "exist together" in the sense that they are both past to some other actual process. One

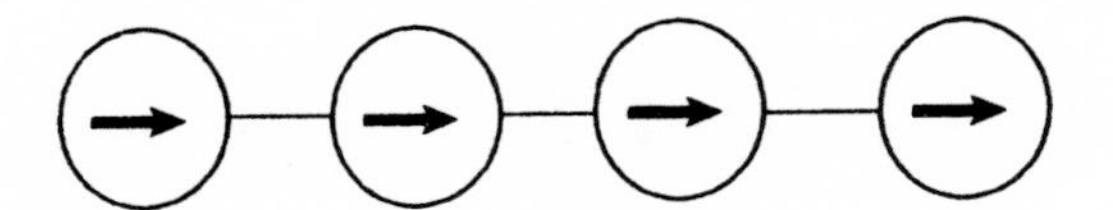

Figure 12.4. Amodal serial chain of events.

could interpret such diagrams as confirming the traditional view that processes and relations are all there in an ultimate atemporal reality, through which the ghostly "now" may or may not pass. But they need not be read that way, just as your use of a planning calendar containing entries for both 1 January and 31 December of the current year does not commit you to think that your acts and experiences of both of these days exist in an ultimate atemporal reality. The diagrams represent, just as words do, some structures of processual relationships that can be discussed without deciding on the modalities. But this question has a further interesting ramification. As long as we discuss a singular serial chain of events (Fig. 12.4), it would be possible to make a drawing representing the temporal fact of the matter of the entire system—we could draw all the past events in one color, say black to suggest solidity and ground, draw the one present event in another, say green to suggest growth, and for the future events we could either skip them altogether or draw them with some airy color, say light blue, to suggest that they are only potentialities. For a Whiteheadian branching system (Fig. 12.5), temporal modality will no longer fit into one spatial picture. It would only be possible to map the modality corresponding to one actual process, and its past and future ramifications. Notice that this dependence on one particular event would in fact not be different from what is the case in the more traditional modality map you may draw on Figure 12.4. But drawing such a trivially consistent map of modalities in the chain of events in that monolithic world, you would not encounter the question of "modalities elsewhere," as there is no "elsewhere." In the general case involving an elsewhere, there is no modal question of what is present elsewhere "at the same

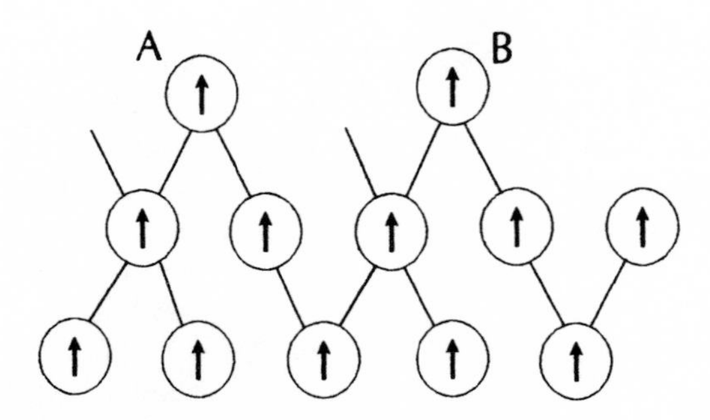

Figure 12.5. General case with partial overlap of universes.

time" as the actual event, because there is no system of time and space at this basic level of modality and causality. Time is constructed out of these relations and depends on more special constraints, regularities, and contexts than discussed so far. If conditions are met for the possibility of constructing space and time, not just one but many time systems can be constructed. Modality is not in time.

There is an interesting similarity between the notion of local temporal fact (reducing 'presence' to a very modest role) and Heidegger's *Zeitlichkeit*. Here too, temporality is more strongly tied to past and future, and clearly in the *local* sense of situatedness and potentiality. The modal reality behind presence is local actuality, the acting of the local process in question; but presence gets hypostatized into a much larger region and even a paradigm of being, while Dasein's own temporality forgets itself in the background. For Heidegger too, this huge region of presence is a construction carried by projects. But Whitehead takes this account of presence much further in two respects. First, temporality is not exclusively or even primarily situated in conscious or practice-involved processes of human existence. And second, the construction of time systems and regions of presence is well grounded in regularities and procedures Whitehead constructs with mathematical rigor.[16]

I argued above that the notion of local rather than global temporal facts is a consequence of Whitehead's categoreal characterization of the process as ontologically primary. Of course, someone may object that we are not obliged to accept a metaphysical principle because it follows from some particular speculative construction. This is obviously true as long as the system has not been established from first principles, or empirically, or however else one might imagine a metaphysics to be grounded. In any case, Whitehead's process metaphysical scheme is not intended as something which is established, or which ought to get established, once and for all. Rather, it is an attempt at contributing to an ongoing enterprise of constructing and refining concepts and grammars, seeking to integrate as many aspects as possible of evidence, experience, and practices in a coherent conceptual structure. Such an attempt should not exclude any practical, scientific, or experiential background as "overseen halves of the evidence," and at the same time it should not just aim at superimpositions of unconnected ideas but at the production of richer and deeper concepts to make meaningful and relevant contrasts out of what appears at first to be contradictory, irrelevant, and/or useless. So this is the gentle tone of voice that should be heard in the process metaphysical suggestion: it is proposed as a conceptual structure that may make coherent sense of vastly different and apparently contradictory fields of language and practice, which is exactly what experienced temporality and scientific time are. And it seeks legitimacy not by pointing to aprioristic authority from some unquestionable fundamentals somewhere else, but rather from turning out to be richer and more useful than other schemes in this kind of situation. So in looking at Whitehead's systematic development of the concept of process, we find not so much authority as instead a particularly rich, radical, and flexible structure.[17]

But shouldn't the metaphysical project be committed to affirming exactly such experiences and notions as presence, co-presence, and simultaneity, which are such a dominant feature in language and experience, and also in most physical descriptions except for the anomalous SR? Yes, it must be committed to showing how and when these structures work, but it is not committed to taking them as fundamental, if that creates conceptual stiffness and closedness. And indeed it can be argued (as I did above when we discussed Prior's "Thank God it's over") that the notion of local temporal fact covers our concrete experience and language at least as well as the classical notion of global temporal fact. Furthermore, as I mentioned, Whitehead has given a detailed account of the way time systems can be constructed given certain kinds of regularities in processual relations, so that a common time system can be well established in practice, accommodating all the well-known uses of chronology and chronologically determined simultaneity.[18] Also, Whitehead argues that this construction of simultaneity is not just a scientific procedure. We have a very vivid immediate impression of being in a space of presence, Whitehead argues, because such construction—ongoing and spontaneous—is always involved in human and probably many other complex processes. What happens, he argues, is that systems with a sufficiently "complex" mode of experience will take reliable processual regularities into account and thus project influences and impressions into a scheme of the kind of extension—e.g., the dimensionality—supported by all processes in a sufficiently large region ("cosmic epoch"). We are participants in a rhythm of the epoch, and through this rhythm, Whitehead accounts for the complex fact that we continuously sense ourselves and everything else in a very real and unlimited space of co-presence, even while everything sensed is causal past. So in the end, on the process account, we are well justified in finding our friends and all kinds of other things in the present, but this is a practical presence that need not and cannot be supported by global temporal facts in the fundamental modal/causal sense. Practical simultaneity is of course not just a question of sensation, it is very much a question of coordination between organisms and other processes sharing partial pasts and futures—a question of what Whitehead termed "unison." Unison and rhythm, two musical metaphors, have the advantage of showing this kind of accordance to be something produced by participants.

The speed of light is very high compared with all of the processes and movements relevant for our projects. Therefore, in practical earthly engineering, the zone of "elsewhere" (which is completely out of reach of the causal trees of any given event) is temporally very thin. A very precise practical synchronization can be achieved without worrying about the choice of inertial system—any choice will do; the difference will be microscopic. But the speed of light is not part of Whitehead's incorporation of "relativity" into the metaphysics of process. Whitehead implies there is a difference between physical and metaphysical relativity. Besides being interesting and important as a physical theory, SR became a provocation—providing both a possibility and a necessity of making implicit a principle of even greater generality: a basic feature of temporality and causality altogether. Physical relativity is much more specific and detailed, but it includes and implies the metaphysical principle Whitehead explicated as significant in understanding things also in fields, with little relation to geometry and electromagnetic radiation. If Whitehead is right about this, it will be because local processuality is a useful grammar also in understanding history, thought, music, etc.

In contemporary discussions of the issue of simultaneity and relativity, it is frequently argued that relativity's challenge to classical temporalism has vanished because the currently favored cosmological solutions to the equations of Einstein's *general* theory of relativity imply the existence of an inertial system with a uniquely simple relation to the overall cosmological evolution and is objectively definable (by the isotropy of the 3°K background radiation). It may or may not be true that a more or less context-independent chronology can be defined thus, but I think the challenge of SR has a deeper character than what can be resolved by this response. First, because the conditions expressed in the equations of relativity have to be more fundamental than the features of the cosmological models *satisfying* GR, whether or not it is true that all reasonable solutions possess the appropriate kind of symmetry (all solutions corresponding to an inhabitable cosmos or a cosmos in other respects similar to what we observe). And, more generally, this is not a satisfactory way of handling metaphysical problems. True, the history of metaphysical questions and answers has always been deeply involved with scientific developments. If Whitehead is right this must be so, because the metaphysical project is really a necessary ongoing improvement of concepts and grammars to take into account new elements of practice, culture, and science. But metaphysical development requires the production of deeper coherence, not just the concatenation of results.

Therefore, if one fundamental theory of science is in conceptual conflict with what we take to be fundamental to everyday time, this is a serious metaphysical problem even if another scientific theory gives some hope that the conceptual conflict will not turn into a direct empirical conflict. In other words, adopting this response as a defense of classical temporalism is accepting that the question of passage-becoming is simply a yes-no question depending on findings of physical cosmology. But I suspect most adherents of classical temporalism would not accept a negative conclusion from physical cosmology regarding the existence of a privileged time system as evidence simply dismissing temporality, just as Prior would not accept such a consequence to follow from SR.

Theological End Note: Where Does God Stand on the Simultaneity Issue?

Seen in the light of the proposed renaturalized understanding of presence as local, the global presence implied in notions of universal time is the ideal of transportation and mediation so perfect that it becomes independent of incarnation in real-world transporting and mediating agencies. The ideal, positive, and unmediated existence of a universal present may be the continuation, into apparently post-theological thinking about time, of a deeply ingrained, Western theology of omnipotence and omniscience. It still makes it so very difficult for us to imagine processuality as concrete and local, and it necessitates a very systematic and explicit metaphysics of process to start shifting this deep habit of thought. It is a very interesting dimension of Whitehead's suggestion that it is coupled with a revision rather than a denial of the connection between theology and concepts of time. If structures of a traditional "distant ruler" God inhabit understandings of time even in supposedly nonreligious modern thought, then a constructive theological shift of divine virtues from omnipotence and omniscience toward participation and wisdom could be more radical and far-reaching. A series of discussions of God, relativity, and the simultaneity issue indicate that even process theologians have yet to fully affirm the strength of their own tradition at this particular, important point.[19] Elsewhere I have given some suggestions of how this could be done.[20]

As Einstein indicates in the famous words of comfort quoted at the beginning of this chapter, developments in fundamental physics can enable deep metaphysical insights into the nature of things. But Whitehead indicated a way of grasping SR's lesson on temporality as even deeper and more radical than Einstein thought.

Notes

1. Einstein to Vero and Mrs. Bice, 21 March 1955. Einstein Archive, reel 7-245; reprinted in *Albert Einstein-Michele Besso Correspondence 1903–1955* (Harmann, Paris, 1972), pp. 537–538.

2. H. Putnam, "Time and Physical Geometry," *Journal of Philosophy*, no. 64 (1967): 240–247.

3. N.V. Hansen, "Modern Physics and the Passage of Time," in *Time, Creation and World Order*, ed. M. Wegener (Aarhus: Aarhus University Press, 1999), 121–138.

4. I use the term *antiscientistic* not in the sense that the attitude necessarily involves hostility toward the natural sciences—it may or may not—but only in the restricted sense that it rejects the "scientistic" idea that the results of natural science provide the answers to fundamental ontological/metaphysical questions. Antiscientism is the view that science, whatever other virtues it may possess, teaches us nothing about such questions.

5. H. Bergson, *Creative Evolution*, trans. A. Mitchell (New York: Macmillan, 1984).

6. H. Bergson, *Durée et Simultanété: A Propos de la Théorie d'Einstein* (Paris: Alcan, 1923).

7. Milič Čapek "What Is Living and What Is Dead in the Bergsonian Critique of Relativity," *Boston Studies in the Philosophy of Science*, 125 (1991): 296–323.

8. M. Heidegger, *Sein und Zeit* (Tübingen: Neimeyer, 1993), 417.

9. N.V. Hansen, "Interpretations in the Historicity of Objects," *Philosophia 25*, 3–4 (1996): 83–113; available from http://hjem.get2net.dk/niels_viggo_hansen/work/.

10. Bruno Latour, *We Have Never Been Modern* (Cambridge: Harvard University Press, 1993); Andy Pickering, *The Mangle of Practice—Time, Agency and Science* (Chicago: Chicago University Press, 1995); Andy Pickering, ed., *Science as Practice and Culture* (Chicago: Chicago University Press, 1992).

11. A.N. Whitehead, *Process and Reality* (New York: The Free Press, 1978). The central idea of a transformation of the metaphysics of time and change based on a notion of local concrete temporal fact is the book's main project. The explicit suggestion of locality of temporal fact is stated from the beginning in The Categoreal Scheme in Part I, particularly the 4th and 5th Category of Explanation, explicitly named "The Principle of Relativity."

12. A. Prior, "Some Free Thinking About Time," in *Logic and Reality*, ed. Copeland (Oxford: Oxford University Press, 1995).

13. See Einstein's "Introduction" in Max Jammer, *The Concept of Space* (Cambridge: Harvard University Press, 1954).

14. In close parallel to Schelling's statement that the past is *aufgehoben* in the present. One may also note the striking structural resemblance with Russell and Whitehead's set theoretical reconstruction of the sequence of numbers.

15. Of course there may be the kind of practical representations of coordinated timing instantiated in SR by the choice of an inertial system, but there is no context-independent ontologically warranted sequence.

16. Whitehead, *Process and Reality*, 294ff, for an exposition of "extensive abstraction."

17. Ibid., 3–18, for a programmatic presentation of philosophy's role and aim as one of ongoing development of deeper (more coherent) and wider (more relevant and inclusive) conceptual structures.

18. See note 16.

19. Charles Hartshorne, "Bell's Theorem and Stapp's Revised View of Space-Time," *Process Studies* 7, no. 3 (1977): 183–191; David R. Griffin, "God and Relativity," *Process Studies* 21, no. 2 (1992).

20. N.V. Hansen, "Time, Change and Construction: On Some Contributions to a Reconstruction of the Metaphysics of Time," available from http://hjem.get2net.dk/niels_viggo_hansen/work/.

13

The Individuality of a Quantum Event: Whitehead's Epochal Theory of Time and Bohr's Framework of Complementarity

YUTAKA TANAKA

Whitehead's Epochal Theory of Time: Genetic Analysis and Coordinate Analysis

The distinction between two modes of analysis of an actual occasion, i.e., genetic and coordinate, is fundamental to Whitehead's "epochal" theory of time. Genetic analysis divides the "concrescence" (the process of becoming concrete), and coordinate analysis divides the "satisfaction" (the concrete thing). The concrete is in its "satisfaction," and the "concrescence" is the passage from real potentiality to actuality. Both can be objects for analysis but under the different perspectives. Whitehead states:

> Physical time makes its appearance in the 'coordinate' analysis of the 'satisfaction.' The actual entity is the enjoyment of a certain quantum of physical time. But the genetic process is not the temporal succession: such a view is exactly what is denied by the epochal theory of time. Each phase in the genetic process presupposes the entire quantum, and so does each feeling in each phase. The subjective unity dominating the process forbids the division of that extensive quantum which originates with the primary phase of the subjective aim. The problem dominating the concrescence is the actualization of the quantum *in solido*.[1]

The above passage seems to have annoyed many commentators on *Process and Reality*. The genetic analysis of an actual occasion divides the concrescence into primary, intermediate, and final phases, which, according to Whitehead, are not "in" physical (i.e., coordinate) time. One phase of genetic division must be prior to another: but what sort of priority is this? William

Christian discusses and rejects four possible ways of interpreting this type of priority: (1) priority in physical time, (2) the logical priority of a premise to a conclusion, (3) a whole-part relation, and (4) a dialectical process of the Hegelian development of an idea. Then he says, "though genetic priority may have some analogies with other sorts of priority, we must accept it as something of its own kind, but he does not analyze further the *sui generis* character of genetic divisions."[2] Charles Hartshorne also questions the validity of genetic analysis, and proposes to accept only the succession of phases in physical time.[3]

What I will show in this chapter is the importance of the distinction between genetic and coordinate analysis and its relevance to the interpretation of quantum physics, especially the relation of Heisenberg's indeterminacy principle to temporality, the Bohr-Einstein debates, and the recent experimental refutation of the Bell inequality.

If we take into consideration the impact of quantum physics on the emergence of Whitehead's metaphysics, as Lewis Ford shows in detail in his book,[4] we naturally expect that the epochal theory of time has something to do with the quantum 'jump,' or the discontinuous transition from potentiality to actuality. But we need to be cautious. The references to quantum physics in *Science and the Modern World* are mainly to the initial stages of quantum theory in the early 1920s, and there is no textual evidence concerning whether Whitehead knew the final stages of quantum physics established by Bohr, Heisenberg, Schrödinger, and other contemporary physicists. The composition of *Process and Reality* began at the Gifford Lectures in 1927, and the same year was memorable in the history of quantum physics: Bohr stated his principle of complementarity and stressed the "individuality" of quantum events in his Como Lectures; and Heisenberg published his paper on the indeterminacy principle in *Zeitschrift für Physik*. Only two years later, in 1929, *Process and Reality* was published. Although Whitehead did not mention Bohr's principle of complementarity nor Heisenberg's indeterminacy principle, there is a striking correspondence between Whitehead's metaphysical analysis of an actual occasion on the one hand, and Bohr's and Heisenberg's physical analysis of quantum events on the other hand.

The purpose of this chapter is not to confirm or disconfirm the historical influence of Bohr or Heisenberg's ideas on Whitehead's metaphysics. That is an interesting study in itself, but will remain only a conjecture. Rather, I will consider first the problem of temporality in the interpretation of Heisenberg's indeterminacy principle, and then discuss Bohr's concept of the "individuality" of quantum events, both under the Whiteheadian perspective. I will show that Whitehead's distinction between genetic and coordinate analysis of an actual occasion proves to be relevant to the interpretation of the delayed-choice experiment in quantum physics. This experiment is about the indeterminate past, which will tantalize process thinkers who take the determinate past for granted, and think that only the future is indeterminate.

Lastly, I will present a new approach of quantum logic to analyze Bohr's concept of the "individuality" of a quantum event. This approach uses the concept of "divisibility" of an event by another event, and defines the concept of "commensurability" of events. Then I will characterize the classical world by saying that all events are commensurable with each other, whereas the quantum world is characterized by saying that some events are incommensurable with each other. This analysis may be interesting to Whiteheadian scholars because it will teach us that the concept of the "individuality" of a quantum event *denies atomism* insofar as atomism presupposes the divisibility of a complex event into atomic component events. Many scholars think of Whitehead's epochal theory of time as "temporal atomism" and arbitrarily conjecture the existence of a temporal atom with a very minute duration. Once we accept the quantum logical analysis and apply it to the epochal theory of time, we will understand the key concept is the "individuality" of an actual occasion, and not "atomism" of any kind.

Heisenberg's Indeterminacy Principle and Time: Is There the Indeterminate Past?

In *The Physical Principles of the Quantum Theory*, Heisenberg insists on the essential contingency of the future in his famous principle of indeterminacy. This principle states that we cannot exactly predict the future state of a physical system on the basis of knowledge of the past. What we can predict is only a probabilistic future that is testable, not by a single experiment, but only by statistical ensembles. The contingency of the future in this sense is a common notion of physicists today who have accepted the Copenhagen interpretation of quantum physics.

Consider the future behavior of a photon impinging on a half-silvered mirror (beam splitter): if it reflects from the mirror, it will travel along route α, and if it goes through the mirror, it will travel another route, β. We do not know which route the photon will choose in advance, but can confirm afterward its choice by using the photodetector: the exchange of energy between the photon and the photodetector will demonstrate the particle-character of a photon, which cannot travel two different routes simultaneously. If the photodetector E at the end of route α clicks, then the photodetector F at the end of route β does not click, and vice versa. The classical description of a particle presupposes determinism in principle, without any reference to an observer. The use of statistics only shows *our* ignorance of the physical system. There is nothing indeterminate in the system itself both in the past and in the future.

Against such a standpoint, the Copenhagen interpretation of quantum physics proposes the indeterminacy principle. It claims statistical analysis is essential because we cannot predict the future behavior of a particle precisely in the individual case because of the indivisible relation of an observer and observed. Heisenberg refrains from talking about the "reality" of any intermediate states of a physical system between actual observations.

When applied to future events, the indeterminacy principle *seems* reasonable to common sense because common sense well knows future contingency. But what about past contingency? If the indeterminacy principle stresses the indivisible connection between the observer and the observed, does the same principle apply not only to the future but also to the past? Not being omniscient about the past, we often need to retrodict, i.e., conjecture about what happened in the past on the basis of present data. That the past is determinate in all its aspects without any observer is the postulate of physical realism, which cannot be accepted unconditionally by the Copenhagen interpretation. Then must we accept the idea of the indeterminate past in some sense in quantum physics?

Indeed, Heisenberg was aware of this question and discussed it in *The Physical Principles of the Quantum Theory*:

> The Indeterminacy Principle does not refer to the past: if the velocity of the electron is at first known and the position then exactly measured, the position for times previous to the measurement may be calculated. Then for these past times $\Delta p \Delta q$ is smaller than the usual limiting value, but this knowledge of the past is of a purely speculative character, since it can never (because of the unknown change in momentum caused by the position measurement) be used as an initial condition in any calculation of the future progress of the electron and thus cannot be subjected to experimental verification. It is a matter of a personal belief whether such a calculation concerning the past history of the electron can be ascribed any physical reality or not.[5]

In the above citation, Heisenberg did not reject the idea of the indeterminate past, but thought that such an idea was of a "purely speculative" character and a "matter of a personal belief," because it cannot be subjected to experimental verification. To Heisenberg in the 1920s, only the prediction of the future was important, and the mathematical theory assisted him to calculate the probability of the end-state given the initial state: the description of the intermediate development of the system between two objectively recorded or recordable states did not seem to correspond to physical reality.

On the other hand, Einstein, as a critic of quantum physics, did not admit Heisenberg's standpoint, especially that the indeterminacy principle does not refer to the past. In the paper "Knowledge of Past and Future in Quantum Mechanics," Einstein proposed an imaginary experiment, in which "the possibility of describing the past path of one particle would lead to predictions as to the future behavior of a second particle of a kind not allowed in the quantum mechanics."[6] So Einstein concluded that "the principle of the quantum mechanics must involve an indeterminacy in the description of past events which is analogous to the indeterminacy in the prediction of future events."[7]

This should be understood in the context of Einstein's argument against

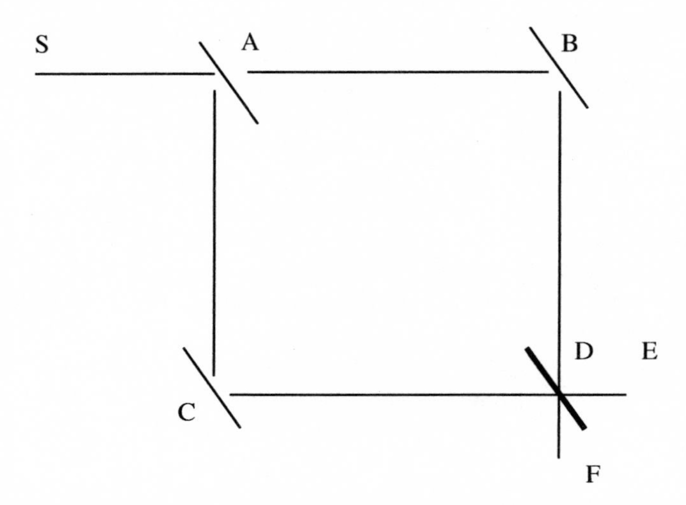

Figure 13.1. Schematic diagram of Wheeler's delayed-choice experiment.

the completeness of quantum physics in the same way that the EPR argument showed that the completeness of quantum physics led to absurdity. In other words, Einstein did not positively assert the existence of indeterminate past events, but only intended to deduce it as the necessary conclusion of the completeness of quantum physics.

The problem of the indeterminate past reappeared about 50 years later in J. A. Wheeler's discussion of the "delayed-choice" experiment (Fig. 13.1). This experiment is not imaginary but is an actual one that uses one particle (say, photon) instead of two particles as in Einstein's case.[8] Laser light incident on a half-silvered mirror *A* divides into two beams: one is along the path *ABD*(α), and the other is along the path *ACD*(β). In this experimental arrangement the detection of a given photon by either by *E* or *F* suffices to determine which of the two alternative routes the photon traveled. This is the particle mode of the experiment. The photon travels either route α or route β.

If a second half-silvered mirror *D* is inserted at the crossing points, the two beams are recombined, part along the route into *E*, and part along the route into *F*. This will cause wave interference effects, and the strengths of the beams going into *E* and *F* respectively will then depend on the relative phases of the two beams at the point of recombination. These phases can be altered by adjusting the path lengths, thereby essentially scanning the interference pattern. This is the wave mode of the experiment. The photon travels in some way along both routes, α and β, at the same time.

Now the crucial point is that the decision of whether or not to insert the second half-silvered mirror *D* can be left until a given photon has almost arrived at the crossover point. Thus one decides whether the photon "shall have come by one route, or by both routes" after it has "*already done* its travel."

After confirming the fact that what we can say of past events is decided by (delayed) choices made in the near past and now, Wheeler discusses the possibility that the phenomena called into being by the present decision can reach backward in time, even to the earliest days of the universe:

> To use other language, we are dealing with an elementary act of creation. It reaches into the present from billions of years in the past. It is wrong to think of the past as "already existing" in all detail. The "past" is theory. The past has no existence except as it is recorded in the present. By deciding what questions our quantum registering equipment shall put in the present we have an undeniable choice in what we have the right to say about the past.[9]

The interpretation of the indeterminacy principle will be altered if we accept the concept of past indeterminacy. Heisenberg originally considered this principle as the limit of the exactitude of two incommensurable quantities at the simultaneous measurement. But the indeterminacy of past events that have not been recorded has a connection, not with their simultaneous measurability, but rather with the definability of their historic routes. That the definition of the past route or history of a particle depends on the present choice of an experimenter is the meaning of the indeterminate past.

Bohr's Framework of Complementarity

Bohr's principle of complementarity is more closely connected with the "individuality" of a quantum event than with the indeterminacy principle. His emphasis is mainly on the definition of a quantum process, and not on the unavoidable "disturbance" or "physical influence" of the observer on the observed. His arguments rest on the insight that in quantum physics we are presented with the inability of the classical frame of concepts to comprise the peculiar feature of indivisibility, or "individuality," characterizing the elementary process. The quantum paradox arises from the apparent contradiction between the exigencies of the general superposition principle of the wave description, and the feature of individuality of the elementary atomic processes. Stressing the impossibility of any sharp separation between the behavior of atomic objects and the interaction with measuring instruments that serve to define the conditions under which the phenomena appear, he writes:

> The individuality of the typical quantum effects finds its proper expression in the circumstance that any attempt of subdividing the phenomena will demand a change in the experimental arrangement introducing new possibilities of interaction between objects and measuring instruments which in principle cannot be controlled. Consequently, evidence obtained under different experimental conditions

> cannot be comprehended within a single picture, but must be regarded as complementary in the sense that only the totality of the phenomena exhausts the possible information about the objects.[10]

Bohr distinguishes two modes of the description of a quantum process, which are "complementary but exclusive": one is "the space-time coordination," and the other is "the claim of causality." The two modes of description, though united in the classical theories, should be considered as "complementary but exclusive features of the description" in quantum physics.

Heisenberg summed up the framework of complementarity in the following diagram:[11]

Space-Time Coordination		*Causal Relationship*
Quantum process is described	Alternatives	Expressed by mathematical laws
in terms of space and time	related	but
but	statistically	**physical description of phenomena**
indeterminacy principle.		**in space-time is impossible.**

Now we are ready to compare Bohr's (physical) framework of complementarity with Whitehead's (metaphysical) distinction between coordinate and genetic analysis of an actual occasion. The structurally similar arguments really characterize the theory of concrescence and space-time coordination in *Process and Reality*. Whitehead's discussion of causality, efficient or final, belongs to the genetic analysis of an actual occasion in his *Theory of Prehension*. The internal development of concrescence of an actual occasion is the theme of this analysis. But according to Whitehead, this internal process itself does not occur in physical time. Physical time makes its appearance in the coordinate analysis of the satisfaction. Each phase in the genetic process presupposes the entire quantum—that is the point of the epochal theory of time.

The distinction between genetic and coordinate analyses has a bearing on the divisibility of the space-time region. The region of an actual occasion is divisible according to the coordinate analysis, but is undivided in the genetic growth. The epochal theory of time stresses the genetically indivisible character of an actual occasion.

Interpreters of Whitehead's metaphysics, as far as I know, do not seem to have grasped fully the "individuality" of an actual occasion. The epochal theory of time was usually discussed through Zeno's paradox of motion and change, and considered as the metaphysical postulate that makes it possible for us to talk about becoming at all. I do not say it is wrong, but only that such a metaphysical postulate is not sufficient to the concrete analysis of becoming and its relation to space-time. For example, the epochal theory of time is often characterized as "temporal atomism."[12] But *atomism* is not a happy word in the sense that it has a connotation of a mechanical worldview. Whitehead's standard usage is "the cell theory of actuality," and not the "atomic theory of actuality." Moreover, an "individual" quantum event is not necessarily microscopic. The simultaneous correspondence of the EPR experiment shows us the "individuality" of a quantum process at a long (macroscopic) distance. The delayed-choice experiment shows us that the individual quantum process may have the indeterminate past according to the coordinate divisions of space-time. So the region of an individual quantum process may have an arbitrary size with respect to space-time coordinates.

In the next section I will analyze the concept of "individuality" by using a quantum logical analysis and show that the concept of completeness, which Einstein presupposes in his criticism of quantum physics, is irrelevant to the quantum world.

A Quantum Logical Analysis of the Indivisibility or Individuality of an Event

There is a hidden presupposition when we apply classical logic to the empirical world, namely, that all events are divisible with each other, or that all events are commensurable. This presupposition fails to be the case in the quantum world, so that one of the most fundamental laws of classical logic cannot claim universal validity. In order to define the "divisibility" and "commensurability" of events, some preliminary discussions are in order.

Suppose we have predictions of two events, a and b:

a: The wind will blow tomorrow. b: Rain will fall tomorrow.

The two statements that predict the weather tomorrow are, strictly speaking, not propositions. Truth-values of propositions must be eternal in the sense that they must be fixed independent of the time when any speaker states them. We cannot fix truth-values of the above statements today unless we are determinists. Tomorrow we can verify these predictions, but today we are not able to assign truth-values to these contingent statements. The truth-table approach of classical logic is irrelevant to the contingent world of quantum physics.

I will give an introduction to quantum logic, which is wider than classical logic in its application in the sense that it can analyze contingent events in addition to determinate propositions. Readers need not follow the logical apparatus below to understand the overall argument, which is laid out in the text.

We define the divisibility of events as follows:

$$\begin{gathered} aDb\text{: The event } a \text{ is divisible by the event } b \\ aDb \Leftrightarrow_{def} a = (a \cap b) \cup (a \cap \neg b) \end{gathered} \tag{1}$$

The right-hand side of this definition is the equivalence that we implicitly assume in our everyday talk. "The wind will blow tomorrow" is equivalent to "The wind will blow and rain will fall tomorrow, or the wind will blow and rain will not fall tomorrow."

We can use Venn's diagram to visualize the divisibility of events.

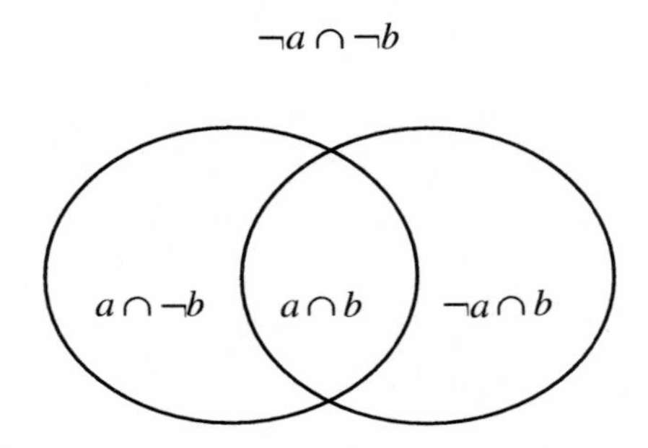

Divisibility of events shown in this Venn diagram presupposes Boolean algebraic structure that characterizes classical logic. It is noteworthy that both Wittgenstein's *Tractatus* and Russell's *Logical Atomism* implicitly presuppose that all events are divisible with each other: their analysis depends on the truth-table. Classical logic with the Boolean algebraic structure is the background of their worldview. If we denote $p_1 = a \cap b$, $p_2 = a \cap \neg b$, $p_3 = \neg a \cap b$, and $p_4 = \neg a \cap \neg b$, we may call p_k $(1 \leq k \leq 4)$ atomic events because they make up the nonoverlapping and exhaustive set of events. So we can decompose a, $\neg a$, b, $\neg b$ into atomic events p_k. In such a logical atomism, the world divides into facts, and the facts consist of atomic events (*das Bestehen von Sachverhalten*)—if we use Wittgenstein's phrase in the *Tractatus*. The set of atomic events is called the "logical spectrum of the world."

Note that the logical spectrum of the world may be relative to our descriptive language. It may be simple and rough (as in the above case) or it may be very fine (as in the case of a professional weather forecaster). Atomic events may be decomposed by another logical spectrum, and we will get the description of the world in more and more detail. The point of logical atomism is not the existence of absolutely atomic events, but the divisibility of any event into more atomic events.

The usual formulation of quantum logic does not use the truth table, to say nothing of Venn's diagram. Quantum logical analysis is rather difficult in its usual formulations, so I will give a very simple and understandable semantical definition of quantum logic by using the concept of divisibility.

We define the concept of "commensurability" of events:

$$aCb\text{: the event } a \text{ is commensurable with the event } b$$
$$aCb \Leftrightarrow_{def.} aDb \ \&\ bDa \tag{2}$$

Note that "&" is the symbol of semantics (metalanguage), and must be distinguished from "$\cap$" of the object-language.

Now we can define the classical world in the following way.

$$\text{The world } W \text{ is classical} \Leftrightarrow_{def} (\forall a \forall b)\{(a \in W \ \&\ b \in W) \rightarrow aCb\} \tag{3}$$

In the classical world all events are commensurable with each other, i.e., all events are divisible with each other. Let 0 denote the null event, which is stipulated by the contradictory statement $a \cap \neg a$, and let 1 denote the all-inclusive event, which is stipulated by the tautology $a \cup \neg a$. Then we can give a natural meaning to the concept of the "complete" description of the classical world by using the logical spectrum defined as follows:

The set of nonoverlapping and exhaustive events $B = \{b_k\}$ is called a "logical spectrum,"

$$\text{i.e., } i \neq j, \rightarrow b_i \cap b_j = 0 \text{ and } \bigcup_k b_k = 1$$

In the classical world, any event a can be decomposed into the logical sum of atomic events with respect to the logical spectrum B as $a = \bigcup_k (a \cap b_k)$.

More generally, if there are multiple logical spectra $B^1, B^2, B^3, \ldots B^l$, a can be decomposed as $a = \bigcup_{k(l)} (a \cap b^1_{k(1)} \cap b^2_{k(2)} \cap \ldots \cap b^l_{k(l)})$, where $b^l_{k(l)}$ is a member of B^l and $\bigcup_{k(l)}$ is the sum of all possible combinations of $b^l_{k(l)}$.

If we confine ourselves to the classical world that contains a finite number of events, then we can easily stipulate the condition of the "complete description" of the classical world.

Let us use the abbreviation $w_m = b^1_{k(1)} \cap b^2_{k(2)} \cap \ldots \cap b^l_{k(l)}$

Then we may say that the whole set of logical spectra $\{B^1, B^2, B^3, \ldots B^l\}$ gives the complete description of the classical world if $a \cap w_m = 0$ or $a \cap w_m = w_m$ for any event a. That is, any event can be decomposed to the logical sum of w_m, and we need no additional logical spectrum. Each w_m may be called an (absolutely) atomic event, and logical atomism is a suitable characterization of the classical world.

If we assign equal probability for every atomic event w_m, then we can calculate the *a priori* probability of the event a.

$$P(a) = \sum_{m(a)} P(w_{m(a)}) = \frac{n}{N}$$

where N is the total number of atomic events, and n is the number of $m(a)$ such that $a \cap w_m = w_m$.

The above description of the classical world is the logical basis of classical physics. The divisibility of any event $a = (a \cap b) \cup (a \cap \neg b)$ is always presupposed, and we may say that one aim of physicists is to invent a new kind of logical spectrum so that we may get nearer to the ideal of the complete description of the world.

Quantum physics tells us that the divisibility formula $a = (a \cap b) \cup (a \cap \neg b)$ does not always hold. In other words, there is a case in which the event a is indivisible by the event b. For example, let a be "the spin of the electron is up along x-axis," and b be "the spin of the same electron is up along y-axis." Then a is indivisible by b, because of the indeterminacy principle.

Note that the indivisibility of an event a does not mean that a is indivisible by any other event, but that there are some events by which a is indivisible. So we can define the individuality of an event as follows:

$$\text{The event } a \text{ has the character of "individuality"} \Leftrightarrow_{def.} (\exists x)(\sim aDx) \quad (4)$$

As there exist quantum events with the character of "individuality," we cannot use the probabilistic formula $p(a) = p(a \cap b) + p(a \cap \neg b)$ when a is indivisible by b. As explained later, this is the reason why the Bell inequality breaks down in quantum events.

The next task is to define the quantum world through the concept of divisibility. The quantum world is the world of quantum logic, which has the algebraic structure of an orthomodular lattice. It is known that the orthocomplemented lattice is orthomodular if and only if the divisibility relation aDb is symmetrical.[13] So we can define the quantum world as follows:

The world W is a quantum world if the relation of divisibility is symmetric, and there exist incommensurable events in W.

$$W \text{ is a quantum world} \Leftrightarrow_{def.} \{(\forall a,b)(a \in W \;\&\; b \in W) \rightarrow (aDb \rightarrow bDa)\} \;\&\; (\exists a,b)(a \in W \;\&\; b \in W \;\&\; \sim aCb) \quad (5)$$

The distinction between classical and quantum worlds is analogous to the situation in relativity physics in which Riemannian geometry holds as a generalization of Euclidean geometry. Some theorems of Euclidean geometry applied to empirical data are not necessarily true in the relativistic world, especially when gravitational effects are strong. In a similar way, some laws of classical logic applied to empirical data are not necessarily true in the quantum world, when we observe "incommensurable" events.

Although we can get classical logic by adding the condition of "commensurability" to quantum logic, quantum logic should not be considered a "weird" or "strange" logic invented by logicians. Rather, quantum logic is more faithful than classical logic to concrete experimental contexts, because it does not presuppose the dogmatic thesis of divisibility of all events. Classical logic is embedded in the deterministic world of classical physics where all propositions have the established truth-values independent of observers. An atomistic view of events is implicitly presupposed in classical physics. On the other hand, in quantum physics, where irreducible contingency appears in the context of observation, the existence of an indivisible event is established. The individuality of a quantum event can involve macroscopic spatiotemporal extension. It can extend over two temporally distant locations that involve "the indeterminate past" in the coordinate division, as in the delayed-choice experiment.

In the next section, we discuss the indivisibility of a quantum event that extends over two spatially distant locations, i.e., the so-called EPR correlation.

Bell's Inequality as an Analytical Result in the Classical World

The experimental test of Bell's inequality, which the French physicist Alain Aspect conducted in 1982, attracted the attention of those who were interested in philosophical problems of quantum physics. This experiment manifested one of the most paradoxical characteristics of quantum systems: the nonseparability of two contingent events, involving the correlation of polarized photon pairs at a distance. Both philosophers and physicists were reminded of the celebrated debate between Bohr and Einstein about the completeness of quantum mechanics in the 1930s. The imaginary experiment, which Einstein used in his polemics against the alleged completeness of quantum mechanics, became real through the progress of technology. The combination of conceptual analysis and experimental tests revived the controversy about the philosophical status of quantum physics. The test of Bell's theorem became a starting point for refreshed research into the nature of quantum events.

In this section, I will deduce the generalized Bell's inequality as an analytical (necessary) result in the classical world, which I have defined in the previous section. The classical assumption about the divisibility of an event into the atomic components causes Bell's inequality. We need only an elementary theory of probability and information, and no additional physical knowledge to derive Bell's inequality.[14]

Suppose that $A = \{a_i\}$, $B = \{b_j\}$ are two logical spectra, e.g., the set of observable values in physical experiments. According to information theory, if we measure A and B and get the value a_i *and* b_j, the newly acquired information is

$$I(a_i) = -\log p(a_i) \qquad I(b_j) = -\log p(b_j)$$

The joint information of a_i *and* b_j is

$$I(a_i \cap b_j) = -\log p(a_i \cap b_j)$$

Similarly, the conditional information of a_i given b_j is

$$I(a_i \mid b_j) = -\log p(a_i \mid b_j)$$

According to Bayes' theorem,

$$I(a_i \cap b_j) = I(a_i \mid b_j) + I(b_j) = I(b_j \mid a_i) + I(a_i)$$

The mean value of information of A and B is, respectively,

$$H(A) = \sum_i p(a_i)I(a_i) \qquad H(B) = \sum_j p(b_j)I(b_j)$$

Let the expectation of the joint information of (A and B) be $H(A \cap B)$.

$$H(A \cap B) = \sum_{i,j} p(a_i \cap b_j)I(a_i \cap b_j)$$

The conditional information of A given b_j is

$$H(A \mid b_j) = \sum_i H(a_i \mid b_j) = \sum_i p(a_i \mid b_j)I(a_i \mid b_j)$$

The conditional information of A given B is the mean value of $H(A \mid b_j)$.

$$H(A \mid B) = \sum_j p(b_j)H(A \mid b_j) = \sum_{i,j} p(a_i \cap b_j)I(a_i \mid b_j)$$

The new formulation of Bayes' theorem is

$$H(A \cap B) = H(A \mid B) + H(B) = H(B \mid A) + H(A)$$

Let $I(a_i;b_j)$be the correlation information between a_i and b_j.

$$I(a_i;b_j) = I(a_i) - I(a_i \mid b_j) = I(b_j) - I(b_j \mid a_i)$$

Though the value of this correlation information may be positive or negative, its mean value must be non-negative according to Gibb's theorem.[15]

$$H(A;B) = \sum_{i,j} p(a_i \cap b_j)I(a_i;b_j) = \sum_j p(b_j)\left(\sum_i p(a_i \mid b_j)\log\frac{p(a_i \mid b_j)}{p(a_i)}\right) \geq 0$$

So we can get the fundamental inequality

$$H(A \mid B) \leq H(A) \leq H(A \cap B) \tag{6}$$

Now we can deduce the generalized Bell's inequality from (6)

Suppose α and β are two separated physical systems. α has two logical spectra A and A'. β has also two logical spectra B and B'. Let a_i, a'_k, b_j, b'_l be discrete observable values of A, A', B, and B'.

In the classical world where all events are divisible with each other, there exists the joint probability for every possible combination such as $p(a_i \cap a'_k \cap b_j \cap b'_l)$.

So we can generalize the fundamental inequality (6) as follows:

$$\begin{aligned} H(A \mid B') + H(B') &= H(A \mid \cap B') \leq H(A \mid B \cap A' \cap B') \\ &= H(A \mid B \cap A' \cap B') + H(B \mid A' \cap B') + H(A' \mid B) + H(B') \end{aligned} \tag{7}$$

Using $H(A \mid B \cap A' \cap B') \leq H(A \mid B)$, $H(B \mid A' \cap B') \leq H(B \mid A')$, we subtract $H(B')$ from both sides of (7), and get the generalized Bell's inequality as follows:

$$\text{Bell-I: } H(A \mid B') \leq H(A \mid B) + H(B \mid A') + H(A' \mid B') \tag{8}$$

We can rewrite (8) in the symmetrical form by using the concept of information distance introduced by Schumacher.[16] Defining the information distance $\delta(A, B)$ between A and B as

$$\delta(A, B) = H(A \mid B) + H(B \mid A)$$

we get the symmetrical representation of the generalized Bell's inequality.

$$\text{Bell-2: } \delta(A, B') \leq \delta(A, B) + \delta(B, A') + \delta(A', B') \tag{9}$$

Note that the inequality (7) presupposes commensurability between A and A' and between B and B'. So this inequality loses its meaning in the case of "incommensurable" observables.

On the other hand, the inequality (8) is meaningful in the quantum world because only "commensurable" pairs of observables (A and B), (A and B'), (A' and B), and (A' and B') appear.

So we can empirically test Bell's inequality (8) to decide whether our world is classical. The results of empirical tests by Aspect and others clearly show Bell's inequality does not hold. This experiment is analogous to the "crucial" experiments of general relativity, which tell us that our world is non-Euclidian, i.e., not "flat" space-time. There is an experimentarily verified sense in which we say *our world is not a classical world.*

Then what about the completeness of quantum physics against which

Einstein protested? Does the disconfirmation of Bell's inequality prove the completeness of quantum physics? As I argue elsewhere, the valid conclusions of the EPR arguments and Bell's theorem, even if we accept classical presuppositions, is the nonlocality of an indivisible quantum event, and not the incompleteness of quantum physics.[17] Einstein's concept of completeness of a physical theory implicitly presupposes the classical world where the relation of divisibility holds. In the quantum world where incommensurable (not mutually divisible) events exist, the very concept of completeness does not hold. Therefore, we must say that quantum physics is neither incomplete nor complete in the classical sense.

The breakdown of Bell's inequality shows that the seemingly plausible logic that all events are divisible does not hold in the quantum world. The crucial point, however, is how we should understand the key concept of divisibility. We have analyzed the concept of individuality in Bohr's sense in terms of quantum logic to elucidate the difference between the classical and the quantum worlds. The "individuality" in question does not assert the existence of *atomic* events, each with a minute spatiotemporal extension. Rather, quantum physics denies the worldview of classical logic, which we may characterize as "logical atomism" after Wittgenstein's *Tractatus*, because the quantum world contains "incommensurable" events which classical logic a priori excludes.

What Bohr understands as "individuality" is not a spatiotemporal atomism but the complementary descriptions of quantum phenomena, i.e., causal relationships and space-time coordinations. Physical atomism implies a mechanistic worldview in that the description of the whole can be reduced to the descriptions of its elementary parts. Logical atomism presupposes that all propositions describing events have fixed truth-values and are commensurable with each other. Both kinds of atomism are not relevant to quantum physics today, and to the philosophy of organism in *Process and Reality.*

Notes

1. A.N. Whitehead, *Process and Reality* (New York: Free Press, 1978), 283.

2. W.A. Christian, *An Interpretation of Whitehead's Metaphysics* (New Haven: Yale University Press, 1967): 80–81.

3. C. Hartshorne, "Whitehead and Ordinary Language," *Southern Journal of Philosophy*, 7 (1969): 437–445.

4. L. Ford, *The Emergence of Whitehead's Metaphysics: 1925–1929* (Albany: State University of New York Press, 1984).

5. W. Heisenberg, *The Physical Principles of Quantum Theory* (Chicago: University of Chicago Press, 1930), 20.

6. A. Einstein, R.C. Tolman, and B. Podolsky, "Knowledge of Past and Future in Quantum Mechanics," *Physical Review* 37 (1931): 780–781.

7. Ibid.

8. J.A. Wheeler, "Law Without Law," in *Quantum Theory and Measurement*, eds. J.A. Wheeler and W.F. Zurek (Princeton: Princeton University Press, Princeton Series in Physics, 1983), 182–213.

9. Ibid., 194.

10. N. Bohr, "Discussion with Einstein on Epistemological Problems in Atomic Physics," in *Quantum Theory and Measurement*, eds. J.A. Wheeler and W.F. Zurek (Princeton: Princeton University Press, Princeton Series in Physics, 1983), 18.

11. Heisenberg, *Physical Principles*, 65.

12. Ford, *Emergence*. "Temporal atomicity" is Ford's usage, with which I disagree, though I owe much to his insights into the compositional history of *Process and Reality*.

13. M. Nakamura, "The Permutability in a Certain Orthocomplemented Lattice," *Kodai Mathematics Series* 9 (1957): 158–160.

14. This argument is based on the probability theory of S. Wantanbe, *Knowing and Guessing: A Quantitative Study of Inference and Information* (New York: Wiley, 1969). Braunstein and Caves show that Watanabe's formulation of non-Boolean information theory is closely related to Bell's theorem in S.L. Braunstein and C.M. Caves, "Writing out Better Bell Inequalities," *Proceedings of Third International Symposium of Foundations of Quantum Mechanics* (1989): 161–170.

15. Watanabe, *Knowing*, Chapter 1, Theorem 1–1.

16. B.W. Schumacher, "Complexity, Entropy and the Physics of Information," *Santa Fe Institute Studies in the Sciences of Complexity*, no. 9 (1990).

17. Y. Tanaka, "Bell's Theorem and the Theory of Relativity—An Interpretation of Quantum Correlation at a Distance Based on the Philosophy of Organism, *The Annals of the Japan Association for the Philosophy of Science* 8, no. 2 (1992): 1–19.

14

Physical Process and Physical Law

David Ritz Finkelstein

I would like to thank the conveners for this chance to recall some of my debts to Whitehead. I am no disciple, but I was certainly influenced by his thought, more than I realized at the time. I will tell the story in roughly chronological order.

Quantum Logic

In high school I read science fiction. One of the novels of L. Sprague de Camp had intriguing symbolic logic equations (only in the original pulp versions, the hard-cover editions are bowdlerized). The idea of an arithmetic of ideas intrigued me and sent me to the *Principia Mathematica* of Whitehead and Russell, probably my first contact with Whitehead's thought.

As a high school student, I couldn't get into the central stacks of the New York Public Library at first, but the Harlem branch library had two volumes of the *Principia*, which I plowed happily through.

It took me some years to recover from the chronic logicism I thus contracted in Harlem. Nowadays I think that making logic fundamental is a mistake, because it separates it from dynamics. This split exists in classical thought, but quantum theory fuses logic and dynamics into one simple algebraic unity that should not be divided.

Even at the time, I felt let down by the *Principia* and the other books on logic I tried. They were all obviously right, merely common sense. I had expected to find that logic too had undergone some spectacular imagination-straining upheaval, just as geometry had under Einstein. It seemed that it had not.

In fact it had, but I did not find out until I stumbled across von Neumann's quantum logic (von Neumann, 1955) later in college. Since so much of physics still used the archaic classical logic, I felt at once that I had found my life's work. Whitehead's aphorism, "Geometry is cross-classification," helped me to understand von Neumann's quantum logic. One can still say that quantum logic is projective geometry interpreted as cross-classification of quantum systems.

Whitehead did not make a sharp distinction between mathematics and logic. That would have been against the main aim of *Principia*: to reduce mathematics to logic. Von Neumann's point was that there is a physical logic just as there is a physical geometry, and it is not a part of mathematics any more than physical geometry is. I am not sure Whitehead would have accepted this view, since he preserved a flat space-time in his theory of gravity.

Quantum Process

Whitehead's *Process and Reality* also made it easier for me to understand some of modern physics. Whitehead had set out to found a cosmology not on the intuitions of Newton but on the latest state of our information about the world, which means the basic ideas of relativity and quantum theory. I find these ideas all through his work.

One problem is that while relativity swept the field within a few years after its enunciation, quantum theory was split up into dialects. Different people describe the same experiences in remarkably different languages. This is confusing even to physicists. How can you tell when physicists agree if they sound so different? It is certainly confusing to the general public. I think this Tower of Babel leads many people still to think of quantum theory as a problem to be solved rather than the solution of a problem.

The pioneering wave theory of Count Louis de Broglie, who never accepted Heisenberg's radical quantum view, still has great influence. Wigner promulgated a non-Copenhagen way to express quantum theory that he termed "orthodox." Orthodox quantum theory drops the quantum wholeness and objectlessness that Bohr and Heisenberg emphasized, and treats the wave function as the fundamental variable object of quantum theory. It speaks just as if the system under study were a wave, but one that "collapses" when we make measurements on the system.

There are systems whose states are wave functions. They are called waves. A quantum is not a wave. Wave functions represent acts of preparing the quantum and measuring its properties. There are two such processes in every experiment, input at the beginning and outtake at the end. We do one, then we do the other. Many teachers of quantum theory mistake one process following another for one object collapsing into another.

Quantum Actuality

Describing the process and not the product is still a new thing in physics. In classical physics, one represses the observer and the act of observation and talks naively about "things as they are," as if we had direct knowledge of "things as they are." I heard someone here too say that physics is about "things as they are." I hate to disagree, but physics today is about what goes on, not about what is; about actuality, not reality. This usage of *actuality* is one of my favorite Whiteheadisms.

Heisenberg formulated the quantum theory in the same city and decade in which Kandinsky coined the phrase *non-objective art.* I assume that Heisenberg borrowed from Kandinsky when he called quantum theory "non-objective physics." The main idea of quantum theory is to talk about what you do, not about "things as they are"; to represent the whole process and use the theory to estimate whether it will happen. Heisenberg set up an algebra of observables in which the fundamental elements are operators, and could just as well be called "processes." They represent what we do, not what "is."

Input and outtake processes seem to be special cases of Whitehead's general concept of process. I suppose that the operational aspect of modern physics, which is so powerful in relativity theory and quantum theory, as well as the precedent of existentialism, encouraged Whitehead to take the idea of process as fundamental. In modern physics we learn to do this. The "event" of relativity theory is a tiny process, and the experiments of quantum theory are extended processes.

Quantum Law

Our philosophy and theology influence our physics. To understand the development of physics, we should take seventeenth-century theism into account.

Newton is an anomaly in this respect. In the first place, he did not believe in a closed physical theory, a fixed governing law. Newton was no Newtonian. He took it for granted that God steps in now and then to keep things working in a reasonable way:

> Since Space is divisible ad infinitum and matter is not necessarily in all places, it may also be allowed that God is able to create particles of matter of several sizes and figures and in several proportions to Space and perhaps of different densities and forces and thereby to vary the laws of nature and make worlds of several sorts in several parts of the universe. At least I see nothing of contradiction with all this. (Newton, 1730, Query 32)

This implies that God does experiments to learn, and so that God is unfinished and develops, a heresy of the Old Unitarians. Newton went to some pains to conceal this belief, unsuccessfully attempting to destroy all copies of the above quotation.

What puzzles me most about the theory of a variable dynamical law is that if such creativity prevails in the universe, how do we account for all the non-creativity that seems to prevail? If the elements of nature are creative, how come so much of nature goes by rote? Why does the apple fall every time we drop it? I get back to this question later.

I have gone through four stages of separation from the concept of fixed universal law:

1. *Polynomy*: When I began to teach physics, I told my students that physics was the search for the laws of nature.
2. *Mononomy*: After I read more of Einstein, I taught that physics was the search for *the* Law of Nature. I thought this was an inspiring insight.
3. *Anomy*: Then I suspected that there is no law.
4. *Panomy*: Now I think that there actually is a law, an evolving law like Newton's rather than an absolute one like Laplace's, and all there is is that law.

For the first hundred years or two after Newton, most physicists assumed mononomy, one fixed, absolute law. Most of the other fundamental absolutes of classical physics have been relativized since then. First Newton's absolute space fused with his absolute time, and in the process both were relativized. In today's physics, one of the few survivors of the mechanical doctrine is still the idea that there is a fixed law. Unified field theory, grand unified theory, "theory of everything," and string theory may still incorporate this faith.

In quantum theory, one becomes doubtful about the doctrine of absolute law. If we observe the law, we change it. If we cannot observe it, who needs it?

By *kinematics* I mean the theory of the description of motion. Kinematics is the linguistics of physics. It sets up a formal language, a syntax, usually dreadfully mathematical, and a semantics, something like a dictionary relating the terms of the language to the experiences that go on or do not go on, including all the measurements you could make and all the experiences you could have in principle. Then a dynamics, an absolute law, is supposed to be what selects a subclass of these kinematic possibilities. Finally, the boundary data complete the process of selection.

Then what in actuality singles out the actual from all that is possible? Whitehead is explicit about this—it is the aim of the organism.

But organisms change their aims. So this sounds as if Whitehead too is giving up the idea of an immutable natural law.

Certainly C. S. Peirce, a profoundly evolutionary thinker, rejected the idea of an absolute law. He insisted that the law had to evolve like everything else, and he even described the process by which he thought it evolved. He began with a "first flash" where today we speak of the big bang. He invoked a "habit-forming tendency" in nature much as we speak of Bose-Einstein statistics today.

I was gratified at this meeting to discover something that probably everyone here knows, that Whitehead indeed imagined the law itself as mutable. Here on page 27 of the program for this meeting is question 10, the last, in the final examination that Whitehead gave at Harvard in 1927:

> Potency refers to the continuum of nature, Act to the community of atomic creatures. Potency is the character of the creativity due to the creatures. Explain this doctrine, pointing out its bearing (*a*) on the logical inconsistency in the notion of a continuous becoming of an extended continuity of creatures, (*b*) on the doctrine of an evolution of laws of nature, (*c*) on the meaning of 'future.'

So Whitehead too had believed that there is no definite fixed law of nature. That law also is part of the creative process. In modern times, J. A. Wheeler is eminent among physicists who insist that the laws of nature are mutable. As he puts it, "There is no law but the law of averages." To which I would add "which is not a law." It does not tell you what is going to happen on any occasion.

As I have already mentioned, dynamics is kinematics plus law. This separation is a modern remnant of the old distinction between space and time.

The original formulation of quantum theory deals with operations at one moment. It is fundamentally nonrelativistic therefore; as has been pointed out from the start by such founding fathers of quantum theory as Bohr, von Neumann, and Wigner. Kinematics is the structure of the instant, dynamics is about how to get from one instant to another.

A true network theory of the kind that Geoffrey Chew shared this morning might describe processes in a way that does not introduce a distinction between space and time, or kinematics and dynamics. Actual processes or events or transitions do not connect events at the same time but take time to cover space. It describes a physical process by giving its fine structure from input to outtake. Each such description then assigns an amplitude to every other description. There seems no place in such a theory for a separate law of nature. The description of the process suffices. Perhaps the process is the law.

That is my main point. The search for a separate law, which has evoked considerable criticism within physics already, may be a relic of another century.

Whitehead saw that earlier than most.

Updates

My students and I at Georgia Tech are looking at a new algebraic formulation of quantum physics. For us, the law is not only a variable, it is the only variable; just as the geometry is not only a variable in Einstein's unified field theory, it is the only variable.

Like Whitehead's theory of organism, ours is a cellular theory of actuality. This raises at the start the question of how the elementary quantum processes of the universe compose into the whole. What is their statistics?

First we tried iterated Fermi-Dirac statistics, then Bose-Einstein, and now we have settled on an iterated Clifford statistics that seems most promising (Baugh et al., 2001).

In Clifford statistics the collective is described by a Clifford algebra with one generator for each element of the collective. The Clifford elements can be interpreted as permutations. This gives a new meaning to spinors, the substratum supporting this algebra. A spinor too describes a collective, not an individual.

Von Neumann's concept of quantum logic has serious limitations. It is not operationally simple, it lacks the higher levels of logic, and it is based in irreversible processes of filtration.

Clifford algebra is operationally simple. It can be used to represent not irreversible filtrations but reversible beam-splittings. It can be iterated to represent the higher levels of logic. It is extensional, providing a class for each predicate. As an algebraic logic it evades the three problems I mentioned with von Neumann's.

This work has been greatly forwarded lately by a belated recognition of a guide to relativization put forward by Irving Segal (Segal, 1951) and by Inönü and Wigner in the early 1950s. They provide an idol-detector, a way to spot false absolutes of the kind that Francis Bacon called "idols" (of the tribe, cave, marketplace and theater), and an idol-smasher, a specific way to break each idol that it detects.

The absolute universal law is among the highest of these idols. The constant imaginary i of quantum theory and the space-time of field theory are also idols ripe for smashing. We break all of these at once in our current exploration of Clifford statistics and feed them into the variable law.

Why has the belief in a single law worked so well? The corresponding question for geometry is answered, "Because the gravitational constant in Einstein's law of gravity is so small." This makes space-time stiff in the classical theory. Probably the same constant accounts for the stiffness of natural law in the classical theory, because the geometry of space-time is just the dynamical law for a small probe. In both cases, I suppose that the quantum structure of space-time or law admits a classical model because of a cosmic quantum condensation that forms the present vacuum.

In my talk, I said an impolite thing about the concept of a wave function of the universe. I accused it of being non-operational. I have edited this out because in the meantime I have seen the error of my ways, thanks to Segal's criterion. The experimenter too is an idol in the usual quantum theory. Smashing this idol lets in quantum descriptions of the experimenter and probably of the cosmic process. Since time is a variable of the experimenter, not the system, we must smash the idol of the classical experimenter to get to the quantum structure of time and the chronon.

From the original quantum-theoretic viewpoint, "the wave function of the universe" is a solecism. It rejects the quantum idea of describing just the acts of the experimenter, and pretends to describe what actually goes on everywhere in the universe.

Now I believe that this is the right thing to do. We must go beyond operationalism if we wish a cosmology that takes into account the facts of quantum theory.

To be sure, applying quantum theory to the universe implicitly puts an all-powerful, all-knowing observer outside the universe looking in at it. But Laplace used just this metaphor in classical cosmology without disaster, and it seems just as useful in quantum cosmology. We still describe the process, not the product, but now it is the cosmic process. We can draw inferences about what we experience from the theory of what a Cosmic Experimenter would experience.

I am happy to learn that Bohr, who was originally infuriated by the notion of a quantum universe, softened to it by 1936, expounded a doctrine of universal relativity, and called for a theory that took into account the quantum nature of the entire experiment, as many do today (Bohr, 1936).

References

Baugh, J., D. R. Finkelstein, A. Galiautdinov, and H. Saller. "Clifford algebra as quantum language." *Journal of Mathematical Physics* 42: 1489 (2001).

Bohr, N. "Causality and complementarity." *Philosophy of Science* 4: 293–4 (1936).

Newton, Isaac. *Opticks*, 4th edition, Wm. Innys, London (1730). Republished, Bell, London (1931).

Segal, I. E. "A class of operator algebras which are determined by groups." *Duke Mathematics Journal* 18: 221 (1951).

Von Neumann, J. *Mathematical Foundations of Quantum Mechanics* (trans. R. T. Beyer). Princeton University Press, Princeton (1955).

15

Dialogue for Part III

The dialogue presented below occurred in response to presentations by David Finkelstein and Yutaka Tanaka.

John Lango: The question I have is about the relevance to Whitehead's metaphysics. Whitehead is primarily influenced by relativity theory, a subject he worked on, and only tangentially influenced by quantum theory. Many of the radical developments of quantum theory were going on in the 1920s while he was writing *Process and Reality* [see Bibliography: Whitehead, 1929] so it's understandable why he would be less influenced by them. It seems to me that there are real problems with squaring his metaphysics with quantum theory. Your interpretation of Heisenberg is that he talked about processes, the process of interrogating nature or experimenting with nature, whereas *Process and Reality* reflects the metaphysical vision of Plato's Timaeus where the whole universe consists in processes, and what physicists do in questioning nature through experimentation is just one subset of those processes. So it seems to me that what Heisenberg is doing is more naturally understood in terms of the idealist tradition of Hume and Kant and that it squares much less with what I think are more objectivist ancient Greek-like approaches to the process of Whitehead.

Finkelstein: Certainly Heisenberg formulated his quantum theory in terms of very special processes—measurements—but I'm sure that what Geoffrey said earlier, that measurements are not that different from other processes, they're not that special, is something that most physicists take for granted, including Heisenberg.

Stapp: It's certainly true that Whitehead was a great scholar in relativity theory and he wrote books on it and had his own variation. But I think it's probably likely that he was aware of quantum jumps. It was 1915 when Bohr came up with that idea, and I think that fed into his idea that actual occasions are event-like, with a jump, and of things happening. A world based on happenings probably was given a lot of support by his awareness of the rudimentary developments in quantum theory even before the late 1920s. There was a lot of quantum theory in the air, and he must have known about it and been influenced. And he does mention quantum theory now and then.

Klein: I think you are correct that there are discrepancies between Whitehead and modern quantum theory, and I recommend to anybody a chapter in Penrose's book by Abner Shimony [see Bibliography: Shimony, 1997], who's a big Whitehead fan but who advocates an augmented Whiteheadianism. There are a lot of similarities that Henry has pointed out. David has pointed out further similarities with potentialities, etc. There are these very close matches, but it's missing something, and what Shimony is talking about is an augmented Whiteheadism and maybe that's what we should be talking about. Let's not take Whitehead of the 1920s too seriously, but imagine an augmented Whitehead, as if he were around now with knowledge of Bell's theorem and modern issues. Would it just be augmented or would it require a fundamental transformation of some or all of his categories?

Clayton: It has often been said that Whitehead didn't take into account much of quantum theory, understood at least according to the Copenhagen interpretation. Are you trying to augment or transform Whitehead and/or rethink some of the standard sayings of quantum theory to bring them together? Can you clarify the common ground in the two projects that you're involved in?

Stapp: John Lango convinced me that Whitehead's idea of platonic ideals is maybe not a good thing. You don't need platonic ideas. The Aristotelian idea of forms as potentialities is much more useful for the quantum view of the universe, so I think that you probably do want to sort out what you want to take out of Whitehead. Certainly as a physicist, he didn't have the benefit of a lot of experience that we have—you don't regard him as "the" authority.

Clayton: And at the same time you're breaking with some standard interpretations of quantum mechanics when you bring in dialogue with this new Whitehead.

Stapp: Well, what is standard? There are so many interpretations, and in most interpretations physicists want to keep mind out—that mind is something we better get out of the system. On the other hand, if you want an ontological interpretation, you can take the simplest view of von Neumann's idea that there is a state of the universe that collapses and has certain structures. That is, ontologically speaking, a very simple way to develop quantum theory. From that simplest view you automatically get these things. The observer is pushed out somehow and his or her questions have to remain.

Finkelstein: I tried to line up the ideas of Whitehead with those of quantum theory, and it doesn't work completely [see Bibliography: Finkelstein, 1997].

The concept of the concrescence of occasions is a problem. Whitehead indicates that an occasion can be resolved into concrescences of other occasions in a non-unique way, it's variable, that sounds very much like the resolution of a vector into components in several different ways, and that makes concrescence sound like quantum superposition. But then there's nothing anywhere in the structure that sounds like the other important operation in quantum theory, which is multiplication—the combining of a proton and a neutron to make a deuteron or the building up of an atom, etc. And of course that's necessary.

Lango: Between concrescence and quantum superposition, I think with concrescence there's final causation. I don't think there's final causation in quantum superposition, and so that leads me to think that there's a radical discrepancy between Whitehead and a quantum picture of the world. In fact, during this whole discussion I never heard the term *final causation* mentioned, and that's an essential ingredient in Whitehead's metaphysics. It's a strange ingredient and it's one thing that prevented Whitehead's philosophy from being the philosophy of the twentieth century. That's another example of where it's difficult to square quantum theory with Whitehead without transforming Whitehead.

Finkelstein: By all means, I would like to transform him. I'm sure that Whitehead is the last person who wants to stop the process. And in quantum theory there is something that could be called final and initial causation. In Dirac's language they're called "kets" and "bras"; the ket represents the process by which you begin an experiment like putting a photon into a polarizer, and the bra represents the action you carry out at the end of the experiment when you have an analyzer and see whether the photon gets through to your eye. That ends the experiment, so it's quite final.

Tanaka: Whitehead wrote the book *The Principle of Relativity* [see Bibliography: Whitehead, 1922]. The meaning of Whitehead's principles of relativity in his physical works is similar to Einstein's but with slight differences. In *Process and Reality*, the principle of relativity has a metaphysical meaning. It means a thoroughgoing relationality between actual occasions. The physical world is not absolute. It can change from cosmic epoch to cosmic epoch, and dimensions of spacetime are also changeable. The present four dimensions of spacetime are an accidental fact to Whitehead's cosmology. So to Whitehead the relationality of actual occasions is basic in a metaphysical sense, and the physical roles must be deduced from the metaphysical relationalities. But the physical role itself cannot be confirmed on empirical grounds because empirical testing requires a uniform sense in the subjective world.

I want to emphasize the meaning of the experiential aspect of actual occasions. Many physicists deny the experiential aspect of elementary events in the physical world. But I think that such events and objects are abstractions from actual occasions, more exactly the relationality of actual occasions. So I am very much inspired by Professor Stapp's ideas of the experiential aspect and then to deduce physical objects so that subjectivity is present also in the microscopic level. Whitehead only stated this idea in the general, metaphysical sense. He did not construct a concrete physical theory—that is the physicist's role.

The question I have for David Finkelstein is that I think he must stress the relativization of classical logic. Your main idea for the interpretation of quantum physics is that the classical logic of classical physics cannot hold in the most basic description of the world because it does not accept the relationality of quantum events. Einstein rejects the completeness of quantum physics because, in part, he assumes classical logic.

I ask Professor Finkelstein, when you relativize absolute dynamic roles, what is your own background logic—is it classical logic or quantum logic?

Finkelstein: First of all, thank you for your kind remarks about my old work, but I no longer believe it. When I first came across von Neumann's lovely idea of quantum logic, first in his book and then in a paper with Birkhoff, on the revision of logic involved with quantum theory, I was still rather fresh from going through Whitehead and Russell, and Whitehead, when they wrote *Principia*, was a thoroughgoing logicist. He really thought logic was fundamental, and so anything that changes logic must be very important. And it's still true, yes it's important, but that's not the best way to look at it. I think, from the point of view either of organism or of quantum theory, logic is sort of an epiphenomenon, a very high level description of the actual dynamical processes that go on. In fact, the von Neumann logic is peculiarly timeless, it's not enough for physics, you can't build a physical theory on Boolean algebra alone. You certainly can't build physics out of a Boolean algebra of things at just one moment, and you certainly can't do it out of the von Neumann modifications of Boolean algebra. You need a deeper theory, and I think that dynamics or kinematics (I no longer distinguish between the two) is that deeper theory. Logicism is just an early stage that everybody has to go through.

Tanaka: A difference between Bertrand Russell and Alfred North Whitehead, in their attitude towards mathematics, was that Russell wanted to reduce mathematics to logic, but Whitehead on the contrary reduced mathematics to algebra. Whitehead was more of an applied mathematician; he majored not in pure mathematics but in applied mathematics and theoretical physics. He constructed a new theory of gravitation that is different from Einstein's theory. Algebra is more to the center for him than logic, so I think there is commonality between Whitehead's ideas of mathematics and Finkelstein's ideas of quantum logic.

Finkelstein: I hadn't realized that. Thank you very much for pointing it out.

Malin: The idea that the laws of nature are habits of nature is something which Whitehead believed in, and it is certainly very attractive. Now, if it's true, the way we would expect it to manifest is to see that the constants of nature are variable. So far as I know there is a very precise measurement that they are not variable to 1 in 10^{12} or something like that—very accurate. How do we deal with that? Well, one way is to say that Whitehead is wrong. Another way is to say that they do vary but they have not detected it yet. And another possibility is that there is an interconsistency between the different laws of nature that we haven't realized yet, which brings about the fact that they cannot change, some self-consistency condition. So what do you think about that?

Finkelstein: I associate the idea that natural law is just a habit with Peirce more than Whitehead. Did he share that idea also?

Malin: Oh, yes.

Finkelstein: Thank you. The idea that things or happenings tend to increase their probability of happening again, we don't call it habit anymore in physics, of course, we call it Bose-Einstein statistics. That's a guess that natural law is a condensation phenomenon. I'm very much inclined to the view that Geoffrey put forward earlier that really it's all in the vacuum. I suspect that the things we call laws today are simply phenomenological descriptions of the quantum fine structure of the vacuum. That I guess is a condensation phenomenon. The limited number of possible laws is the result of a limited number of possible phases. And so the apparent constancy of law arises because there are jumps between them, just as there are between the various crystalline forms of ice. You don't have a smooth variation from one to the other. It takes a real disaster.

Fagg: How congenial are your ideas on this with Wheeler's talking about mutability of law, law without law, and so forth?

Finkelstein: He simply came to this conclusion before I did. I have no problem with it at all.

Fagg: If you're dealing with the entire universe, then where is the observer? It's outside the universe, for which there is no physical outside. I'm not up to date on Jim Hartle and Murray Gell-Mann, but the last I remember reading about their work, they're talking about a coarse-graining kind of quantum mechanics in which they are able to average over parts of the wave function of the universe that are insignificant. They invoke the idea of decoherence and so forth, and they seem to feel that they can proceed with this without the necessity of an observer.

Finkelstein: At the beginning of one's encounter with quantum theory, there really are two roads, probably more than two, but I want to confirm one particular junction. If you look at Kandinsky's paintings, you can say there are no objects on the canvas because he's a bad painter, or you can say he didn't think there are objects and he wanted to express this in his work. Likewise, if you look at the quantum theory of Bohr and Heisenberg, you can say they left out an ontology because they're not great physicists, they had to get the experiments working and so on, they just wanted the answers. Or, if you actually read what they say, you see it's a matter of principle that they didn't give an ontology. They really thought in terms of process as primary. It's not that something is left out. Heisenberg called it nonobjective physics because he didn't think there are objects. Now, if you take that road, then you work algebraically with the processes themselves, but if you insist on objects, if you insist on an ontology, rather than a kinematics, you'll have to run around desperately for something these operators can act on. And Schrödinger provided wave functions for this. You have to ignore the fact that von Neumann is very careful about indicating that a quantum system does not have a wave function.

A quantum system does not have a state. When we take up quantum theory, we make a list of the questions we can ask of a quantum system. The question "What is your state?" is not in that list. You can ask about the energy, the momentum, etc., but nowhere are you allowed to ask, "What is the state?" So in an important sense there is no quantum variable corresponding to the state. You put one in because it's suggested by the mathematics and because we're desperate for an object. And if you do that, then you have to wonder about the crazy things this "state" does. To conceal the fact that it doesn't exist, you have to make it "collapse" at critical times. The whole business of collapsing states obviously comes from clutching the idea that a quantum system has a state. Remember that in von Neumann's book the wave function is something that describes not one quantum system but what he calls a "reiner Fall," which translates as a "homogeneous case." For a homogeneous case you just need to have some way of making quantum systems, and you make a lot of them all the same way. So as Belinfante, Dirac, and others say, Ahah! it's the way you make the quantum system that's described by the wave function, or the psi vector, or what have you. It's the process.

I like that very much, but then you don't pull one member out of the ensemble and say it has a state which somehow tells you what the whole ensemble is. You cannot look at a quantum system and see how it was made. It doesn't have enough information in it. You can ask it any question, but you will not find out from it how it was made. So it's totally nonoperational, totally counterexperiential to introduce the idea that a quantum system has a state. There's a huge variety of so-called states actually representing questions we choose to put, and the system will come up with either a "yes" or a "no" depending on, in part, its past and, in part, on a decision that's made on the spot.

That's the other choice that Henry Stapp speaks of. There really are two choices. We decide the question, and nature or the system or a collective or God decides "yes" or "no." But all you get from the experiment is one bit. You put in a huge amount of information, rotate a polarizer or tune a particle detector, and all you get out is a "yes" or "no." To imagine that the system is carrying an infinite amount of information besides this "yes" or "no" doesn't agree with our experience.

Eastman: In contrast to experiments that can be linked rather directly to experience, Whitehead also introduced the "extensive continuum." Can someone comment on this abstract concept?

Tanaka: Whitehead distinguished between spacetime and an extensive continuum, the latter being a proto-spacetime. Since Whitehead discusses the extensive continuum, he drops the dimensionality of spacetime. Dimensionality is a physical concept; we are located in three dimensions of space and one of time, a physical, contingent fact. In the extensive continuum there is mutual immanence of pre-events despite the fact of temporal order in the physical world. Whitehead explicitly says that every event is immanent in other events regardless of the causal order. My paper is a tentative formulation of Whitehead's extensive continuum as a proto-space-time.

Finkelstein: Since I may be involved a little in getting you started on this project, I feel that I should point that there's a limit to how much satisfaction you can get out of quantum logic of this kind. If you think of the letters A, B, and so on as representing processes, then a simple way to combine two processes is to do one after the other. That goes on in the laboratory all the time. You put one polarizer after another, for example. But that's not what occurs in this kind of quantum logic. Here we deal with $A \cap B$ and $A \cup B$. To verify that $A \cap B$ equals C, it really takes an infinite number of experiments of the kind I just mentioned, the serial composition type. That doesn't mean anything here is wrong or inadequate, it just suggests that it can't be fundamental.

A more fundamental thing is the dynamical composition of processes in terms of which the $\cap$ and $\cup$ can then be expressed. There is another unsatisfying feature of this quantum logic for me, which again doesn't mean it's wrong, it just means that more work has to be done. With classical logic there's an important kind of extensionality. Each property corresponds to a set of individuals, those having that property, and each set corresponds to a property, the property of belonging to that set. (I understand there are problems when you get to infinite sets and problems of self-reference. We now know how to take care of these nowadays, in fact, von Neumann is one of the pioneers in doing this, but I point out that he only did this for classical logic.) Extensionality cannot really be asserted for this system of quantum logic. There isn't the quantum set theory to go with this quantum logic to make it an extensional theory. To make a set theory requires a further algebraic structure that is still lacking here. You require it to go to the higher-order logic. This is all first-order logic, and you can't do physics with a first-order logic. You need the whole apparatus, the hierarchy of statements about statements about statements, and that's still lacking.

Eastman: You're saying that current quantum logic is intrinsically incomplete?

Finkelstein: At any rate, the quantum logic that we have seen used today is a small part of what is needed. [I am avoiding the word *incomplete* only because of its technical meanings.]

Tanaka: I think that quantum propositional logic is rather simple. It has the modular orthocomplemented structure of a projective lattice. There are many difficult problems with quantum predicate logic, so I cannot resolve these problems, especially as one applies quantum logic to infinite structures. Classical logic is very straightforward and we can easily use it as a tool of infinities, but quantum logic cannot be used so easily as a tool for infinities. So when we deduce many things we easily use the classical logic, but I want to say that the fundamental principle of classical logic does not hold necessarily in this world—the world we live in. So the conclusion of my quantum logic paper is that in the quantum logical world, the analytical truths of the classical world do not necessarily hold, so we can say that we live in the quantum physical world after the experimental disproof of Bell's inequalities.

Finkelstein: I agree: We need a more quantum logic.

Valenza: I would like to ask Professor Tanaka to repeat his argument against this linear ordering underlying actual events as, for example, in the Bell Aspect experiments, followed by a response by Professor Stapp.

Tanaka: For three events E1, E2, and E3 at locations l1, l2, and l3, if all events lie in a well-ordered sequence of occurrence as Professor Stapp assumes, there must be an unambiguous temporal order between events E1 and E2. One of the two events must be prior to the other. A difficulty of the above picture is that there does not seem to be any experimental apparatus to really determine which event is prior, E1 or E2.

Stapp: It is precisely because there is no empirical way to determine the order in which the actualizations located in spacelike separated regions occur that there is freedom to make the ordering at the ontological level go either one way or the other. The basic ontology goes beyond what the empirical tests reveal.

Tanaka: So your full theory operates at the ontological level and not just the physical level. Your model is very welcome to Hartshorne's type of process metaphysics, so I try to give another version more faithful to a Whiteheadian, ontological framework.

Stapp: Hartshorne certainly liked the idea that occasions occur in a well-defined order. I think that the Whiteheadian idea of an actual occasion is that an actual occasion, in its coming into being, fixes where it is. In my understanding of Whitehead, he talks about this actual world for an event, which I interpret as being the events that have already occurred in the backward light cone of that event, and that are fixed and settled. This is fine if you have a well-defined idea of coming into being in which the space-time location of the actual occasion is already fixed. Then you know its backward light cone, and its actual world, the world upon which the occasion can draw causally, in the Whiteheadian point of view. But if you have two possible events and you're not saying which event comes into being first, then you don't know enough. This event might decide to locate itself in the backward light cone of the other one. So you seem to arrive at some sort of logical problem in making the Whiteheadian idea work, if you don't say that they occur in a certain sequential order. What is fixed and settled if the order of coming into being is not definite? That is just reading Whitehead literally. Now I heard you say something about each event being immanent in the other, and how that was going to get you out of this difficulty, but I did not understand that notion well enough to comment upon it.

Eastman: Jorge Nobo has referred to a certain range, width or duration, to include space-like separated regions other than ones in the backward light cone—space-like separated regions that give rise to this dilemma that we're now referring to. Perhaps there is some duration that is necessarily part of this. Does the concept of duration help in this discussion?

Stapp: Whitehead, in order to deal with this sort of problem, gets into the idea of 'presentational immediacy.' That is, when you experience some distant event, it has the appearance of happening "now" even if you have access only to your past, and are causally influenced only by the past. Nonetheless, you perceive the faraway event as happening now, even though you don't really have causal access to the region where it seems to be happening. You have an indirect access because of your access to the past of that faraway event, and that past will propagate into the location of that event. This is somewhat like how the human brain works. You have processes in the brain that automatically account for delays and that create a nice picture as though the things are all happening now. But that picture is constructed from various things that actually happened in the past. So it's some sort of an illusion that experience is in the now. But I don't think that this sort of shuffling, which can be accounted by classical processing, can account for causal influence of the type that the dis-confirmation of Bell's theorem seems to require. If you have another idea of Whitehead according to which two events can be immanent in each other, within an evolving universe, then I will be glad to listen. But there certainly is a problem reconciling the correlation data of quantum theory with any theory that allows causal effects only in the forward light cone of their causes.

Part IV

Metaphysics

16

Whitehead's Process Philosophy as Scientific Metaphysics

FRANZ G. RIFFERT

Why Whitehead and Bunge?

This chapter tries to explicate some parallels between analytic philosophy and process philosophy. It is intended to show that process thought and analytical convictions may be quite consonant on a formal-methodological level. But since there is not just *one* process philosophy and even less *one* analytic philosophy, we first have to adjust our task, limiting our undertaking to two concrete exponents of these two traditions.

Rescher correctly pointed out that under the heading of "process philosophy" there are "distinct approaches to implementing its pivotal idea of pervasiveness and fundamentality of process, ranging from a materialism of physical processes (as with Boscovitch) to speculative idealism of psychic processes (as in some versions of Indian philosophy)" (*IPP* 8). Whitehead's particular process philosophy is used here because he attempted to connect with and build upon scientific results. Prigogine and Stengers, for instance, confess that they are "truly fascinated by his [Whitehead's] unusual resoluteness to reach comprehensive consistency" (*DN* 101, my translation). And they go on: "Whitehead's cosmology is so far the most ambitious attempt of such a philosophy. Whitehead saw no basic contradiction between science and philosophy. . . . The aim was to formulate a minimum of principles with the help of which, all physical existence—from stone to men—could be characterized" (*DN* 102, my translation). A similar reaction came even much earlier, in 1951, from the logician and philosopher Bochenski: "His [Whitehead's] work is the most complete philosophical treatment of the results of (natural) sciences we own" (*EPG* 106, my translation). Many similar quotations could be added. It is no question that Whitehead is one of the outstanding figures in the fields of logic and metaphysics in the twentieth century. This is the reason why his work will be the exponent of process philosophy for the following comparison.

However, not only process philosophy is multifaceted. Even more so is the analytic tradition in philosophy. It has developed into several branches or schools. One can at least distinguish between logical positivism, ordinary language philosophy and critical rationalism.

An attempt to compare Whitehead's metaphysics to the first two mentioned branches of analytic philosophy only can yield an abyss of fundamental differences. Logical positivism, as developed in the Vienna Circle—for instance by Rudolf Carnap (*LAW*), Moritz Schlick (*MV*), and Alfred J. Ayer (*LTL*)—in its very essence is of a strict antimetaphysical attitude. The so-called criterion of meaning was formulated as the verification principle: "The meaning of a proposition is the method of its verification" (*MV* 148). Since metaphysical propositions are not verifiable, they are claimed to be meaningless. Metaphysics, in the eyes of the positivists, was reduced to bare nonsense. That one will not be able to find any substantial positive relations between Whitehead's thought and the logical positivists' rejection of any kind of metaphysics whatsoever does, of course, not need any further demonstration.

The same also holds for the relationship between Whitehead's philosophy and the so-called Oxford Model (Strawson, *IDM*; Tugendhat, *VES*) within the analytic tradition. Ordinary language philosophy is the attempt to get to truths by the mere analysis of ordinary languages and their (deep) grammatical structures. Whitehead opposes this approach quite openly in the following passage: "There is an insistent presupposition sterilizing philosophic thought. It is the belief, the very natural belief, that mankind has consciously entertained all the fundamental ideas that are applicable to its experience. Further it is held that human language, in single words or in phrases, explicitly express these ideas" (*MT* 173). Whitehead has called this belief "The Fallacy of the Perfect Dictionary" (*MT* 173). He was convinced that pure *descriptive* analysis of natural languages in the end would be barren. His alternative approach could—according to Strawson (*IDM* 9)—be termed revisionary or constructive: "Words and phrases must be stretched towards a generality foreign to their ordinary use" (*PR* 4). This was the revisionary procedure Whitehead used when he states that an actual entity is a "feeling"—a clear and deliberate metaphor. Only by revisionary procedures—just like in sciences—can metaphysicians hope to push the limits of knowledge. (For a detailed discussion on that topic see McHenry, *DRT* and Haak, *DRM*.) So here again we arrive at a deep difference: Whitehead's process approach and the Oxford Model seem to be incompatible. Any attempt to find parallels between these two approaches seems to be hopeless from the start.

I now turn to the last-mentioned school of analytic philosophy: critical rationalism. Sir Karl Popper founded the critico-rationalist position. He held neither that there is any clear-cut demarcation line between fruitful sciences and senseless metaphysics, nor that pure analysis and description of the grammatical structures of natural languages is the only—even less the best—way of attaining fundamental and innovative philosophical insights.

All (re-)formulations of the criterion of meaning have proven deficient, being either too narrow and thereby excluding some type of science (for instance the formal sciences or even all factual sciences, Popper, *LF* 11), or too wide and thereby including metaphysics (Stegmüller: *HGP* 425–28). Hence Popper (*CR*, *LF*) and his followers (Bunge, *TBP*; Albert, *TKV* 48; Weingartner, *GM*; Lenk, *PFF*) rejected these different formulations of the positivist criterion of meaning. Metaphysics, though in Popper's view still unrefutable, had its merits at least in (a) providing useful and heuristic insights for scientists in their efforts to develop new theories and (b) by presenting solutions for nonscientific questions such as ethical and/or existential ones. There can be no doubt that metaphysical systems have been of important influence on the course of scientific developments (see Koyre, *EG* 174ff). Alternative metaphysical approaches enable scientists to look at problems from different perspectives and thereby to focus on features of certain fields of problems hitherto simply neglected. In this way, metaphysical systems are useful by proposing alternative views and enhancing theory pluralism, which is the presupposition of theory competition (*TKV* 25, 49).

This positive valuation of metaphysical theories opens the possibility to compare critical rationalism to Whitehead's metaphysics. Among the critical rationalists, philosopher and physicist Mario Bunge (*FP*) developed this positive view of metaphysics furthest. In several books and articles (see for instance *MMM*, *TBP*, *SMP*) he established formal methodological criteria that metaphysical systems have to meet in order not to be dogmatic but critical or, as he put it, scientific. He even tried to construct such a metaphysical theory that is sensitive toward these criteria, his so-called thing-ontology. Therefore his formal methodological account of metaphysics, not his substantial account, will be chosen for a comparison with Whitehead's process approach. On the substantial level the two approaches are quite different.

So first we turn toward Bunge's criteria, which have to be adequately combined with a metaphysical approach in order to qualify as scientific (below). Next we shall examine whether Whitehead's process metaphysics qualifies as such a scientific metaphysics.

Bunge: Scientific Metaphysics

The critical rationalist Bunge is very clear in stating that sciences are (implicitly or explicitly) influenced by ontological theories—for good or for bad. This being the case, we have to do our best in making our presupposed (though often unconsciously held) metaphysical positions explicit and thereby open them to criticism. This is the chief source for Bunge to establish certain criteria; metaphysical approaches have to provide for some enhancement of criticism.

Bunge distinguishes between three types of metaphysics: plain, exact, and scientific. These three types are ordered hierarchically inasmuch as each later one is a more fully developed kind of metaphysics. The term *plain metaphysics* encompasses all approaches—traditional and contemporary—that neither use formal sciences (logic and mathematics) explicitly as helping tools, nor take care of results of scientific research, but at least make an appeal to any kind of rational argumentation. Neglecting the formal and the factual sciences renders this type of metaphysics highly deficient in Bunge's view. Since the development of physics in the seventeenth century and the extraordinary progress of formal sciences, especially in the twentieth century, these deficits have to be taken very seriously, and hence leave these approaches unacceptable for today's scientific community.

Nevertheless, plain metaphysics still plays a heuristic role in metaphysical research. They are "a rich mine of problems and insights" (*MMM* 145). Certainly these valuable concepts, that may inspire metaphysical thinking, have to be refined and accommodated to the requirements of scientific metaphysics.

The next and more advanced type of metaphysics Bunge calls "exact metaphysics." This set of metaphysical theories are "built with the explicit help of logic or mathematics" (*TBP* 8). Because of its use of formal sciences, this type of metaphysics is an advance compared to plain metaphysics. But still there is a disadvantage: it neglects—at least to great extent—the research results of the nonformal (factual: natural and social) sciences. Exponents of exact metaphysics are, for instance, Leibniz, Bolzano, and Scholz (see *TBP* 8). Bunge even mentions Whitehead as a representative of this metaphysical type. Discussing the main possibilities to generate scientific metaphysical theories, Bunge states: "*Overhauling theories in exact metaphysics*. Example: revising Whitehead's theory of space and time to render it consistent with relativity physics and manifold geometry, and freeing it from phenomenalist ingredients" (*MMM* 147).

The scientific metaphysical approach is characterized by a strong effort to take into full consideration the results of scientific research.[1] This is very ambitious, to say the least. Bunge has to admit: "As for scientific metaphysics it is still largely a program" (*TBP* 8). Nevertheless, in his *Treatise on Basic Philosophy* (8 volumes), Bunge has undertaken an extensive and fascinating attempt to realize this program of building a scientific metaphysics according to his own criteria. What now are these criteria? In his work *Method, Model, and Matter* he has given a list of six formal methodological criteria, which are described in the following sections.

Generality

In contrast to scientists, each dealing with a specific domain of discourse, a metaphysician has to search for "the most general features of reality and real objects" (*MMM* 145). Here Bunge quotes Charles Sanders Peirce (*CPP* 5). This difference in the generality of the domains of discourse between sciences and ontologies is the only difference.[2] There is no difference in method; both science and metaphysics use the same retroductive method in research.[3] This leads Bunge to maintain that "*ontology is general science and the factual sciences are special metaphysics.* In other words, both science and ontology inquire into the nature of things but, whereas science does it in detail . . . metaphysics is extremely general" (*TBP* 16). Within this point of view, the search for a so-called criterion of meaning, as the representatives of the Vienna Circle did, has become meaningless.

Generality is a criterion for all three types of metaphysical approaches. So this first criterion does not distinguish between plain, exact, and scientific metaphysics.

Systemicity

As just mentioned, a scientific metaphysics is nothing but a very general scientific theory, differing from specific sciences only in its wide range of application. According to this position, a metaphysical approach is like any other science. Hence, it has to be undertaken in a systematic way. The requirement of systemicity requires that all basic notions of a metaphysical approach have to be connected with each other in order to form the basic relations, i.e., axioms, of the theory. In Bunge's own approach these basic notions are attribute, thing, part, change, event, process, life, mind, and value, to mention the most important. This axiomatic basis is the starting point for further deductive extensions. Such a system functions as a hypothetico-deductive frame.

A system has certain advantages over simple aggregations of basic ideas: a system allows for better testing and so enhances criticism. The basic concepts being distributed among the axioms secure a logical connection between the axioms. The whole system is tested (indirectly) even if only one part (i.e., one axiom) is tested explicitly. So basic notions and axioms that might otherwise not be testable gain indirect support by corroboration from other directly tested axioms (see *SR* 308).

Bunge mentions some more specific criteria for scientific systems. Since ontologies are general sciences, they have to fulfill these requirements too. These conditions can be split into two classes: syntactical and semantical ones. The first-mentioned set of conditions, because of their formal character, will be dealt with when criterion three (explicit uses of formal sciences) is discussed. The semantical conditions are discussed below.

1. First, Bunge demands that a systematic approach has to be built on a "*common universe of discourse*" (*SR* 392). This requirement states that a realm of objects has to be indicated. In other words, a single set of objects of reference has to be introduced explicitly. This criterion is a prerequisite of deducibility.

2. Second, "semantical homogeneity" (*SR* 392): all predicates used within the axioms of a system have to be members of the same semantical family. Since it is not easy to decide whether semantical heterogeneity is simply the result of lacking insight into a deeper unity (*SR* 393), this criterion is of minor importance in our metaphysical context for metaphysical systems, which by definition aim at transcending the narrow boundaries (i.e., narrow semantical families) of special scientific disciplines.

3. Bunge's third sub-criterion is called "semantical closure" (*SR* 393): it is introduced to call a halt to the smuggling in of new concepts and ad hoc hypotheses in order to save the system from a breakdown. Usually such new concepts and ad hoc hypotheses are brought into play in order to save a theory by explaining otherwise unacceptable consequences. This procedure would lead to an ongoing theoretical readjustment of the theory in question and would render empirical refutation impossible.

4. The last one is called "conceptual connectedness" (*SR* 394): The purpose of this criterion is to prevent the axioms from falling apart unconnected. It is fulfilled, "if and only if no primitive concept occurs in a single axiom of the set" (*SR* 394). If the concepts are not distributed among the axioms, the axioms are not connected with each other. Hence no deduction can take place.

The content of these four subcriteria will be discussed when turning to the topic of systemicity in Whitehead's approach.

Exactness

Metaphysical systems, like scientific systems, have to meet the requirement of exactness. This criterion for Bunge is best satisfied by explicit application of formal sciences, especially logic and mathematics. That means that the basic concepts and axioms have to be defined in a logical manner, using logico-mathematical symbols and their background theories. The explicit use of formal sciences contributes to the enhancement of systemicity by making the interconnections between the concepts clearer and more definite. Thereby it purifies them (to a certain extent) from the host of ambiguities of ordinary languages. Finally they enable "blind" formal procedures in deducing consequences from the axiomatic basis.

At this point, we present Bunge's more distinct formal-syntactical subcriteria. Bunge mentions four desiderata in his book *Scientific Research*:

1. First, he calls for "formal consistency" (*SR* 436): No contradictions whatsoever are to follow by correct deductions from the axiomatic matrix. This requirement is important, since from an inconsistent

theory everything follows: a statement (p) and its contradicting statement ($\neg p$). Hence an inconsistent theory becomes irrefutable and at the same time is refuted: the hypothetico-deductive method breaks down.[4]

2. Second, "external consistency" (*SR* 437) is required: the system in question has to be related in a way free of contradictions to other noncompetitive, yet well-confirmed (scientific) theories. Bunge elaborates this criterion to one of the main criteria for scientific metaphysical approaches. It will be discussed in the next section.
3. Third, "primitive independence" (*SR* 439): the independence of the basic concepts (primitives) among themselves is demanded. This means that no basic term should be definable by other basic concepts of the same theory.
4. Finally, "axiom independence" (*SR* 441): it is required that none of the axioms of the system in question can be deduced from other axioms making up the axiomatic matrix. If this were the case, the derivable "axiom" would instead be a theorem. We shall return to these syntactical subcriteria when discussing the criterion of "exactness" in the Whiteheadian context.

Connectedness to Sciences

So far we have only dealt with necessary conditions for plain and exact metaphysical systems. Criterion 1 (generality) is met by plain metaphysics; criteria 2 (systemicity) and 3 (exactness) are necessary for exact metaphysical approaches. The three requirements still left are necessary conditions for scientific metaphysics.

As already mentioned in the last section (syntactical subcriterion 2), scientific metaphysical systems have to be connected to well corroborated and noncompetitive scientific theories. It is not quite clear if Bunge holds a weak or a strong interpretation of "connectedness." The strong interpretation of connectedness is indicated by the term *coherence* (*TBP* 16; also see *MMM* 158). Through such strong connectedness, the scientific and metaphysical systems would be related in an implicative way. The weak interpretation of connectedness may be confirmed by expressions such as *consonance* (*TBP* 15) or being *in tune with* (*MMM* 145). These terms suggest the less ambitious relation of compatibility between metaphysical and scientific systems.

The weak interpretation can be conceived as the minimal condition, whereas the strong interpretation can be understood as an ideal that has to be aimed at in order to develop the scientific metaphysics fully. It requires that the relation of deducibility should hold between the metaphysical system and scientific theories. Of course, specifying conditions and auxiliary hypotheses have to be added in order to be able to deduce the basic concepts or axioms of scientific theories from a metaphysical system. Hence the two systems (metaphysical and scientific) are related in a hypothetico-deductive way.

Since metaphysical systems aim at extreme generality, their concepts can hardly be confronted with particular facts and events. Nevertheless, they can be tested empirically. This test is executed by using intermediate theories of a lesser generality, which are the sciences. They establish an indirect contact with empirical facts for metaphysical theories. "The evidence for a theory in scientific metaphysics consists of judgements about its ability (a) to cohere with science. . . . In short, the test for scientific metaphysics is science" (*MMM* 158). This relation can be formalized in the following (oversimplifying) way:

$$(T_M \wedge C_{S1}) \rightarrow T_S$$
$$(T_S \wedge C_{S2}) \rightarrow P_X \cong E_Y$$

Formula 1: Indirect connectedness of metaphysical theory to empirical observation, where T_M = metaphysical theory; T_S = scientific theory; C_{S1}, C_{S2} = specifying conditions, auxiliary hypotheses; P_X = prognosis X (expressed in scientific language); E_Y = empirical evidence Y (expressed in observation language); $\rightarrow$ = logical implication; $\wedge$ = logical conjunction; $\cong$ = translation (from observation language to scientific language and vice versa). We shall return to this formula when discussing Whitehead's approach. The following two criteria can be conceived as specifications of criterion 4 (connectedness to science).

Elucidation of Philosophical and Scientific Key Concepts

Elucidation of "key concepts in philosophy or in the foundations of science" (*MMM* 145) is required from scientific metaphysical systems. As mentioned before, scientific theories and more specific philosophical theories, such as anthropology and ethics, rest on and are influenced by metaphysical presuppositions. Hence, to clarify the metaphysical presuppositions implies a clarification and elucidation of the basis of scientific and specific philosophical theories. So, for instance, a metaphysical system should shed some light on the role of mathematics by determining its metaphysical status.

Capacity of Transforming Metaphysical Approaches into Scientific Theories

This criterion specifies the way a metaphysical system is connectable to more specific scientific theories. It is so by "being enriched with specific assumptions" (*MMM* 146). The same is true for (general) factual scientific theories:

they can only be tested empirically by adding auxiliary hypotheses and specifying conditions (*MMM* 158). Newton's theory can only be tested empirically by adding auxiliary hypotheses and/or specifying conditions, including idealizations of any kind. So his theory of gravitation, for instance, can only be tested by supposing that "the *only* significant gravitational sources are the sun, the earth . . ." etc. There is no "empirical meaning" of Newton's Law of Gravitation by itself apart from specifying conditions. So this criterion specifies the way general factual theories, be they metaphysical or scientific, can be tested empirically. If we add specific conditions to a metaphysical system, we may arrive at a new scientific theory. We now turn to Whitehead's approach.

Whitehead's Process Metaphysics as a Scientific Metaphysics

In this section I examine whether Whitehead's metaphysical approach meets the above Bungean criteria. We need to analyze some of Whitehead's works because he reflected his methodological commitments nowhere extensively.

The Metaphysical Domain of Discourse

This criterion has to be met by any metaphysics, be it plain, exact or scientific. There is indeed no difficulty finding passages in Whitehead's philosophical writings that support how his metaphysics was intended as an extremely general theory. "Metaphysical categories are . . . tentative formulations of the ultimate generalities" (*PR* 8). For Whitehead, as for Bunge, it is this generality that distinguishes metaphysical systems from scientific ones and not any basic methodological differences (*PR* 5), for they both are preceded by the same methodology that Whitehead once called—long before Popper used the same term (*LF*)—the "logic of discovery" (*FR* 67). This leads us straight to the second criterion.

Hypothetico-Deductive System

The basic, most general concepts of a metaphysical approach have to be linked in order to build a system. "*Speculative Philosophy is the endeavour to frame a coherent, logical, necessary system of general ideas*" (*PR* 3, italics mine). Here Whitehead leaves no doubt that he did not just present an array of loosely connected insights but tried to build a metaphysical *system* by connecting the basic concepts.

Whitehead formulates his set of axioms in the so-called categoreal scheme of *Process and Reality*. McHenry (*AMW*), for instance, has convincingly argued that the categoreal scheme is built parallel to the axiomatic core of the *Principia Mathematica* and therefore can be conceived as an "axiomatic matrix" (*AMW* 176). Hence it is a "matrix from which true propositions applicable to particular circumstances can be derived" (*PR* 9). Whitehead becomes even more clear when he discusses the advantages such systems offer. "The use of such a matrix is to argue from it boldly and with rigid logic" (*PR* 9). More consequences can be deduced from interconnected basic assumptions than from isolated notions. Hence, more and better tests can be executed on the basis of a metaphysical system! A systematic connection of basic notions "gains the character of generating ideas coherent with itself and receiving continuous verification" (*FR* 69f).

Thus, Whitehead sees the same advantages in system building as does Bunge. We now examine Bunge's more detailed semantical subcriteria.

1. The common universe of discourse: Whitehead indeed has done justice to this subcriterion by explicitly introducing one set of (fundamental) entities: namely actual entities. Of course there are eight categories of existence introduced in the categoreal scheme, but the actual entities are the most fundamental ones because all other forms of existence are derivatives or aspects of actual entities. Even the concept of God is no exception (*PR* 19). Whitehead himself stressed the fact that his approach is one "of one type of actual entities," a "one-substance cosmology" (*PR* 19).

2. Semantical homogeneity: this criterion loses its raison d'être in the metaphysical domain and has to be restricted to the fields of special sciences. Whitehead argues that ideally the general metaphysical theory should encompass all sciences. "Its [the metaphysics'] business is not to refuse experience but to find the most general interpretative system. Also it is not a mere juxtaposition of various sciences. It generalizes beyond any special science, and thus provides the interpretative system which expresses their interconnection" (*FR* 86). Providing a general interpretative system of differing sciences implies an effort to transcend the specific domains and their corresponding narrow semantics. Otherwise, there would be no difference from old systems and hence no progress could be expressed. Whitehead has frequently defended his unusual exaggerated use of ordinary notions by referring to this need of expressing novel insights in order to transcend limitations of traditional thought systems. "The technical language of philosophy represents attempts of various schools of thought to obtain explicit expression of general ideas presupposed by the facts of experience. It follows that any novelty in metaphysical doctrines exhibits some measure of disagreement with statements of the facts to be found in current philosophical literature. . . . It is, therefore, no valid criticism on one metaphysical school to point out that its doctrines do not follow from the verbal expression of the facts accepted by another school" (*PR* 12). This may suffice to show

that Whitehead was fully aware of that issue and why he did not cling to this criterion in the metaphysical context.

3. Semantical closure: Whitehead's position toward this subcriterion is not explicit, but it is reasonable to assume that he agreed with it. In the opening passage of section II "The Categories" of *Process and Reality* he states: "Every entity should be a specific instance of one category of existence, every explanation should be a specific instance of categories of explanation, and every obligation should be a specific instance of categoreal obligations" (*PR* 20). This implies that if these ideals were fully realized, which Whitehead of course doubted (*PR* 20), no ad hoc introduction of new concepts or hypotheses would be necessary and hence semantical closure would be obtained. Thus, the above quotation indicates that Whitehead tried to keep his system semantically closed.

4. Conceptual connectedness: Whitehead again meets this subcriterion when, in discussing coherence, he writes that "the fundamental ideas, in terms of which the scheme is developed, presuppose each other so that in isolation they are meaningless" (*PR* 3). This somewhat obscure formulation becomes fully clear if we turn to another passage where Whitehead takes up the discussion of coherence: "Incoherence is the arbitrary disconnection of first principles. In modern philosophy Descartes's two kinds of substance, corporeal and mental, illustrate incoherence. There is, in Descartes's philosophy, no reason why there should be a one-substance world, only corporeal, or a one-substance world, only mental" (*PR* 6). The category of mental substance has no connection to the category of corporeal substance; hence Descartes's system falls to isolated pieces. Descartes's approach has fallen short of conceptual connectedness. If there is an axiom that does not even contain one concept that another axiom contains as well, the system ceases to be a system. Hence Whitehead can maintain: "The requirement of coherence is the great preservative of rationalistic sanity" (*PR* 6).

So far we have shown that Whitehead's approach meets all semantical subcriteria that are—according to Bunge—relevant to scientific systems. We now turn to the relation between sciences and metaphysics.

The Role of Formal Sciences in Metaphysics

This criterion may be the most controversial one. First it may be astonishing to notice that an outstanding logician, as Whitehead surely was, never tried to apply explicitly the mathematical and logical instruments in any of his late philosophical writings. This is amazing since the coauthor of the *Principia Mathematica* in his days surely was among those few who were most qualified to do so. This fact—in connection with a few misleading passages in his later works—may have given rise to the opinion that Whitehead thought that the task of formalization in the domain of metaphysics was impossible. The analytical philosopher (critical rationalist) Peter Kirschenmann in one of his essays describes "Bunge's ideas about an exact metaphysics, which will contrasted with Whitehead's conception that speculative metaphysics cannot be exact" (*IES* 86). Metaphysical approaches according to Whitehead—so Kirschenmann continues—cannot be exact because they are "not characterizable by a definite mathematical structure" (*IES* 95). Even some Whiteheadians share this conviction (see, for instance, L. Armour, *LEW* 203).

I now present a few indications that Whitehead is in agreement with this Bungean criterion even though he never undertook an attempt to achieve this goal.

1. In his early so-called logical phase (see *UW* 117–299), Whitehead wrote a treatise that attracted almost no attention at the time it was published, although Whitehead himself thought it to be one of his masterpieces (see *PSM* 33). It is "On Mathematical Concepts of the Material World" (*MCM*), published in 1905 during his cooperation with Bertrand Russell. In this article Whitehead outlines five different cosmologies, from each of which Euclidean geometry, i.e., its axioms, should be derivable and which thereby should prove to be able to "represent the most general characteristics of time, space and entities in time and space" (*WPR* 343). Whitehead uses formal logic extensively in this essay. Thus, he did use formal logic as a tool in theory construction in one of his earliest philosophical papers. Of course, that does not necessarily imply that the late Whitehead still believed in the usefulness of such attempts.

2. In his magnum opus, *Process and Reality*, he demands that a metaphysics has to be a logical system. "The term 'logical' has its ordinary meaning, including *'logical' consistency*, or *lack of contradiction*, the *definition of constructs in logical terms*, the exemplification of general logical notions in specific instances, and the *principles of inference*" (*PR* 3, italics mine). For the characteristic of "logical consistency" Whitehead essentially postulates Bunge's subcriterion of formal consistency. The demand to define the basic notions in logical terms and to apply the rules of inference clearly implies the task of formalization. Whitehead entertained such formalizations in his essay "On Mathematical Concepts of the Material World." Later, Whitehead stated that "Logic implies symbols" (*MT* 61).

3. Another indication that Whitehead held formalization of metaphysical systems possible, and even necessary, lies in the fact that the structure of the core of *Process and Reality*—the categoreal scheme—equals a logical calculus. The basic concepts are defined and the axioms are explicitly stated. This again supports the assumption that he believed the formalization of this core to be possible. How else could his demand be understood to state the scheme "with the utmost precision and definiteness" (*PR* 9)? Only precisely formulated systems allow one to infer from it "with rigid logic" (*PR* 9).

4. Further, Whitehead also was well aware of the advantages of formalization. He stressed its usefulness in enhancing economy of mental efforts: "By relieving the brain of all unnecessary work, a good notation sets it free to concentrate on more advanced problems, and in effect increases the mental power of the race." Again, it is hard to see why this advantage of economizing our rational powers should end at the entrance to the field of metaphysics.

5. Of special importance in this context is an utterance that Whitehead once made when talking to the logician R. M. Martin, which is mentioned by Martin in his essay "An Approximate Logical Structure of Whitehead's 'Categoreal Scheme." It demonstrates very clearly that Whitehead appreciated any attempt to formalize the core of his metaphysical system: "Whitehead himself would have welcomed it [formalization of the categoreal scheme]. On one occasion, in fact, in conversation . . . he commended [me] to this effort, adding that he would have attempted such an account himself had he had the time, but it was essential to make the intuitive sketch first in the few remaining years allotted him for philosophical writing" (*ALS* 25). Hence, the fact that Whitehead did not formalize the axiomatic core of his system is not due to his conviction that such formalization is impossible or useless, but to the lack of time that was left for him. We must keep in mind that he was 63 years old when he became a professional philosopher at Harvard after decades of working as a logician and mathematical physicist. R. M. Martin has undertaken a first promising attempt to formalize the categoreal scheme (see *ALS*). Another very promising direction in formalizing central parts of Whitehead's metaphysics was taken by Granville Henry (*FC*). He applied PROLOG programming structures on Whiteheadian key notions such as 'actual entity' and 'eternal object' and thereby hoped "to gain some new perspective on the structure of concrescence of actual entities" (*FC* 17; see also *PCP*).

These five references should make clear that any attempt to characterize Whitehead as a philosopher with an antimathematical or antilogical attitude must be mistaken. On the contrary, Whitehead thought formal sciences to be important tools for any kind of research. "Mathematics is the most powerful technique for the understanding of pattern, and for analysis of relationships of patterns. . . . If civilization continues to advance, . . . the overwhelming novelty in human thought will be the dominance of mathematical understanding" (*ESP* 84). Since metaphysical theories investigate the most general patterns of reality, it is hard to see why Whitehead should have excluded it from being a field of application for mathematics.

I now examine Whitehead's thesis that formal sciences presuppose metaphysics. This is of interest here because it differs strikingly from Bunge's account. While, according to Whitehead, "logic presupposes metaphysics" (*MT* 107), Bunge maintains, "The relation between logic and ontology is that of presupposition or logical priority: any cogent metaphysics presupposes logic" (*TBP* 14).

From Whitehead's point of view, a metaphysical system that does not provide a place of interpretation for formal sciences is not comprehensive (see *PR* 3). Indeed, this seems to be a severe problem in Bunge's approach. For instance, L. Apostel states: "'How can it be explained that there exists a system producing mathematics?' . . . is one of the main questions of regional (*not* general) ontology and the question 'What *in* reality makes mathematics applicable to it, while mathematics is not created in constant interaction with the external world?' is another basic question of general ontology" (*SRO* 4). And Apostel reminds Bunge that "these two matters should be clarified if work is to progress in a satisfactory manner" (*SRO* 4). As we have mentioned, Whitehead has discussed the status of mathematics and mathematical (as well as logical) entities (particulars, variables, predicates, inferences . . .) within his metaphysical approach (see also Code's detailed investigations on that topic: *OO*, especially chapter II). After this extensive discussion of the formal criteria, we now turn to the last three criteria that render a metaphysical system scientific.

Relations Between Metaphysical and Scientific Systems

To recapitulate, Bunge's view is that metaphysical systems have no direct contact with empirical facts because of their generality. "True, ontological theories, whatever their degree of scientificity, cannot be tested empirically" (*TBP* 21). The test of metaphysical theories is an indirect one, mediated by theories of less general scope: scientific theories.

We now show that Whitehead conceived the relation between metaphysics, sciences, and empirical fact in essentially the same way: "A special scheme [of science] should either fit in with the general cosmology, or should by conformity to fact present reasons why the cosmology should be modified" (*FR* 76). How does Whitehead conceive this "fitting together" of the metaphysical and the scientific systems? "Also it [metaphysics] is not a mere juxtaposition of various sciences. It generalizes beyond any special science, and thus provides the interpretative system which expresses their interconnection" (*FR* 86). Whitehead had in mind that a metaphysical system comprises all basic notions of single sciences and thereby integrates them into a unity (*PR* 116). But the basic concepts of the sciences do not follow necessarily from cosmological systems. To be deducible, additional specifying conditions and/or auxiliary hypotheses are needed. Otherwise the more specific and complex content of scientific patterns could not be deduced from the more general theory of less content. Whitehead leaves no doubt on that point when writing about the relation between his process metaphysics and what he terms the "principles of psychological physiology" (*PR* 103): "These principles are not necessitated by this cosmology; but they seem to be the simplest principles which are both consonant with that cosmology, and also fit the facts" (*PR* 103).

We have shown that Bunge conceived the relation between both kinds of systems in a hypothetico-deductive manner. Hence, the victory of any scientific theory that is consonant with a certain metaphysics does not imply the victory of the metaphysics. This would mean to commit the fallacy of the affirmation of the consequence.[5] On the other hand, the falsification of a scientific theory that is consistent with a certain metaphysical theory could entail its refutation. However, things are more complicated, of course, for in order to be able to deduce from a scheme—as we have seen—it is necessary to add additional specifying conditions, and hence the falsification of the consequence only entails that at least one of the statements of the antecedents must be wrong. Things become still more complex if we take into account that deductions from scientific theories need additional auxiliary and specifying conditions too. So indirect corroborations and/or falsifications are far from being definitive. Whitehead was well aware of that fact: "The only logical conclusion to be drawn, when a contradiction issues from a train of reasoning, is that at least one of the premises involved in the inference is false" (*PR* 8).[6]

This may suffice to make clear that Whitehead held a hypothetico-deductive method between metaphysics and sciences while on the other hand he was well aware of the problems of this procedure in the metaphysical context.

Elucidation of Philosophical Concepts and Scientific Assumptions

Whitehead held the opinion that the basic notions of the more specific schemes—be it sciences or specific philosophic disciplines—in general have to be derivable from metaphysical principles. Thereby metaphysical systems function as interpretative schemes of the sciences. Hence it follows that in cases where this is possible, the metaphysical system sheds light on these less general systems. Whitehead conceived this task of metaphysics as important: "It has been a defect of modern philosophies that they throw no light whatever on any scientific principles. Science should investigate particular species, and metaphysics should investigate the generic notions under which those specific principles should fall" (*PR* 116). On several occasions Whitehead tries to make clear that his approach does not fall short of that demand. So he maintains, for instance, that his process metaphysics is able to elucidate the basic principles of the physics of his day. "But the general principles of physics are exactly what we should expect as a specific exemplification of the metaphysics required by the philosophy of organism" (*PR* 116). Here once again the kind of connection between metaphysical and scientific system becomes clear. But Whitehead expresses the relation even more distinctly at the end of a section titled "The Quantum Theory" in his *Science and the Modern World.* "The point illustrated by this example is that the cosmological outlook, which is here adopted, is perfectly consistent with the demands for discontinuity which have been urged from the side of physics" (*SMW* 136). On the other hand Whitehead, of course, was fully aware that metaphysical principles cannot simply be applied to the special fields of sciences. Specifying conditions have to be added: "Scientific descriptions are, of course, entwined with the specific details of geometry and physical laws, which arise from the special order of this cosmic epoch in which we find ourselves" (*PR* 116).

This relation of process metaphysics to physics, frequently mentioned by Whitehead in his works, has been investigated quite often.[7] There have also been a series of books and articles published that explore the relationship of process philosophy to biology and psychology,[8] but less frequently to chemistry.[9] So Whitehead's approach, indeed, is one of the few philosophical approaches of the twentieth century that can be explicitly and comprehensively related to scientific research results.

Concerning the task of throwing light on philosophical subdisciplines such as ethics and anthropology, Whitehead again offers promising hints. Only one shall be mentioned here: In the categoreal scheme (categories of obligations viii and ix, *PR* 27f) Whitehead explicitly identifies the source of the concept of freedom within his system. He also mentions the relevance of that concept for ethics. Referring to category ix of the categories of obligations, he writes: "The point to be noticed is that the actual entity, in the state of process during which it is not fully definite, determines its own ultimate definiteness. This is the whole point of *moral responsibility*" (*PR* 255, italics mine). We shall not pursue that topic any further here. However, we hope that it has become clear that Whitehead fulfills the demand of elucidating more specific philosophical concepts outside the domain of metaphysics as well as basic scientific notions.

Metaphysics as Presupposition of and Candidate for Science

The aforementioned has shown how both Bunge and Whitehead viewed metaphysics as general systems that could be connected to the more special systems by adding more specific characteristics that allow for implicative reasoning. In that sense, a metaphysical system is a presupposition of a special science. It is a candidate for a special science in as far as it can—by adding specifications—mutate into a science.

Concluding Remarks

What has been said so far should suffice to substantiate that Whitehead's process philosophy meets Bunge's formal methodological criteria or was at least designed to be able to meet them (explicit formalization). This implies that Whitehead's metaphysics can be termed a *scientific metaphysics*. His process metaphysics hence is a serious alternative to Bunge's materialist 'thing ontology.' Since, according to the critical rationalists, the rivalry of theories cannot be dispensed without the sacrifice of progress in every field of human thinking, one can only hope that members of this branch of philosophy will dedicate more intensive study to Whitehead's process approach.

However, it was the founder of critical rationalism, Sir Karl Popper, who qualified Whitehead in one of his most influential books, *The Open Society and Its Enemies*, as "one of the two most influential irrationalist authorities of the present" (*OSE* 247) that "represented his philosophy without argument" (*OSE* 450). Popper's critique of Whitehead's philosophy may have been one of the main reasons for the fact that Whitehead's process metaphysics was hardly ever mentioned by exponents of critical rationalism and even less often discussed. In the light of the outcome of the investigations presented in this chapter, this judgment must be qualified as a severe misunderstanding or even ignorance of Whitehead's work.[10] More than 30 years later, Popper seemed to have changed his opinion concerning Whitehead's way of philosophizing. In his book (coauthor, Sir John Eccles) *The Self and Its Brain*, he stated: "Today the universe does not appear as an assemblage of things, but as a multitude of interactions and processes (like especially Whitehead had stressed)" (*SB* 7). On one occasion, in fact when Popper was rewarded with the title Doctor Honoris Causa at the Catholic University of Eichstätt, the author talked to him on that topic. When asked if he had changed his judgment about Whitehead's metaphysics since the publication of *The Open Society and Its Enemies*, and if he would value it positively today, his answer was "Yes. You can express it that way."

In line with Popper's later positive view of Whitehead's metaphysics, the analytic philosopher Marx Wartofsky wrote that building a metaphysical system "should be attempted only with the clear-headed awareness of mutual relevance of scientific theory and discovery on the one hand and critical metaphysics on the other." And he continued by referring to Whitehead's approach: "There has been only one such grand design in this century, whose scope, rigour and depth earn for it such a title, as far as I can tell; and whose difficulty has left it, for the present, without significant heuristic force for science. Perhaps it remains to be rediscovered" (*MHS* 170; see also Simons, *MS* 378).[11] So one can only hope with Wartofsky that other critical rationalists will follow Popper and at least draw more critical attention to Whitehead's work in particular and to the process approach in general.

Abbreviations

AEC Michel Weber. "The Art of Epochal Change." Eds. F. Riffert and M. Weber. *Searching for New Contrasts. Whiteheadian Contributions to Contemporary Challenges in Psychology, Neurophysiology and the Philosophy of Mind.* Vienna: Lang, 2003 (in press).

ALS Richard M. Martin. "An Approximative Logical Structure for Whitehead's Categoreal Scheme." Ed. R. M. Martin. *Whitehead's Categoreal Scheme and Other Papers.* The Hague, 1974, 1–26.

AMW Leemon B. McHenry. "The Axiomatic Matrix of Whitehead's Process and Reality." *Process Studies* 15/3 (1986), 172–180.

APE David R. Griffin (Ed.). *Archetypal Process Evanston.* Illinois: Northwestern University Press, 1989.

CAP D. S. Clark. "Whitehead and Contemporary Analytic Philosophy." *Process Studies* 16/1 (1987), 26–34.

CBF Marcel Kinsbourne. "Consciousness: The Brain's Private Psychological Field." Eds. F. Riffert & M. Weber. *Searching for New Contrasts: Whiteheadian Contributions to Contemporary Challenges in Psychology, Neurophysiology and the Philosophy of Mind.* Vienna: Lang, 2003 (in press).

CCC A. Karim Ahmed. "Causality, Chaos, and Consciousness." *Process Studies* 27/3–4 (1998), 255–266.

CMP Granville C. Henry & Robert Valenza. "The Concept of Mass in Process Theory." *Process Studies* 27/3–4 (1998), 292–307.

COP Avraham Schweiger. "The Common Origin of Perception and Action: A Process Perspective." Eds. F. Riffert and M. Weber. *Searching for New Contrasts: Whiteheadian Contributions to Contemporary Challenges in Psychology, Neurophysiology and the Philosophy of Mind.* Vienna: Lang, 2003 (in press).

CPP Charles S. Peirce. "Scientific Metaphysics." Eds. Ch. Hartshorne and P. Weiss. *Collected Papers of Charles Sanders Peirce.* Vol. 6, Harvard: The Belknap of Harvard University Press, 1960.

CR Karl R. Popper. *Conjectures and Refutations. The Growth of Scientific Knowledge.* London: Routledge & Kegan, 1963.

CRE Jean Piaget. *La construction du reel chez l'enfant.* Neuchatel: Delachaux et Niestle, 1937.

CWQ Stuart Hameroff. "Consciousness, Whitehead and Quantum Computation in the Brain: Panprotopsychism Meets the Physics of Fundamental Spacetime Geometry." Eds. F. Riffert & M. Weber. *Searching for New Contrasts: Whiteheadian Contributions to Contemporary Challenges in Psychology, Neurophysiology and the Philosophy of Mind.* Vienna: Lang, 2003 (in press).

DN Ilya Prigogine and Isabell Stengers. *Dialog mit der Natur.* Munich: Piper, 1979.

DRM Susanne Haack. "Deskriptive versus revisionäre Metaphysik: Strawson und Whitehead." *Conceptus* 12 (1978), 80–100.

DRT Leemon B. McHenry. "Descriptive and Revisionary Theories of Events." *Process Studies* 25 (1996), 90–103.

DST Richard Gale. "Disanalogies Between Space and Time." *Process Studies* 25 (1996), 72–89.

ECA J. M. Burgers. *Experience and Conceptual Activity: A Philosophical Essay Based Upon the Writings of A.N. Whitehead.* Cambridge: The MIT Press, 1965.

EG Alexandre Koyré. *Études Galileennes.* Paris: Hermann, 1966.

EOS Granville C. Henry and Robert Valenza. "Eternal Objects at Seal." *Process Studies* 30/1 (2001), 55–77.

EPG Josef M. Bochenski. *Europäische Philosophie der Gegenwart.* Bern: Francke, 1951.

ESP Alfred N. Whitehead. *Essays in Science and Philosophy.* New York: Rider and Company 1948.

ETI Lawrence Fagg. "Electromagnetism, Time and Immanence in Whitehead's Metaphysics." *Process Studies* 26/3–4 (1997), 308–317.

FC Granville C. Henry. *Forms of Concrescence—Alfred North Whitehead's Philosophy and Computer Programming Structures.* London: Bucknell University Press, 1993.

FCM Jason Brown. "Foundations of Cognitive Metaphysics." *Process Studies* 27/1–2 (1998), 79–92.

FP Mario Bunge. *Foundations of Physics.* Berlin: Springer, 1967.

FR Alfred North Whitehead. *The Function of Reason.* Princeton: Princeton University Press, 1929.

GM Paul Weingartner. "Der Gegenstandsbereich der Metaphysik." Ed. Th. Michels. *Heuresis.* Salzburg: Otto Müller Verlag, 1969, 102–140.

HGP Wolfgang Stegmüller. *Hauptströmungen der Gegenwartsphilosophie vol. I.* Stuttgart: Körner, 1978.

HPP Anderson Weeks. "Hippogriff of Physics and Psychology or Important Metaphysical Knowledge?" Eds. F. Riffert and M. Weber. *Searching for New Contrasts: Whiteheadian Contributions to Contemporary Challenges in Psychology, Neurophysiology and the Philosophy of Mind.* Vienna: Lang, 2003 (in press).

IAP George Shields. "On the Interface of Analytic and Process Philosophy." *Process Studies* 25 (1996), 34–54.

IDM Peter F. Strawson. *Individuals: An Essay on Descriptive Metaphysics.* London: Methuen & Co., 1959.

IES Peter Kirschenmann. "Some Thoughts on the Ideal of Exactness in Science and Philosophy." Eds. J. Agassi and R. Cohen. *Scientific Philosophy Today—Essays in Honor of Mario Bunge.* Boston: Reidel, 1982, 85–99.

IPP Nicholas Rescher. *Process Metaphysics—An Introduction to Process Philosophy.* Albany: SUNY Press, 1996.

IRP Franz Riffert. *Whitehead und Piaget—Zur interdisziplinären Relevanz der Prozessphilosophie.* Vienna: Langer, 1994.

ISW Elisabeth Kraus. "Individualism and Society: A Whiteheadian Critique of B.F. Skinner." Ed. J. R. Roth. *Person and Community: A Philosophical Exploration.* New York: Fordham University Press, 1975, 103–132.

L Wesley C. Salmon. *Logik.* Stuttgart: Reclam, 1983.

LAW Rudolf Carnap. *Der logische Aufbau der Welt.* Berlin, 1928.

LEW L. Armour. "Logic and Experience in Whitehead's Metaphysics." *Process Studies* 21/4 (1992): 203–218.

LF Karl R. Popper. *Logik der Forschung.* Vienna: Springer, 1935.

LTL Afred J. Ayer. *Language, Truth and Logic.* Oxford, 1936.

MAN Jason Brown. *Mind and Nature—Essays on Time and Subjectivity.* London: Whurr, 2000.

MCM Alfred North Whitehead. "On Mathematical Concepts of the Material World." *Philosophical Trans. Royal Society of London.* Ser. A (205) (1906), 465–525.

MHE Hilary Putnam. "Meaning Holism and Epistemic Holism." Eds. K. Cramer et al. *Theorie der Subjektivität.* Frankfurt a. M.: Suhrkamp, 1987, 251–296.

MHS Marx Wartofsky. "Metaphysics as Heuristic for Science." Eds. M. Wartofsky and R. Cohen. *Boston Studies in the Philosophy of Science.* Vol. III, New York: Humanities Press, 1967, 123–172.

MMM Mario Bunge. *Method, Modell and Matter.* Boston: Reidel, 1973.

MP T. J. Regan. "The Matrix of Personality: A Whiteheadian Corroboration of Harry Stack Sullivan's Interpersonal Theory of Psychiatry." *Process Studies* 19/3 (1990), 189–198.

MPC Roger Sperry. "Mental Phenomena's Causal Determinants in Brain Function." *Process Studies* 7/3 (1975), 173–183.

MS Peter Simons. "Metaphysical Systematics: A Lesson from Whitehead." *Erkenntnis* 48/2–3 (1998), 377–393.

MT Alfred North Whitehead. *Modes of Thought.* Cambridge: Cambridge University Press, 1938.

MV Moritz Schlick. "Meaning and Verification." *Philosophical Review* 45 (1963), 148.

OAM Luca Vanzago. "The One and the Many. Reflections on Whitehead's Notion of Personal Identity." Eds. F. Riffert and M. Weber. *Searching for New Contrasts: Whiteheadian Contributions to Contemporary Challenges in Psychology, Neurophysiology and the Philosophy of Mind.* Vienna: Lang, 2003 (in press).

ONP Franz Riffert. "On Non-Substantialism in Psychology." *International Journal of Field Being* Special Issue: Process Philosophy and the Sciences (2002).

OO M. Code. *Order and Organism—Steps to Whitehead's Philosophy of Mathematics and the Natural Sciences.* Albany: SUNY Press, 1985.

OP David Finkelstein and W. Kallfelz. "Organism and Physics," *Process Studies* 26/3–4 (1997), 279–292.

OSE Karl R. Popper. *The Open Society and Its Enemies* (2 vols.). London: Routledge & Kegan, 1965.

PAP Avraham Schweiger. "The Process Approach in Perception." *Salzburger Theologische Zeitschrift* 3/2 (1999), 207–213.

PAW A. H. Johnson. "The Psychology of Alfred North Whitehead." *The Journal of General Psychology* 32 (1945), 175–212.

PBO George Wolf. "The Place of the Brain in an Ocean of Feelings." Eds. J. B. Cobb and F. I. Gamwell. *Existence and Actuality.* Chicago: University of Chicago Press, 1984.

PCP Granville C. Henry and Michael G. Geertsen. "Whiteheadian Philosophy and Prolog Computer Programming," *Process Studies* 15/3 (1986), 181–191.

PD Norwood R. Hanson. *Patterns of Discovery—An Inquiry into the Conceptual Foundations of Sciences.* Cambridge: Cambridge University Press, 1972.

PFF H. Lenk: "Philosophie als Fokus und Forum," Ed. H. Lübbe. *Wozu Philosophie?* Berlin: deGruyter, 1978, 35–69.

PMP Franz Riffert and John B. Cobb. "Process Metaphysics and Psychology—On Some Basic Issues and Pioneer Works." Eds. F. Riffert and M. Weber. *Searching for New Contrasts: Whiteheadian Contributions to Contemporary Challenges in Psychology, Neurophysiology and the Philosophy of Mind.* Vienna: Lang, 2003 (in press).

PP David Roy. "The Promise of Process Psychology." Eds. F. Riffert & M. Weber. *Searching for New Contrasts: Whiteheadian Contributions to Contemporary Challenges in Psychology, Neurophysiology and the Philosophy of Mind.* Vienna: Lang, 2003 (in press).

PPM David R. Griffin. "Panexperiential Physicalism and the Mind-Body Problem." *Journal of Consciousness Studies* 4 (1997), 248–68.

PPO Charles Hartshorne. *The Philosophy and Psychology of Sensation.* Chicago: University of Chicago Press, 1934.

PPP Nicholas Rescher. "The Promise of Process Philosophy." *Process Studies* 25 (1996), 55–71.

PPS David R. Griffin. *Parapsychology, Philosophy, and Spirituality: A Postmodern Exploration.* Albany: SUNY Press, 1997.

PR Alfred North Whitehead. *Process and Reality: An Essay in Cosmology.* Cambridge: Cambridge University Press, 1929.

PSI Clive Sherlock. "The Presumptive Self—Implications of Process Philosophy for Psychology." Eds. F. Riffert and M. Weber. *Searching for New Contrasts: Whiteheadian Contributions to Contemporary Challenges in Psychology, Neurophysiology and the Philosophy of Mind.* Vienna: Lang, 2003 (in press).

PSM Wolfe Mays. *Whitehead's Philosophy of Science and Metaphysics.* The Hague: Nijhoff, 1977.

PSP George Wolf. "Psychological Physiology from the Standpoint of a Physiological Psychologist." *Process Studies* 11/4 (1981), 274–291.

PTE Liliana Albertazzi. "A Psychology for the Ecosystem." Eds. F. Riffert and M. Weber. *Searching for New Contrasts: Whiteheadian Contributions to Contemporary Challenges in Psychology, Neurophysiology and the Philosophy of Mind.* Vienna: Lang, 2003 (in press).

PTN Timothy Eastman. "Process Thought and Natural Sciences" *Process Studies* 26/3–4 (1997), 239–246.

PTS C. Papatheodorou and Basil Hiley. "Process, Temporality and Space-Time." *Process Studies* 26/3–4 (1997), 247–278.

PUW Julian G. Pilon. "On Popper's Understanding of Whitehead." *Process Studies* 8/3 (1978), 192–195.

QPR Shimon Malin. *Nature Loves to Hide: Quantum Physics and Reality, a Western Perspective.* Oxford: Oxford University Press, 2001.

QW Leemon McHenry. "Quine and Whitehead: Ontology and Methodology." *Process Studies* 26/1–2 (1997), 2–12.

RLM Willard v. O. Quine. "Response to Leemon McHenry." *Process Studies* 26/1–2 (1997), 13–14.

RW George Lucas. *The Rehabilitation of Whitehead—An Analytic and Historical Assessment of Process Philosophy.* Albany: SUNY Press, 1989.

SAP Joseph Rosen: "Response to Hartshorne Concerning Symmetry and Asymmetry in Physics." *Process Studies* 26/3–4 (1997), 318–323.

SB Karl R. Popper and John C. Eccles. *The Self and Its Brain.* New York: Springer, 1977.

SCC Franz Riffert. "On Scientific Confirmation of Causal Efficacy." Eds. F. Riffert and M. Weber. *Searching for New Contrasts: Whiteheadian Contributions to Contemporary Challenges in Psychology, Neurophysiology and the Philosophy of Mind.* Vienna: Lang, 2003 (in press).

SCN Jay Schulkin: "Evolving Sensibilities of Our Conception of Nature." *Process Studies* 27/3–4 (1998), 241–254.

SMP Mario Bunge. "Is Scientific Metaphysics Possible?" *Journal of Philosophy* 68 (1971), 507–520.

SMW Alfred North Whitehead. *Science and the Modern World.* New York: Macmillan, 1925.

SNC Franz Riffert and Michel Weber (Eds.). *Searching for New Contrasts: Whiteheadian Contributions to Contemporary Challenges in Psychology, Neurophysiology and the Philosophy of Mind.* Vienna: Lang, 2003 (in press).

SPP Nicolas Rescher. "On Situating process Philosophy." *Process Studies* 28/1–2 (1999), 37–42.

SR Mario Bunge. *Scientific Research I: The Search for System.* New York: Springer, 1967.

SRO Leo Apostel. "Some Remarks on Ontology." Eds. J. Agassi and R. Cohen. *Scientific Philosophy Today—Essays in Honor of Mario Bunge.* Boston: Reidel, 1982, 1–44.

SS George Lucas. " 'The Seventh Seal'—On the Fate of Whitehead's Proposed Rehabilitation." *Process Studies* 25 (1996), 104–116.

T John B. Cobb. "Therapy." Eds. F. Riffert and Weber. *Searching for New Contrasts: Whiteheadian Contributions to Contemporary Challenges in Psychology, Neurophysiology and the Philosophy of Mind.* Vienna: Lang, 2003 (in press).

TBP Mario Bunge. *Treatise on Basic Philosophy: Ontology I—The Furniture of the World.* Vol. III, Dordrecht: Reidel, 1977.

TKV Hans Albert. *Traktat über kritische Vernunft.* Tübingen: Mohr, 1968.

TPP David Roy. *Toward a Process Psychology: A Model of Integration.* Fresno, California: Adobe Creations Press, 2000.

UW Victor Lowe. *Understanding Whitehead.* Baltimore: Hopkins, 1966.

UWK David R. Griffin: *Unsnarling the World-Knot—Consciousness, Freedom, and the Body-Mind Problem.* Berkley: University of California Press, 1998.

VES Ernst Tugendhat. *Vorlesungen zur Einführung in die sprachanalytische Philosophie.* Frankfurt a. M.: Suhrkamp, 1976.

VWP Reto L. Fetz. "Für eine Verbindung Whiteheadscher und Piagetscher Ansätze." Eds. H. Holz and E.W. Gazo. *Whitehead und der Prozessbegriff/Whitehead and the Idea of Process.* Munich: Alber, 1984, 220–239.

WCF Pierre Rodrigo. "What Is Called 'Feeling'? Lure and Certainty in Whitehead and Descartes." Eds. F. Riffert and M. Weber. *Searching for New Contrasts: Whiteheadian Contributions to Contemporary Challenges in Psychology, Neurophysiology and the Philosophy of Mind.* Vienna: Lang, 2003 (in press).

WCP Reto L.Fetz. "Whitehead, Cassirer, Piaget: Drei Denker—ein gemeinsames Paradigma." *Salzburger Theologische Zeitschrift* 3/2 (1999), 154–168.

WD Karl Raimund Popper. "What is Dialectic?" *Mind* 49 (1940), 312–336.

WIP John A. Jungerman. *World in Process: Creativity and Interaction in the New Physics*. Albany: SUNY Press, 2000.

WLP Isabelle Stengers. "Whitehead and the Laws of Physics." *Salzburger Theologische Zeitschrift* 3/2 (1999), 193–206.

WNP Daniel Athearn. "Whitehead as Natural Philosopher: Anachronism or Visionary?" *Process Studies* 26/3–4 (1997), 293–307.

WOV Jason Brown. "A World of Value." Eds. F. Riffert & M. Weber. *Searching for New Contrasts: Whiteheadian Contributions to Contemporary Challenges in Psychology, Neurophysiology and the Philosophy of Mind.* Vienna: Lang, 2003 (in press).

WP Claude Dumoncel. "Whitehead's Psychology." Eds. F. Riffert and M. Weber. *Searching for New Contrasts: Whiteheadian Contributions to Contemporary Challenges in Psychology, Neurophysiology and the Philosophy of Mind.* Vienna: Lang, 2003 (in press).

WPR Granville C. Henry. "Whitehead's Philosophical Response to the New Mathematics." *Southern Journal of Philosophy* (1969), 341–349.

WPS Robert Palter. *Whitehead's Philosophy of Science*. Chicago: University of Chicago Press, 1960.

WRP William Seager. "Whitehead and the Revival(?) of Panpsychism." Eds. F. Riffert and M. Weber. *Searching for New Contrasts: Whiteheadian Contributions to Contemporary Challenges in Psychology, Neurophysiology and the Philosophy of Mind.* Vienna: Lang, 2003 (in press).

ZD Joseph Bochenski. *Zeitgenössische Denkmethoden.* Munich: Francke 1980.

ZDE W. v. O. Quine. "Zwei Dogmen des Empirismus" In *Von einem logischen Standpunkt*. Frankfurt a. M.: Ullstein, 1951, 27–50.

ZST P. Duham. *Ziel und Struktur physikalischer Theorien.* Reprint: (1978) Hamburg: Meiner, 1908.

Notes

1. The term *scientific metaphysics* was first introduced by Charles Sanders Peirce (*CPP*).

2. For the purpose of this chapter, it is not necessary to distinguish between the terms *ontology* and *metaphysics*.

3. This research method has been given different names: "retroduction" is the term Hanson (*PD* 85ff) has introduced; Bochenski has called it "reduktiv" (*ZD* 101) and Peirce finally spoke of "abduction" (*CPP* 90).

4. A formal demonstration is given in Popper (*WD*).

5. For details see W. C. Salmon (*L* 57f).

6. Of course this is an oversimplification, since deduction is made possible only by introducing additional specifying conditions and auxiliary hypotheses. Hence either the metaphysical system (i.e., at least one of its laws or hypotheses involved in deduction) or at least one of the additional conditions may be wrong. Good introductions to this problem of "holism of theory testing" are presented by Duham (*ZST*), Quine (*ZDE*), and Putnam (*MHE*).

7. See, for instance, Ahmed (*CCC*), Henry and Valenza (*CMP*), Fagg (*ETI*), Finkelstein and Kallfelz (*OP*), Mays (*PSM*), Eastman (*PTN*), Papatheodorou and Hiley (*PTS*), Malin (*QPR*), Rosen (*SAP*), Schulkin (*SCN*), Jungerman (*WIP*), Stengers (*WLP*), Athearn (*WNP*), and Palter (*WPS*).

8. So far there have been several attempts to elaborate the implications of process philosophy for different fields of psychology. Some of them are: Albertazzi (*PTE*), J. Brown (*MAN, WOV, FCM*), Burgers (*ECA*), J. B. Cobb (*T*), Dumoncel (*WP*), Fetz (*VWP, WCP*), Griffin (*APE, PPS, UWK, PPM*), Hameroff (*CWQ*), Hartshorne (*PPO*), A. Johnson (*PAW*), Kinsbourne (*CBF*), Kraus (*ISW*), Regan (*MP*), Riffert (*IRP, SCC, ONP*), Riffert and Cobb (*PMP*), Riffert and Weber (*SNC*), Rodrigo (*WCF*), Roy (*TPP, PP*), Schweiger (*PAP, COP*), Seager (*WRP*), Sherlock (*PSI*), Sperry (*MPC*), Vanzago (*OAM*), Weber (*AEC*), Weeks (*HPP*), G. Wolf (*PSP, PBO*).

9. For an overview concerning the relations between process philosophy and sciences in general, see Riffert (*IRP* 399–402).

10. For a critical discussion of some of Popper's critical comments on Whitehead's approach in the *Open Society and Its Enemies,* see Pilon (*PUW*).

11. From the side of process philosophy there have been several interesting attempts to determine the relationship between process philosophy and the analytic tradition. See, for instance, Leemon McHenry (*QW*) and Quine's response (*RLM*), Nicolas Rescher (*SPP*), D. S. Clark (*CAP*) and Granville Henry and Robert Valenza (*EOS*). Outstanding are the works of Nicolas Rescher (*IPP*), George Lucas (*RW*), and the special focus on analytic philosophy in *Process Studies* (1996) edited by George Shields (including articles by George Shields (*IAP*), Leemon McHenry (*DRT*), Nicolas Rescher (*PPP*), Richard Gale (*DST*), and George Lucas (*SS*). Shields in his article gives the most comprehensive overview available at the moment about published works on the mutual relationship between process philosophy and analytic philosophy.

17

Whitehead and the Quantum Experience

JORGE LUIS NOBO

Toward the end of their 1996 book, *The Nature of Space and Time*,[1] physicists Stephen Hawking and Roger Penrose take opposite sides in a discussion of whether quantum theory should be interpreted epistemologically or ontologically. Of immediate interest is the fact that Penrose, in arguing for an ontological interpretation, considers the likely relevance of an adequate theory of experience. As he puts it,

> in order to explain how we perceive the world to be, within the framework of QM, we shall need to have one (or both) of the following:
> (A) A theory of experience.
> (B) A theory of real physical behavior. (*NST*, 128)

Penrose is himself committed to some theory of real physical behavior. Accordingly, he thinks that those who, like Hawking, fail to support such a theory have an obligation to develop a theory of experience that consistently dispenses with real physical behavior. But theories of experience, he adds, are dangerous to adopt.

Penrose does not explain why it would be dangerous to adopt a theory of experience. Perhaps he thinks any such theory would undermine the objectivity and universality that are the hallmarks of scientific knowledge. Perhaps he believes a theory of experience is likely to erect an impenetrable barrier between apparent and real physical behaviors. Whatever his reasons, Penrose must be prepared to admit that the presumed dangers of a theory of experience can be avoided by any such theory that does not dispense with real physical behavior, does not cut off physical behavior from experience, and, more to the point, essentially links the ontological status of real physical behavior to that of experience in general.

Essentially interrelated theories of experience and of real physical behavior are hallmarks of Whitehead's speculative philosophy. These interrelated theories are aspects of a novel metaphysical theory that seems inherently well-suited to provide a firm philosophical foundation for an ontological interpretation of quantum mechanics. Accordingly, my ultimate goal in this essay is to sketch aspects of Whitehead's metaphysics that are most likely to be of some relevance to the concerns of quantum theorists. But those same aspects bear on broader ontological and epistemological issues that are of interest in science and philosophy alike.[2]

I

The adequate conceptualization of our experience so as to provide a coherent interpretation of its ingredient elements is, for Whitehead, the defining goal of *speculative philosophy*: "the endeavour to frame a coherent, logical, necessary system of general ideas in terms of which every element of our experience can be interpreted" (*PR*, 3). This endeavor is made necessary by the fact that every activity of theorizing begins with human experience, addresses some problem or puzzle or difficulty encountered in human experience, is wholly conducted within the inescapable confines of human experience, and determines its own degree of success or failure by comparing the expectations it yields with the subsequent actual deliveries of human experience. This fact requires the recognition that human experience has primacy for the construction of any theory whatsoever. But truly to recognize the theoretical primacy of human experience is to recognize that human experience must itself be adequately conceptualized and elucidated, and thus must itself be the subject of a theory more fundamental and more comprehensive than any other. Other theories deal only with aspects of, or abstractions from, our experience. But speculative philosophy must address our experience as a concrete whole, and must exhibit all abstractions from it in their proper relationships, each to the others and all to our experience as a whole.

As Whitehead understands it, concrete human experience, *in each of its concrete moments*, is *not* an external relation between someone experiencing and something experienced (*AI*, 233). It is, instead, an *integral inclusive reality* within which, as aspects of itself, there are differentiated, among others, the human organism from his or her environment, the human subjectivity from the human body, the subjective response from the object responded to, the subjectively added from the objectively given, the self-determining from the other-determined, the private from the public, the present from the remembered past or the anticipated future, and, more generally, the experiencing from the experienced. Concrete human experience is not the sum of these contrasting aspects, but their aboriginal context and unity. Thus understood, concrete human experience is the original datum from which all other disciplines abstract their respective subject matters. The elucidation of that datum as one concrete whole is one chief aim of Whitehead's speculative philosophy.

It should be immediately noted that *Whitehead does not equate concrete human experience with human subjectivity; for the latter is only an ingredient aspect of the former.* This fact by itself may allay some of Penrose's fears concerning the dangers of a theory of experience. The immediate datum for all theorizing includes, but also greatly exceeds, anything that can be meaningfully and consistently construed as subjective. It should be noted also that *not just quantum theory, but every theory less comprehensive than speculative philosophy requires for its complete cogency a theory of human experience*; for any theory whose subject matter is an abstraction from our concrete experience cannot be *fully* understood until what its subject matter is abstracted from is itself fully understood.

Whitehead's characterization of concrete human experience and the various contrasts it harbors may seem mistaken to those who consider experiences to be ephemeral states of enduring mental substances or ephemeral processes undergone by enduring human brains. But the theoretical primacy of human experience entails that the nature of concrete human experience cannot be completely explained in terms of any theory whose subject matter or experiential basis is some limited aspect or other of such experience; for the theory that explains the limited aspect can fully explain the whole only if aspect and whole are in all essential respects homogeneous. But since no abstract aspect of experience can be fully homogeneous with the whole of concrete experience, no theory adequate to such an aspect can be fully adequate to such a whole. To think otherwise is to commit what Whitehead termed the "fallacy of misplaced concreteness" (*SMW*, 53–54). Materialism and idealism alike commit this fallacy, each trying to reduce all of human experience to a different preferred abstraction from the concrete wholeness of such experience. Matter-mind dualism embraces both of those abstractions, but fares no better because it forgets that the physical and the mental are alike abstractions from our concrete experience (*SMW*, 58).

The theoretical primacy of human experience must weigh heavily in the framing of any ontology. Whatever else may be said to exist or to be real, our experiences exist and are real. Even if all the various aspects of a concrete moment of experience are not what they appear to be, are in some sense illusory, then, still, as appearances or illusions they exist or are real, and the concrete moment of experience that includes them exists or is real. Hence, there is no getting away from the fact that human experiences belong to that community of realities we call the universe. If we do not know that, we do not know anything. Moreover, if we know, as in fact we do, that other things exist besides our experiences, it must be because our experiences contain within themselves clues to, and hence are *significant* of, all other existents. But our experiences can be thus significant only if all else, including the commerce of other existents among themselves, has commerce with them. One task of speculative philosophy, therefore, is to determine how—*not whether*—all else has in some sense commerce, or communication, with human experience.

The theoretical primacy of our experience must also weigh heavily in the articulation of any epistemology. *The only realities accessible to us are our own experiences, whatever communicates with, or makes a difference in, our experiences, and, as a special case of the latter, whatever is a necessary ontological presupposition of the existence of our experiences.* Therefore, since human knowing is a modality of human experiencing, our knowledge is subject to the same threefold restriction. Neither the range of possible human knowledge nor the criteria for the actual attainment of any portion of that range can exceed or be torn from our experiences, what communicates with them, or their necessary ontological presuppositions. If we appeal to the past, it must be because our experience somehow requires and exhibits the reality of that past. If we plan for the future, it must be because the potentiality for that future is somehow displayed in our experience.

On the other hand, human knowledge is not the only species of knowledge possible, for human experience is not the only species of experience. Indeed, to properly understand human experience is to understand it as a species of something more general and more pervasive than itself. This conclusion follows from the fact that speculative philosophy, in having to be adequate to the reality of human experience, must be adequate to the reality of everything else, including the reality of nonhuman experiences and, hence, of experience in general. Also, human experience is a recent addition to reality. Accordingly, we must conceive of reality as compatible with the emergence, and even the eventual disappearance, of human experience. This means that we must be able to conceive of human experience as significant of *all* reality—reality with or without the concurrent existence of human experience. This can be done most straightforwardly if human experience is thus significant not because it is human, but because it is experience. This means, in effect, that being significant of all reality must be a property of all experience, not just of human experience. But this fact, if it is a fact, must itself be signified by human experience; for only in our experiences can we find the signs of any other reality we may wish to distinguish or posit.

It follows that speculative philosophy must seek to exhibit reality in general and experience in general as interdependent natures. To accomplish this broader task, it must restrict reality to experience, whether human or not, and to whatever communicates with some experience or other, including the necessary ontological presuppositions of every experience whatsoever. For Whitehead, the conceptualization of reality in general as inclusive of, and as signified by, experience in general is the task of *metaphysics*, the most basic major subfield of speculative philosophy. Metaphysics seeks the generalities that must be true of all possible worlds. But a possible world that is knowable in principle can only be a world in which, sooner or later, experience obtains.

Metaphysics has no choice but to derive its ultimate notions from actual human experience and from the one actual world therein revealed. Metaphysical ideas, Whitehead holds, can be obtained only by generalizing from the essential features of all our experiences. This is the method of *imaginative, descriptive*, or *philosophic* generalization: "the utilization of specific notions, applying to a restricted group of facts, for the divination of the generic notions which apply to all facts" (*PR*, 5). The divining of the generic in the specific may begin from any aspect of our experience that serves as subject matter for some discipline or other. But the goal is always the same: to find concepts and principles that, suitably generalized, can illuminate other aspects of any experience or any experience as a whole. Thus, the success of the generalization "is always to be tested by the applicability of its results beyond the restricted locus from which it originated" (*PR*, 5). What is ultimately sought, though, is "a generality transcending any special subject-matter" (*PR*, 10), but also indispensable for the elucidation and interpretation of any concrete moment of experience, whether human or not. The notions derived from the various restricted or specialized loci can attain the required transcendence and indispensability only if they can be coherently correlated and contextualized in terms of the utmost generalizations obtainable by applying the method to human experience in its concrete and integral wholeness. Thus, generalization from the individual wholeness of concrete moments of human experience is indispensable to the creation of any viable metaphysics. It is also the only way to avoid the fallacy of misplaced concreteness.

If any in principle knowable possible world must include *some* world-states in which experiences obtain, it seems plausible to explore the working hypothesis that experiences necessarily obtain in *every* possible world-state of any possible world. Whitehead's metaphysics is explicitly based on such a working hypothesis: that the ultimate individual actualities of the universe have the metaphysical characteristics of concrete moments of human experience, suitably generalized (*AI*, 221). These ultimate individual actualities are technically referred to as "actual entities," or, equivalently, as "actual occasions" or "occasions of experience." Actual entities constitute one of the seven metaphysical genera, or categories, of *proper* existents posited by Whitehead.[3] But the genus of actual entities is metaphysically *basic* because the remaining six genera consist of existents that either are differentiable but inseparable features, aspects, relationships, or states of actual entities, or are eternal ontological presuppositions of *all* actual entities. For example, the genus *nexus* comprises all entities made up of interdependent occasions; and the genus *eternal object* consists of all entities (but not all realities) functioning as ontological presuppositions of all occasions. Nexūs, like their component occasions, are entities that become. On the other hand, eternal objects are entities that do not become, but are presupposed by the becoming of every occasion because each occasion becomes definite through the *ingression* into (or reproduction within) its nature of some set or other of eternal objects. Insofar as they are immanent determinants of our experience, eternal objects are the real, uncreated universals that we encounter as qualia, patterns, and structures: e.g., a color, an emotion, a shape, or a scalar form of energy.

The postulation of actual entities, or occasions of experience, as metaphysically basic is what Whitehead refers to as the "working hypothesis" of his metaphysics. I refer to it as the "postulate of metaphysical experientialism." The postulate is not intended as a self-evident truth. This point bears emphasizing because the old conception of metaphysics as a deductive discipline starting from self-evident truths dies hard (*PR*, 7–10). True enough, metaphysics seeks the ultimate generalities indispensably relevant to the true description of every possible world state. But, as Whitehead says, "the accurate expression of the final generalities is the goal of discussion and not its origin" (*PR*, 8). Accordingly, "metaphysical categories are not dogmatic statements of the obvious; they are tentative formulations of the ultimate generalities" (*PR*, 8). But the postulate of experientialism, although tentative, is not a shot in the dark. It takes seriously the theoretical primacy of human experience for all disciplines; it recognizes the implications of that primacy for ontology and epistemology; it takes at face value the manifest wholeness and irreducibility of each concrete moment of our experience; and it seeks the ultimate nature of reality in the only place it can look for it or hope to find it: in experience's own generic nature.

Whitehead's *metaphysical experientialism*, then, is no more and no less than a plausible working postulate on which to essay the construction of a scheme of metaphysical ideas adequate to the successful interpretation of our experience and of our world. Such an interpretation is what Whitehead means by a *philosophical cosmology*, which is both the goal and the second major subfield of speculative philosophy. Whether it can ground the construction of a successful cosmology is the ultimate test of a metaphysical scheme, a test at once empirical, hermeneutical, and heuristic. The construction of the cosmology is not a deduction from the metaphysics. The cosmological ideas must specify and supplement the metaphysical ones because the actual world, for all its exhibition of generalities and necessities, is inevitably specific, is in some respect or other contingent, and is always in some regard or other singular or unique. It follows that the metaphysical ideas discussed below will require considerable specification and supplementation before they can ground a viable ontological interpretation of quantum physics.

II

In Whitehead's metaphysics, *actual occasions* are the ultimate discrete and indivisible realities constituting the *actual world*. The conception of an actual occasion is reached primarily by descriptive generalization from, first, the paradoxical relationship between any one concrete moment of our experience and the world of which it is part, and second, from the becoming, being, and interdependence of successive concrete moments of our experience. In both cases, the generalization is toward features that are essential to any experience, however basic or primitive, and away from features that are uniquely human, and even away from features that we have reason to think are uniquely associated with what we normally regard as living organisms. For example, neither thought nor consciousness nor sense-perception is a metaphysical, or necessary, feature of occasions of experience. On the other hand, each occasion of experience necessarily involves a process in which the occasion unconsciously grasps the objective reality of earlier occasions as efficient causes of its own existence and as determinants of its own initial ingredient subjectivity. Also, although begotten and partly determined by its correlative actual world, each occasion must be partly self-determining. In fact, for Whitehead, the technical meaning of *actual* cannot be torn from the notion of *self-determination* or, equivalently, of *self-realization.* To be actual is to be, or to have been, self-realizing (*PR*, 222).

Actual occasions "are drops of experience, complex and interdependent" (*PR*, 18). The manner of their interdependence is of special interest. *Actual entities are interdependent because they are mutually immanent.* The doctrine of mutual immanence is a straightforward theoretical interpretation of the paradoxical relationship that obtains between any one concrete moment of our experience and the world of which it is a component. Since, on all plausible accounts, any concrete moment of human experience is construed as embodying a *finite* locus within the world, the said experience and the *remainder* of the world must be *outside* each other. Nonetheless, what is manifested *within* that moment of experience is the entire world and not a finite region of it—at least, it is the world as extending well beyond the *here* and *now* of the experience in question. Thus the paradox: "there is a dual aspect to the relationship of an occasion of experience as one relatum and the experienced world as another relatum. The world is included within the occasion in one sense, and the occasion is included in the world in another sense" (*MT*, 163).

Since the actual world, on this theory, is made up of actual occasions, the same paradoxical relationship necessarily obtains between any two, or more, occasions of experience. "Any set of actual occasions are united by the mutual immanence of occasions, each in the other" (*AI*, 197). This mutual immanence of all occasions is sufficient to define the metaphysical category of *nexus*: "the term Nexus does not presuppose any special type of order, nor does it presuppose any order at all pervading its members other than the general metaphysical obligation of mutual immanence" (*AI*, 201). Since no occasion can exist except as connected to, or in a nexus with, all other occasions, *the actual world is a nexus.* Any nexus less inclusive than the actual world may exhibit additional types of order pervading its respective members. Indeed, any enduring entity is construed by Whitehead as a limited nexus of occasions exhibiting some type or other of contingent order among its members. But all contingent types of order, and even their attendant types of disorder, alike presuppose the generic order of mutual immanence (*AI*, 228).

Given the discreteness of occasions, which entails their mutual transcendence, their mutual immanence is undoubtedly a paradoxical doctrine. But it is a paradox not unfamiliar to quantum physicists, who daily contend with the interconnectedness of quantum phenomena or with the holistic influence of the universe on local events. Thus, quantum physics and ordinary experience alike call out for a resolution of the paradox of mutual immanence (*MT*, 163). The elucidation of this paradox is the basic task Whitehead assigns to his scheme of metaphysical categories: "The world within experience is identical with the world beyond experience, the occasion of experience is within the world and the world is within the occasion. The categories have to elucidate this paradox of the connectedness of things:—the many things, the one world without and within" (*AI*, 228).

The need to account for the mutual immanence, or *connectedness*, of things is the main reason why the postulate of experientialism recommends itself to Whitehead. After noting the impossibility of accounting for the connectedness of things in terms of any ontology—such as Descartes's mind-matter dualism—whose primitive entities are only externally related to each other (*AI*, 220–21), he says: "But if we hold, as for example in *Process and Reality*, that all final individual actualities have the metaphysical characters of occasions of experience, then on that hypothesis the direct evidence as to the connectedness of one's immediately present occasion of experience with one's immediately past occasions, can be validly used to suggest categories applying to the connectedness of all occasions in nature" (*AI*, 221).

Elsewhere (in *WMES*) I have discussed at length how a presystematic analysis of memory suggests many of Whitehead's metaphysical categories and doctrines. Here I must restrict the discussion to a few of the suggested doctrines themselves. One such doctrine is that each actual occasion is in essence a *becoming* ending in a *being* that functions as a *datum* for the becoming of all later occasions. Each actual occasion is *both* its becoming and its being, both the creative activity and the created product, "both process and outcome" (*PR*, 84). The occasion's being is *created*, or *constituted*, or *produced*, by its own becoming (*PR*, 23, 166). As a becoming, the occasion is *an actuality in attainment*; as a being, the same occasion is an *attained actuality*. In deference to the importance of its ingredient, self-determining subjectivity, an actuality in attainment is said to be a "subject" and, in deference to the objective roles it plays beyond itself, the attained actuality is said to be a "superject" (*PR*, 29). Since the attained actuality is the occasion as *fully made*, it is also referred to as the occasion's individual "satisfaction."

The succession of an occasion *qua* becoming by the same occasion *qua* being is only a special case of the fact that each occasion's becoming involves it in an *intra-occasional supersession* of phases of existence. *This supersession of phases is not in physical time*. By *physical time* Whitehead means the time of relativity physics. Such time, he holds, is only a *contingent* feature of the actual occasions forming what he calls our "cosmic epoch," roughly the actual world insofar as we can trace the relevance to us of its history and pervasive features.[4] Physical time is one of the features that become, but physical time is *not* the becoming of the features (*PR*, 283). This is one reason why it is misleading to refer to occasions as "events." Actual occasions, unlike events in relativity physics, are not divisible into earlier and later events, and are not point instants. Nonetheless, the becoming of an occasion "is a process proceeding from phase to phase, each phase being the real basis from which its successor proceeds towards the completion of the thing in question" (*PR*, 215). Moreover, through the ingression of appropriate eternal objects, each "predecessor-phase is absorbed into the successor-phase without limitation of itself, but with additions necessary for the determination of an actual entity in the form of individual satisfaction" (*PR*, 149). In other words, the becoming of the occasion is a *cumulative* process because what each of its creative phases produces remains a determinate feature of that occasion. The occasion as a being, superject, or satisfaction is the final synthesis of all the accumulated determinations.

Supersession is *inter*-occasional as well as *intra*-occasional. "Each occasion supersedes other occasions, it is superseded by other occasions, and it is internally a process of supersession, in part potential and in part actual" (*IS*, 241). In inter-occasional supersession, the terminal, or superjective, phase of any one occasion's existence is immediately superseded by at least one other occasion's initial, or *dative*, phase of existence (*IS*, 240). In general, the converse relation also holds: the initiation of an occasion's becoming immediately supersedes the termination of at least one other occasion's becoming. But two exceptions must be noted: God, who is the primordial instance of becoming and the only unceasing one; and the first set of nondivine occasions. God's initial stage of becoming supersedes only those factors of the universe that are presupposed by, and manifested in, the becoming of *every* actuality, including God. Among those factors are the eternal objects. The initial divine stage is the unconditioned valuation and ordering of eternal objects into *abstractive hierarchies* of sets of eternal objects capable of joint ingression as determinants of occasions (*SMW*, 167–69; *PR*, 31). Abstractive hierarchies are logical constructs (*SMW*, 170). They do not create the eternal objects' capacities for joint ingression; rather, they effect a primordial *reduction* of the infinite possibilities for such joint ingressions (*SMW*, 177–79). The completed initial divine stage is termed "God's primordial nature." It conditions the becoming of all other divine stages and of all other nondivine actualities.[5] The first set of nondivine actual entities is made up of occasions that supersede only God's primordial nature.

The term *supersession* stands *both* for the creative *process* constituting the becoming of an occasion *and* for the peculiar *relations* generated by that process. With equal truth it can be said that the process is supersessional because it generates supersessional relations, or that the said relations are supersessional because they are generated by a supersessional process. The relata of supersession are successively created, and are either successive occasions or successive phases in the becoming of an occasion. In both modes, however, the relation of supersession involves much more than the relation of succession. First of all, in each instance of the relation, whether inter- or intra-occasional, the earlier, or already created, relatum always remains in existence either to partly *determine* (in inter-occasional supersession) or to *condition* (in intra-occasional supersession) the creation of the later relatum.[6] Second, in each instance of either mode, the creation of the later relatum involves a *taking into account* of, and hence constitutes a *reaction* to, the earlier relatum.[7] Finally, also in each instance of either mode, the later relatum *incorporates* within itself the earlier relatum: the later occasion *includes* within its own nature a reproduction of the earlier occasion; and the later phase *absorbs* the earlier phase.[8]

Since the existence of a superject is never abolished by the existence of later occasions, and since earlier occasions are reproduced in later ones, *actuality is both cumulative and reproductive*. The accumulated superjects function as the *absolutely initial data* for the creative processes whereby occasions later than themselves are begotten. But the creative process that begets a new occasion also reproduces within the said occasion all the individual superjects constituting that occasion's correlative metaphysical past. As thus reproduced—or as thus *causally objectified*, to use Whitehead's technical term—the occasions of the past function as *objective data* for the nascent ingredient subjectivity of the novel occasion in the making. In themselves, the past occasions remain *beyond* the new occasion; but as reproduced they are *within* the new occasion, though not within the new occasion's ingredient subjectivity. In this manner, the cumulative and reproductive nature of actuality explains how earlier occasions can simultaneously transcend and be immanent in later occasions.

The causal objectification in a given occasion of all occasions earlier than itself is metaphysically necessary. All such objectifications are produced by the very first phase of the occasion's becoming. Therefore, in regard to inter-occasional supersession, *earlier than* is a metaphysical notion defined as follows: any occasion x is earlier than any other occasion y if, and only if, the becoming of x is *completed* before the becoming of y is *initiated*. Equivalently, the *metaphysical past* of any given occasion is the nexus of all occasions that are causally objectified in it. The occasion's metaphysical past is broader than the occasion's associated backward light cone; for, without regard to any space-like separation, *all* occasions in the occasion's metaphysical past are causally objectified within the occasion's nature and locus. As thus objectified, the past occasions constitute the given occasion's *physical memory* (*IS*, 243). Since causal objectification is a transitive, asymmetric, irreflexive relation, physical memory is the reason why each occasion contains, within its own nature and locus, the *chronology* of inter-occasional supersession leading up to the initiation of its own becoming.

In causal objectification, the particular nature and unique identity of one occasion is reproduced within the particular nature and unique identity of another occasion. This is a revolutionary and much misunderstood doctrine. Despite Whitehead's explicit claim that the doctrine was intended to blur the sharp distinction between universals and particulars, most of his interpreters have been unable to take the doctrine literally. Yet the doctrine makes complete and nonparadoxical sense once it is understood that *each attained actuality is in essence a quantum of information*; for any determinate quantum of information is inherently a *reproducible* reality. Each actuality in attainment is in essence the becoming of a novel quantum of information, but a quantum which, because of the causal objectification in it of its correlative past, subsumes within itself the history of the universe up to the initiation of that quantum's becoming. It should be immediately noted, however, that causal objectification is *not* the transmission of a signal. By the same token, causal objectification is presupposed by the transmission and significance of any signal. I return to this point later.

The analysis of an occasion as the becoming and being of a quantum of information involves all of Whitehead's metaphysical genera, or categories, of existents: occasions, eternal objects, nexūs, subjective forms, prehensions, contrasts, propositions, and multiplicities. The causally objectified occasions and nexūs constitute the information inherited from the past. But every occasion adds to the information it receives from the past. The novel information includes the definiteness of the occasion's subjective responses to its immanent objective data. The particular definiteness of a particular response is what Whitehead terms a "subjective form." Each subjective form results from the ingression of an eternal object, or set of eternal objects, into the locus of the occasion's ingredient subjectivity. Subjective forms include emotions (which include what we normally think of as scalar forms of energy), desires, valuations, purposes, adversions, aversions, moods, and various species of consciousness (*PR*, 24).

The particularity of a subjective form cannot be torn from the particularity of the subject to which it belongs, nor from the particularity of the datum to which it responds and refers. Each subjective form expresses *how* the *subject* ingredient in the experience responds to some *datum* or other ingredient also in the experience. The datum *provokes* the subject's response, but not necessarily the *how*, or nature, of the response. The completed subjective operation of taking into account the provoking datum and responding to it with a definite subjective form is termed a "prehension." The provoking, or objective, datum of a prehension may be any entity belonging to any of Whitehead's metaphysical categories of existents, including another prehension or set of prehensions. A prehension whose complex objective datum is the integration of two or more intra-occasionally earlier prehensions is called an "integral prehension." Insofar as an occasion's *ingredient subject* is self-determining,[9] its process of becoming is an autonomous weaving together of prehensions (*PR*, 108). In this process, simple prehensions serve as data for a variety of integrative prehensions, which in turn serve as data for even more complex integrating prehensions. Ever more encompassing, but progressively fewer, integrating prehensions are generated in the successive phases of the occasion's becoming. The final all-integrating prehension is the completed occasion, or satisfaction (*PR*, 26).

Any synthesis of objective data brought about by an integral prehension is termed a "contrast." The terminal contrast of objective data is termed the "objective datum" of the satisfaction; for it is the complex objective referent of the satisfaction's complex subjective form. But the satisfaction's complex subjective form is itself a terminal contrast; for the category of contrast also includes *any* realized synthesis of entities of whatever type. In fact, contrast is the most inclusive of the categories of noneternal existents, and is an open-ended category in respect to the evolution of contingent species of itself (*PR*, 22).

Since a contrast is a mode of synthesis of various entities in one prehension, it includes the metaphysical category of *proposition* as one of its metaphysically necessary special cases: a proposition is the prehensible contrast of an actual entity, or of a nexus of actual entities, with a set of eternal objects expressing a possible determination of the actual entity or of the nexus. Propositions are data for *propositional prehensions*, and *may* serve as data for very complex integrative prehensions of various kinds, including *judgments* and other *intellectual prehensions*. At least one propositional prehension—the occasion's *subjective aim*—plays a decisive metaphysical role by guiding the occasion's subjective process of self-completion to its final determination. The subjective aim involves a proposition prehended by the subject "with the subjective form of purpose to realize it in that process of self-creation" (*PR*, 25). In other words, some set or other of eternal objects prehended by the subject constitutes the final subjective *definiteness* at which the subject aims. This final definiteness is nothing other than an ideal of superjective existence to be progressively realized by the subject. Accordingly, what is generated in the self-determining phases of an occasion's becoming "is the outcome of the subjective aim of the subject, determining what it is integrally to be, in its own character of the superject of its own process" (*PR*, 241). This is why each "actual entity is at once the subject of self-realization, and the superject which is self-realized" (*PR*, 222).

The relevant possibilities aimed at by the subject are broad and general in the earlier phases of self-completion, but, through the subject's own autonomous decisions, become progressively narrower and more specific in the later phases. *Decision*, in this regard, is a technical term. In general, it does not mean a conscious choice, though it may mean that in regard to the experience of high-grade occasions. The term is intended in its root sense of a *cutting off* (*PR*,

43). What is cut off is always some *possibility* for the determination of some actual entity or other, or of some actual nexus or other. For example, if the subjective forms autonomously realized by a subject be *these*, they cannot also be *those*; if these data are integrated in *this* manner, they cannot also be integrated in *that* manner.

In its technical sense, decision belongs to the essence of actuality. Whitehead captures this feature in his principle of efficient, and final, causation, also termed the "ontological principle" (*PR*, 24), which "asserts the relativity of decision; whereby every decision expresses the relation of the actual thing, *for which* a decision is made, to an actual thing *by which* the decision is made. But 'decision' cannot be construed as a casual adjunct of an actual entity. It constitutes the very meaning of actuality. An actual entity arises from decisions *for* it, and by its very existence provides decisions *for* other actual entities which supersede it" (*PR*, 43). In *immanent* decisions, the actual thing *for which* a decision is made, and the actual thing *by which* a decision is made, are one and the same autonomous subject (*PR*, 88). In *transcendent* decisions, on the other hand, the actual thing *for which* a decision is made is an occasion in the future of the occasion *by which* the decision is made (*PR*, 254).

Every subject makes both kinds of decision because the relevant possibilities it entertains have to do not only with its own final definiteness, but also with the role its causally objectified superject may play in occasions in its relevant future (*PR*, 27). Indeed, the subject's immanent decisions aim at intensity in its own subjective form; and its transcendent decisions aim at intensity in the subjective form of occasions in its relevant future (*PR*, 27). The relevant future consists of those occasions *anticipated* by the present subject "by reason of the real potentiality for them to be derived from itself" (*PR*, 27). Strictly speaking, therefore, the subject can only realize what I term the *immanent component* of its subjective aim; for the aim at intensity in the relevant future—the subjective aim's *transcendent component*—can be realized only by the descendant, or descendants, of the subject. In fact, since the immanent component is itself twofold, even the simplest subjective aim has three components: an immanent component inherited from an immediate ancestor; an autonomously decided transcendent component to be realized by an immediate descendant; and, also autonomously decided, another immanent component mediating between the other two components. More complex subjective aims may involve a series of two or more descendants. In the anticipation of descendants, what the subject prehends are not the descendants as such, but the potentiality from which and by which such descendants will arise. Also prehended as inherent in that potentiality is the necessity that its own superject be causally objectified in, and subjectively *conformed* to, by them (*IS*, 242–44).

Each occasion's generic necessity to anticipate at least one *immediate descendant* is what Whitehead means by *physical anticipation* (*IS*, 243). The initiation and locus of an occasion's becoming is determined by its being physically anticipated by at least one *immediate ancestor.* The anticipating occasion is then, in the strictest sense of the term, *the efficient cause* of the *when* and *where* of its immediate descendant's existence; for it is the completion of the anticipating occasion that immediately elicits the initiation of its descendant's becoming at the anticipated locus. All other occasions in the descendant's metaphysical past will be efficient causes of its *objective content,* but not of its origination *then* and *there.* Thus, "[the] whole antecedent world conspires to produce a new occasion. But some one occasion in an important way conditions the formation of a successor" (*MT*, 164).

What Whitehead calls "a successor," I have been calling "an immediate descendant" to distinguish between those successors that are physically anticipated and those that are not. Conversely, I also distinguish between *immediate ancestors* and mere *predecessors.* All four notions are defined in terms of the relation of causal objectification, which, as said earlier, is transitive, asymmetric, and irreflexive. Thus, occasion *x* is a *predecessor* of some other occasion *y* if, and only if, (1) *x* is causally objectified in *y*, and (2) *x* is not causally objectified in any occasion that is also causally objectified in *y*. On the other hand, *x* is an *immediate ancestor* of *y* if, and only if, (1) *x* is a predecessor of *y*, and (2) *x* physically anticipates *y*. Similarly, occasion *z* is a *successor* of occasion *y* if, and only if, (1) *y* is causally objectified in *z*, and (2) *y* is *not* causally objectified in any occasion that is also causally objectified in *z*. On the other hand, *z* is an *immediate descendant* of *y* if, and only if, (1) *z* is a successor of *y*, and (2) *z* is physically anticipated by *y*.

Physical anticipation is a special mode of *anticipatory objectification*, the latter being the manner in which the metaphysical potentiality for all future occasions achieves *objective reality* within an occasion's experience. An occasion's physical anticipation of a descendant constitutes a transcendent decision by that occasion as to *when* and *where* in that potentiality the descendant in question is to have its locus. The decision is, in part, an autonomous structural determination of the anticipating occasion. Its nature and transcendent effectiveness alike require an appeal to the eternal ontological presuppositions of all occasions. These presuppositions, to be examined below, must also be invoked to account for the manner in which the metaphysical potentiality for an occasion's contemporaries achieves objective reality in the occasion's experience. In a higher-grade occasion, this *presentational objectification* of the potentiality in which its contemporaries are arising is the immanent field whose subsequent diversification by *physical imagination* constitutes the terminus of sense-perception (*IS*, 245). In any concrete moment of human experience, physical imagination is primarily a function of brain activity, but its ultimate data are the physical anticipations made by occasions in the immediate past of that concrete moment.

The doctrine of physical anticipation is also essential for understanding Whitehead's conception of a social nexus and of the actual world as such a nexus. A *social nexus*, also termed a *society*, is a nexus of occasions such that some of its members are metaphysically later than, and *genetically* linked to, other members, and such that all its members exhibit certain features and relations in common by reason of inheriting them from their ancestors in the society. The actual world, insofar as it is made up of earlier and later occasions, is a social nexus.

Occasions in a nexus are genetically linked, or form a *genetic nexus*, if every later member is physically anticipated by, and conforms to, at least one earlier member or other. Here to *conform* means to repeat, re-enact, or reiterate some feature or relation exhibited by the particular nature of the ancestor or ancestors. When the conformation is in respect to subjective form, the later subject conforms the definiteness of its subjective form to some element of the definiteness of the subjective form of an earlier occasion as objectified for it. In any such case, a *signal* may be said to be transmitted from the earlier occasion to the later one, but, for the expression of this doctrine, *reiterated* would be a more appropriate term than *transmitted*. Any restriction of such transmissions to successors must be construed as cosmological rather than metaphysical. Also, two successive occasions, though supersessionally contiguous, need not be extensively contiguous with each other. Time in the supersessional sense is *not* "another form of extensiveness" (*SMW*, 125). The notion of extension to which I am alluding will be explained later. It is one of the eternal ontological presuppositions of all occasions.

A genetic nexus is properly a society only when at least three generations of its members are such that the later generations conform to the earliest generation in respect to the same set of features and relations. Any generation within a society, except the first and the last, is preceded and succeeded by generations exhibiting more or less the same set of features and relations. The particular set of features and relations shared by all generations of a given society constitutes the *defining characteristic* of that society. A society prolongs itself, or continues to exist, for just as long as its defining characteristic continues to be reiterated by the latest generation of the genetic nexus. If conformation fails significantly, or if it shifts to a significantly different set of features and relations, the society will have ceased to exist, even though the genetic nexus may continue to exist, albeit with a different defining characteristic and, hence, as a different society. The continuing existence of a society is compatible with its gaining and losing nonessential, or nondefining, characteristics.

A society may be made up of subordinate societies, and may also include some nonsocial nexūs. Thus, the *actual world* is, for the most part, a society of societies of societies, with its most pervasive features and relations attaching to nearly all members, and the less pervasive ones attaching only to members of one or another subordinate society. As a society that is ever prolonging itself, the actual world always includes occasions that are still in process of becoming. But the term *actual world* is also used by Whitehead to refer to a *determinate* world-state correlative with a given actual occasion. The *occasion's actual world* is then synonymous with the given occasion's metaphysical past. In yet a third sense of the term, *actual world* may signify the entire universe correlative with a given occasion, and thus be understood to include not only the actualities in the past of the given occasion, but all actualities that are *metaphysically contemporary* with it, and also the potentiality for actualities in its *metaphysical future* (*IS*, 242). Two occasions are *metaphysically contemporary* if neither is causally objectified in the other. Similarly, an occasion's *metaphysical future* is the ever-expanding nexus of occasions in which the said occasion is causally objectified. Again, these relationships of contemporaneity and futurity are not in any way dependent on the speed of light and hold regardless of any space-like separation between the occasions involved.

At the lowest level of societal inclusiveness is what Whitehead referred to as a "personally ordered society": a society such that no two of its members are contemporaries with each other. My stream of experience is, in the main, a personally ordered society, but so is an electronic society. My stream of experience is a subordinate component of the complex social nexus constituting my brain. A complex society made up of interacting strands of personally ordered societies is termed a "corpuscular society." Personally ordered societies within a corpuscular society may split and coalesce as part of their interactions. Corpuscular and personal societies may be created and annihilated in fields composed of genetic and nonsocial nexūs.

Societies are the true *enduring entities* of Whitehead's metaphysics. In our experiences, aspects of their defining characteristics constitute the familiar *enduring objects* of our everyday life: e.g., stars, rocks, desks, animals, and human beings. Other such aspects constitute the unfamiliar enduring objects posited by science: e.g., molecules, atoms, and electrons. In each case, however, the enduring object is either the whole or a part of the defining characteristic of a society perceived or conceived in abstraction from the society of which it is the defining characteristic. In both cases, the relevant defining characteristics are ultimately derived from the societies as causally objectified in our experiences. But causal objectification normally abstracts from the qualitative content of the occasion or nexus being objectified. Moreover, in us, additional abstractions and transmutations are involved in the generation of the sensa constituting the qualitative terminus of sense-perception. Thus, as given in our sense-perception, or even as conceived by science based on sense-perception, enduring objects are *extreme abstractions* from the more complex natures of societies. Metaphysical materialism mistakes these abstractions for the ultimate constituents of the universe.

The distinction between an enduring object and an enduring society is metaphysically important. Quantum physics may constitute the first scientific context where concrete societies of occasions, and not just their abstract defining characteristics functioning as enduring objects, have to be taken into account. If so, each enduring object should be construed either as a single ingressed eternal object or as a contrast of ingressed eternal objects. As such, enduring objects are repeatable. Therefore, the same enduring object may be abstracted from several societies. It follows that two enduring objects may be indistinguishable even though they are associated with two perfectly distinct societies. For example, electrons are enduring objects; consequently, any two electrons are exactly alike in all respects, except their situations. On the other hand, any two electronic societies are perfectly distinct. Also, since they are made up of eternal objects, enduring objects are non-temporal and non-spatial, however much we may associate them, in our cosmic epoch, with the spatio-temporal volumes in which they are situated. Thus, enduring objects do not age, whereas enduring societies do.[10] Also, enduring objects can be created, annihilated, and created again. On the other hand, an enduring society has a beginning and may have an end; but, once ended, it cannot begin again.[11]

A human subject experiences itself both as a concrete enduring society and as an abstract enduring object. By reason of the three components of subjective aims, each occasion in a human stream of experience subjectively enjoys itself as realizing an aspect of a complex subjective aim belonging to the personally ordered society of which it is a passing moment. It is then prehending a personal, or enduring, aim that, through the mediating brain, affects and is affected by the enduring body in its interplay with the environment. For this reason, the momentary subject normally identifies with the body and construes itself as a moment in the life of a self-conscious bodily organism. But this identification with the body, though ontologically well founded, can be present in various degrees and may be absent. In any case, though the body changes significantly over time, and though various personal aims of varying endurance succeed one another with great rapidity, certain general characteristics of the body, certain types of subjective form, and certain abstract features of the personal aims remain the same and are integrally prehended by each momentary subject as constituting an enduring self-identical character, or personality, *residing* in all such moments of subjectivity. This resident personality, or enduring object, is objectively real and, insofar as it elicits conformation from moment to moment, is causally efficacious. Nonetheless, it is an *abstract* entity, though only in Whitehead's sense of the term; for referring to it as an abstraction "merely means that its existence is only one factor of a more concrete element of nature" (*CN*, 171). The relevant point is that abstract and concrete entities have different ways of functioning in experience. Not surprisingly, therefore, an amnesiac loses his conscious grasp of much or all of his abstract personality without thereby losing his conscious grasp of his temporal continuity with a unique, concrete personal past currently beyond recollection. The amnesiac also grasps his potential future as possibly the locus for regaining, or recollecting, his old "self" or for constructing a new one. The Western philosophical tradition elevates the abstract self into the status of a concrete substance; whereas the Eastern tradition dismisses it as illusory. On this point, given Whitehead's metaphysics, both traditions are mistaken: the substance-like concreteness is illusory, but the abstract self is real.

III

Enduring things, whether societies or abstract objects, are products of the supersessional process. But "eternal things are the elements required for the very being of the process" (*SMW*, 108). These eternal elements are realities *presupposed by*, and *manifested in*, the becoming and being of every actuality, God included. They are the ontological presuppositions of all experiences or, equivalently, of all ontologically basic acts of becoming. As such they are the only factors in the concrete universe that do not issue from the becoming of actual entities. There are four such factors and they constitute what the universe is by way of *eternal potency*—an inexhaustible power to beget interconnected, self-completing occasions. The four factors are: *eternal objects*, the reproducible ultimate essences that can function as immanent determinants of anything that becomes; *creativity*, the *whereby* of all becoming; *extension*, the *wherein* of all becoming and of all interconnected non-eternal existence; and *envisagement*, the *whereby* of all taking into account and the *wherein* of all eternal objects apart from their ingression into actualities. These eternal factors jointly function as an *envisaging creative prespace* that is sensitive to its holistic states in respect to the determination of each of its loci of actualization.

The eternal factors are presupposed by and manifested in every actuality because each actuality is *an individualization of the integral wholeness of the universe* (*PR*, 225) and, hence, an individualization of its eternal factors as well as of its history of attained actualities (*PR*, 223).[12] Thus, every occasion is a creative taking into account of the world-state that is given for it; and thus, too, every occasion is a *finite extensive region* rendered determinate by the ingression into it of a selection of eternal objects. But the eternal factors of the universe also transcend each such manifestation. Thus, the universe's eternal factors, as well as the actual world correlative with a specified occasion, are alike items in the universe with which the said occasion is at once mutually immanent and mutually transcendent. Accordingly, the doctrine of mutual immanence, stated most generally, is the claim that "every item in the universe, including all the other actual entities, are constituents in the constitution of any one actual entity" (*PR*, 148). Each actuality is a new *incarnation* of the concretely evolving universe.

Whitehead's concrete universe is the eternal universe as inclusive of the determinations that accrue to it through the becoming and being of interconnected actual entities. The accruing determinations constitute the ever-expanding nexus of actual entities that is what the universe is by way of actuality. Thus, the actual universe is a nexus embedded in eternity. In their becoming and being, the actual entities making up that nexus not only manifest the eternal universe, but are also constrained by its eternal properties. In what follows, we first examine the more relevant properties of eternal extension; then we analyze the supersessional history of a particular occasion to explicate the remaining eternal properties of the universe.

Eternal extension, which in itself is *neither temporal nor spatial*, is an infinite, unbounded continuum that is devoid of all quality and is completely indeterminate in respect to dimensions or geometrical axioms. This continuum is infinitely *differentiable*, and in that sense *divisible*, into potential regions of itself exhibiting such topological relations as complete inclusion, partial overlap, mutual externality, and contiguity. Since potential regions have no real boundaries, they are said to be "improper" regions and are one expression of the continuum's indeterminacy. But though in itself indeterminate, eternal extension is determinable, and its determination is effected by the becoming and being of actual occasions. Every actual occasion bounds, structures, embodies, and qualitatively determines a finite region of the eternal continuum, each its own. The embodied region is the *formative locus*, or *formative standpoint*, of the occasion's becoming and being. Because each such locus is determinately bounded and structured, it is said to be a "proper" region.

What Whitehead terms the "separative property of extension" forbids the overlapping of formative loci (*SMW*, 64; *PR*, 309). These loci and the occasions that embody them must be mutually transcendent. It follows that schemes of jointly actualizable proper regions constitute a proper subset of schemes of merely potential regions. The former schemes, but not the latter, are metaphysically relevant. Nonetheless, both kinds of schemes can play a role in the analysis of formative standpoints and of objectified occasions and nexūs.

A second metaphysical property of extension requires that its proper regions retain *everlastingly* any structural or qualitative determination gained from, or crystallized by, the creative activities of the occasions whose standpoints they are. For this reason, the products of each phase in an occasion's becoming, and hence the completed occasion itself, are permanent additions to the concrete universe. Supersession and experience are irreversible processes. What has become does not unbecome. Actuality is necessarily *cumulative*.

Actuality is also necessarily *reproductive*. A third constraining property of eternal extension is its *modality*, or *projectiveness* (*SMW*, 65–74, 91). By reason of this property, all extensive loci are each *projected* into the others and, hence, *as projections*, are each immanent in the others. Modal projection holds for potential loci as well as for proper ones. What is projected into a potential region is the current extensive structure of the universe. But the structure that the region thus gains is only potential. As long as the region remains merely potential, the projection it harbors shifts with every progressive shift in the actual extensive structure of the universe. Indeed, what transforms a potential region into a proper region embodied by a nascent occasion is the determinate *crystallization* of the universal projection the region bears at the initiation of the occasion's becoming. In other words, what is permanently projected into a nascent proper region is the extensive structure—actual and potential—of the universe correlative with the initiation of the becoming of the occasion embodying the region in question. The crystallization is effected by the ingression into the region of a complex contrast of eternal objects establishing the region's boundary and its modal differentiation into proper and improper subregions. By virtue of this crystallization, or permanent projection, the extensive structure of the occasion's formative region forever *mirrors* the extensive structure of the universe-state by which the occasion is begotten.

Relative to the initiation of an occasion's becoming, whatever proper region is *already completely determinate* is projected into the occasion in that character and thus with its particular complement of objective and subjective content. Such projections constitute *proper subregions* of the formative region into which they are projected and are the causal objectifications of all occasions in the metaphysical past of the occasion in question. All other regions, including proper regions embodied by incomplete occasions, are projected into the occasion as merely potential regions—regions that are differentiable but as yet undifferentiated—except that the occasion's own proper region is projected into itself as *bounded* but otherwise devoid of actual structure or content. Thus, what is *self-projected* is only the state of the nascent region relative to the initiation of its becoming. This limited self-projection allows the region to mirror its own extensive location, or *collocation*, relative to the extensive location of all other occasions, actual or potential.

The projection of one region into another is termed the "modal presence" of the former *in* the latter. It is also termed the "modal aspect" of the former *from* the latter. The self-projection of a region within itself is termed its "self-aspect." In turn, the projection of the rest of the universe's extension into the region is termed the region's "non–self-aspect," or "other-aspect." A region's other-aspect is the said region minus its own self-aspect. The other-aspect completely surrounds the self-aspect and is the intra-occasional locus of the world as objectified within experience. In turn, the self-aspect is the intra-occasional locus of what we have been referring to as the "ingredient subject." It is also the locus for all the subjective forms belonging to that subject. By the same token, the other-aspect is the locus of all causal, presentational, and anticipatory objectifications.

The mutual modal immanence of extensive regions is metaphysically more basic than the mutual objectification of actual occasions. For one thing, modal projections of extensive regions provide the loci, or *niches*, for all three modes of objectification.[13] For another, presentational and anticipatory objectifications occur universally only in respect to the potential extensive regions that *might* be embodied by hypothetical occasions contemporary with, or in the future of, the occasion in which the objectifications occur. In other words, since relative to the becoming of a given occasion, say *C*, there are no already existing occasions in its future, and since metaphysically contemporary occasions cannot be causes of one another, the immanence in *C* of occasions later than, or contemporary with, itself can only be in respect to some essential element of their respective natures that precedes their respective existences and that can thereby play a role in *C*. The element in question, for each such occasion, is its extensive standpoint; for the extensiveness of each such standpoint, though not its embodiment and structuring by an occasion, is an eternal feature of the universe. Thus, the extensiveness of any such standpoint can play an objective role in *C's* experience because, *qua* potential region, it is modally immanent in the other-aspect of *C's* own formative standpoint. At the very least, *C* is internally related to the respective extensive regions from which all the other occasions have arisen, are arising, or will arise.

The metaphysical relation between a region and its modal presences in other regions involves both extension and envisagement and is here termed "ontic reference." Ontic reference is a reciprocal relation between a projector region and any of its projections into other regions. But it is that relation as an immediate terminus of envisagement or, derivatively, of experiencing. The projector and its projection *signify* each other because each refers to the other as termini of envisagement. Also, insofar as projector regions are embodied by individual occasions, ontic reference obtains between occasions and their objectifications in other occasions.

To facilitate the illustration of ontic reference and the modal structure of occasions, I introduce the following notational conventions. First, let &*** stand for the infinite continuum of eternal extension, and let any capital letter followed by two asterisks stand for a *finite potential region* of &***. Second, let any capital letter followed by only one asterisk stand for the formative region of an actual occasion, and let an actual occasion be designated by the same capital letter used to denote its formative locus, but without the asterisk. Third, in designating modal presences, let the upper and lowercase versions of an asterisked letter respectively denote a region and its modal projection, and let proper or improper subregions of a proper region be indicated by means of a slash notation such that what is to the left of the slash is a subregion of what is to the right of the slash. Finally, let &**/... stand for the modal projection into the region to the right of the slash of all of &*** that is entirely outside the region in question. I continue to use occasion *C* as our explanandum and frame of reference. Also, I analyze each phase of becoming into a *creative* subphase issuing in a *created* subphase.

The very first creative subphase of *C*'s becoming transforms C** into C* and thereby crystallizes its modal structure of proper and improper subregions. The most important feature of that structure is the contrast between c*/C* and &**/C*, respectively *C*'s self-aspect and other-aspect. Now c*/C* bears the relation of mutual ontic reference with the rest of C*, and &**/C* bears the same relation with all regions of $\&***$ that are entirely outside $C*$. Also, if B and L are predecessors of C, then $b*/\&**/C*$ and $l*/\&**/C*$ are the modal presences in $C*$ of $B*$ and $L*$, respectively. Mutual ontic reference then obtains between each projector and its corresponding projection in $C*$. Moreover, $b**/c*/C*$ and $l**/c*/C*$ are differentiable subregions of *C's* self-aspect that respectively bear mutual ontic reference with $b*/\&**/C*$ and $l*/\&**/C*$. Notice that what is true for individual causal objectifications is true for all the modal presences into which $\&**/C*$ is actually or potentially differentiated.

From the point of view of the ingredient subject embodying $c*/C*$—let us term it c/C—ontic reference is either experientially *transcendent* or experientially *immanent*. In transcendent reference, the projection is an immanent differentiation of the occasion's formative locus, whereas the projector is beyond that locus. The subject c/C prehends the projection as referring to a projector that is beyond its experiential domain, but that is not beyond the envisagement of which that domain is a localized and creative enrichment. Thus, c/C experiences $b*/\&**/C*$ as referring beyond $C*$ to $B*$; but it does not experience $B*$ itself. Equivalently, c/C experiences the contents of $b*/\&**/C*$ as the self-revelation of an occasion, B, whose revealed existence and essence transcends that particular self-revelation, but which, for C, would be indistinguishable from nonentity if it were not so revealed.

The very meaning of *intra-occasional objectivity* depends on this fundamental relation of transcendental reference. Because of it, whatever is contained in $\&**/C*$ is for c/C something other than its own subjectivity, though it may be the subjective definiteness of some other occasion. It is also the reason why the objective immanence of the universe in each finite locus of experience is intuitively taken to be the self-revelation of what is other than, and independent of, the recipient of that self-revelation. From the point of view of the ingredient subject, $\&**/C*$ is the perspectival self-disclosure of *its world*. Subject c/C derives from, is objectively anchored to, and prehends a world transcending its experience. "For, it belongs to the nature of this perspective derivation, that the world thus disclosed proclaims its own transcendence of that disclosure. To every shield, there is another side, hidden" (*AI*, 228). The world disclosed in experience is the projection, or objectification, of the world hidden from experience. But objectification is self-revelation (*PR*, 227). In this manner, Whitehead's categories do "elucidate this paradox of the connectedness of things:—the many things, the one world without and within" (*AI*, 228).

In revealing itself within a given occasion, the world reveals its self-revelation in and for other occasions—occasions earlier than, concurrent with, or later than, the occasion in question: for the reciprocal ontic reference of regions is itself objectified in and for any given occasion. On this theory, then, the modal and referential texture of each occasion of experience ever proclaims the falsehood of solipsism and subjectivism. No experience can be what it is except as anchored to, and significant of, what is other than itself. Likewise, the subject ingredient in an occasion cannot be what it is except as referring to what is other than itself, the world as ingredient in the same occasion. Indeed, the importance of the modal differentiations of an occasion's self-aspect is that the subjective form embodying each such differentiation refers beyond itself to a corresponding region of the occasion's other-aspect. For example, the subjective form of *c/C's* prehension of *b*/&**/C** embodies *b**/c*/C** and thereby bears an essential ontic reference to *B*'s objectification in *b*/&**/C**.

The immanent, or intra-experiential, reference of subjective form to objective datum is self-evident in sense-perception. I fear the dog I see baring its fangs at me. In this concrete moment of my experience, there is not the slightest question regarding what my fear refers to. The fear is a modification of my subjectivity; the dog is an item in the objective content of my experience; and my fear *refers* to the dog. Even if the dog is a hallucination—which means that it fails the test of trans-occasional objectivity within one enduring perceiver or among a community of such perceivers—what I fear is *not* in my mind, but is out there in my objective experience, even if in no one else's. Notice, finally, that on this theory the so-called *intentionality* of consciousness is only a special case of immanent reference, and that immanent reference is itself only a special case of ontic reference. Intentionality is not some mysterious property co-emergent with a mysterious human consciousness; rather, it is a special case of a metaphysical feature of all ultimate actualities.

The importance of the modal structure of an occasion's formative region cannot be exaggerated; for it is the reason why every experience bears an ontic reference to past, present, and future. For example, in the course of *C*'s becoming, the other-aspect of its formative locus is finally differentiated into three potential, or *improper*, subregions: first, *pa**/&**/C**, the relatively well-defined, smallest potential subregion encompassing the modal presences of the formative loci of all occasions in *C*'s metaphysical past, each such locus bounded, embodied, and structured by a causal objectification; second, *pr**/&**/C**, the less well-defined, smallest potential subregion encompassing the modal presences of the formative loci of *C*'s metaphysical contemporaries; and third, *fu**/&**/C**, the remainder of the other-aspect, which thus constitutes a potential subregion encompassing the modal presences of the formative loci of all occasions in *C*'s metaphysical future. The first of these potential subregions is the *field of C's metaphysical past* as modally immanent in *C**. It is the field housing *C*'s physical memory. The differentiation of this field from the rest of *C*'s other-aspect is an aboriginal feature of *C*'s dative phase.

On the other hand, the differentiation of *C*'s other-aspect into separate fields harboring respectively the presentational objectifications of occasions contemporary with *C* and the anticipatory objectifications of occasions in *C*'s future is *not* an aboriginal feature of *C*'s dative phase. Therefore, this further differentiation is to be ascribed to the intermediate phases of *C*'s becoming. But the differentiation will itself be based on information provided by the modal presence in *C* of its causally objectified past. The information in question is a function of the physical anticipations that are necessary ingredients of all occasions (*AI*, 196), including the physical anticipations made by all occasions in *C*'s immediate past. Conditions permitting, the immediate descendants of these immediately past occasions will embody their anticipated loci. Accordingly, if *C* is an occasion of sufficient propositional complexity, this information will allow it to determine, within certain limits of probability, which subregion of its other-aspect is the modal presence of the transcendent extensive region likely serving as the substratum for *C*'s contemporaries. The smallest potential subregion that includes all modal presences rescued in this manner from the otherwise undifferentiated other-aspect constitutes the modal presence in *C* of *C*'s contemporary world. It is *C's field of presentational immediacy*. In turn, the modal presence of *C*'s future world is what remains of the other-aspect after the subtraction from it of *C*'s fields of physical memory and presentational immediacy. This remaining field, *the field of futurity*, is the modal presence in *C* of that potential region of eternal extension wherein the formative locus of *C*'s immediate descendant may be physically anticipated with some probability of success. Thus, if *!d**/fu**/&**/C** stands for the objective datum of *c/C*'s physical anticipation of an immediate descendant, then, conditions permitting, that descendant will be begotten as the embodiment of *D***. In this fashion, an occasion's physical anticipation of an immediate descendant functions as an *imperative* (hence, the exclamation mark in the notation) that, if at all possible, must be *heeded* by the activity begetting the descendant.

Regardless of *C*'s contingent complexity as an occasion, *pr**/&**/C** necessarily becomes the objective datum for one of *c/C*'s bare regional feelings. This "general regional feeling . . . is the whole of . . . [*c/C*'s] direct physical feeling of the contemporary world" (*PR*, 316). The region in question is then the simple terminus of *c/C*'s *perception in the pure mode of presentational immediacy*. On the other hand, if *C* is a high-grade occasion, such as those constituting a stream of human experience, *pr**/&**/C** is differentiated into regions of itself by means of sensa generated by *C*'s physical imagination under the causal influence of brain states. In fact, all sensa—whether they are associated with perception, recollection, waking or dreaming fantasy, or hallucinations—are generated by physical imagination under the causal influence of brain states. What differentiates these modalities of presentational immediacy from each other is the set of causal objectifications providing the information to which the sensa conform. Thus,

> physical imagination has normally to conform to the physical memories of the immediate past: it is then called sense-perception and is non-delusive. It may conform to the physical memories of the more remote past: it is then called the image associated with memory. It may conform to some special intrusive element in the immediate past such as, in the case of human beings, drugs, emotions, or

> conceptual relationships in antecedent . . . occasions: it is then variously called delusion, or ecstatic vision, or imagination. (*IS*, 245)

Since the brain is itself a social nexus harboring the stream of experience as a subordinate social nexus, it should be clear that all modalities of presentational immediacy derive their sensory content and regional differentiations from information provided directly by physical memory and indirectly by physical anticipation. Perception of this information is termed "perception in the pure mode of causal efficacy." When contrasted with sense-perception, it is also termed, somewhat misleadingly, "nonsensuous perception." Whatever its designation, this mode of perception is metaphysically necessary and basic. Through it, every occasion perceives all occasions in its past as causally efficacious in bringing about its own existence and in partially determining its nascent nature. Also through it, every occasion perceives all potential occasions in its future as subject to the causal efficacy of its own final nature. In contrast with the fundamentality and universality of causal efficacy, any modality of presentational immediacy is derivative and may be negligible or even absent.

What in human or animal experience is ordinarily referred to as sense-perception is really, in Whitehead's theory, an ongoing interplay between presentational immediacy and causal efficacy. This interplay is termed "perception in the mixed, or impure, mode of symbolic reference" because the termini of prehensions in either mode are individually or collectively significant of the termini of prehensions in the other mode. In us, but not necessarily in all other animals, conceptual and propositional analyses of symbolic reference provide the objects of conscious and unconscious knowledge. On the other hand, if we abstract presentational immediacy from the interplay of symbolic reference, we are left with Hume's mutually independent sense-impressions and, hence, with a resulting skepticism concerning causation, induction, and knowledge in general. Equivalently, such an abstraction leaves us with what Santayana referred to as "the solipsism of the present moment." Hume's appeal to *human nature* or *custom* and Santayana's appeal to *animal faith* are both unseemly attempts to dodge the unacceptable implications of their theories of experience. The failures of their theories to provide any basis for empirical knowledge is a *reductio ad absurdum* of any attempt to construe sense-perception as the foundation of all experience and knowledge.

Whitehead's contention is that, were causal efficacy not the primitive mode of perception, the sensa of presentational immediacy would be meaningless and uninterpretable. Any such sensum "is silent as to the past or the future. How it originates, how it will vanish, whether indeed there was a past, and whether there will be a future, are not disclosed by its own nature. No material for the interpretation of sensa is provided by the sensa themselves, as they stand starkly, barely, present and immediate. We *do* interpret them; but no thanks for the feat is due to them" (*AI*, 180). Interpretation is possible only on the evidence or information "drawn from the vast background and foreground of non-sensuous perception with which sense-perception is fused, and without which it can never be" (*AI*, 181). Conscious mnemonic imagination provides a good example not only of the dependency of presentational immediacy on causal efficacy, but also of the insufficiency of materialism. The images associated with conscious recollection are generated by physical imagination under the causal influence of brain activity. But the images are conditioned by, and refer to, physical memory. For this reason, conscious recollection can never be completely explained in terms of brain activity. As Whitehead puts it: "The ordinary mechanistic account of memory is obviously inadequate. For a cerebration in the present analogous to a cerebration in the past can, on this theory, only produce an image in the present analogous to the image in the past. But the image in the present is not the *memory* of the image in the past. It is merely an image in the present" (*IS*, 244). The image in the present does copy some aspect or other of the past experience, but "its character of being a copy arises from its comparison with the objectification of the past which is the true memory" (*IS*, 244).

Due to the transitivity of causal objectification and of modal presences in general, physical memory is all-inclusive, extremely redundant, and non-specialized. For that reason, social nexūs of any significant complexity eventually develop derivative systems of selectively germane memories mapped onto the structure and behavior of their subordinate societies. An animal brain involves a complex hierarchy of such derivative memory systems, their various mappings and codes coordinated by the common physical memory that funds them all. By means of such systems, a brain can develop and coordinate constantly updated systems of appropriately encoded representations of the organism, its environment, and their actual and potential interactions. In the context of symbolic reference, the contents of all modalities of an organism's presentational immediacy symbolically code and decode such representations. It must be added that the mutual significance of presentational immediacy and causal efficacy on which symbolic reference rests would be impossible apart from the mutual immanence and mutual ontic reference of the various components of each occasion's modal structure.

The modal structure of an occasion's formative region functions as an *indicative scheme* that not only permanently enshrines the occasion's collocation in the extensive continuum, but also captures the initiation and termination dates of the occasion's becoming, the former date in its physical memory of its immediate ancestor and the latter in its physical anticipation of its immediate descendant (*PR*, 67). The initiation, or *alpha*, date is the *supersessional juncture* between the occasion's dative phase and the superjective phase of an immediate ancestor; whereas the termination, or *omega*, date is the supersessional juncture between the occasion's own superjective phase and the dative phase of an immediate descendant. The chronological order of supersessional junctures is one and the same with the chronological order of *immediate* causal objectifications. This is one reason why the complex supersessional chronology of occasions up to a given occasion is enfolded within the extensive locus of the said occasion; for, in the occasion's dative phase, the causal objectifications of all past occasions are systematically disposed in their relative supersessional order, according as one is, or is not, immediately included in another (*PR*, 293).

An occasion's alpha and omega dates are particular supersessional relations the occasion bears to an immediate ancestor and an immediate descendant, respectively. Nonetheless, each such relation also marks, or indicates, a more general supersessional relation between the occasion and its immediately past, or immediately future, universe—a supersessional relation that the occasion may share with other occasions. For an occasion's alpha date also marks the occasion's supersession of its immediate metaphysical past, and its omega date also marks the supersession of the occasion by its immediate metaphysical future. It then follows that two or more occasions may supersede the same immediate past or may be superseded by the same immediate future. In other words, the initiation of two or more occasions may supersede the same determinate state of the universe; and, conversely, the same determinate state of the universe may supersede the termination of two or more occasions.

It thus becomes apparent that the alpha and omega dates of particular occasions mark supersessional junctures between occasions and states of the universe. The relations thus marked will be referred to as "universal instants" in the supersessional chronology of occasions. Two occasions, say *b* and *c*, have alpha dates that mark the same universal instant if, and only if, (1) every predecessor of *b* is a predecessor of *c*, and (2) every predecessor of *c* is a predecessor of *b*. Two alpha dates are said to be equivalent if they mark, or indicate, the same universal instant in the supersessional chronology of occasions. Similarly, the omega dates of *b* and *c* are said to be equivalent if, and only if, (1) every successor of *b* is a successor of *c*, and (2) every successor of *c* is a successor of *b*.

An occasion's alpha and omega dates jointly define its locus in the supersessional chronology of occasions. The unique combination of an occasion's extensive and supersessional loci is termed its "position." No two occasions can share the same position even if they have equivalent alpha dates and equivalent omega dates. Hence, an occasion's position is a sufficient expression of its unique particularity. Moreover, because it is instituted with its dative phase, an occasion's position grounds its self-identity through its successive phases of becoming. For the same reason, no genetic fallacy is committed when we identify the occasion *qua* subject with the occasion *qua* superject. Also, an occasion's position is derivatively the position of its self-aspect. Finally, because each causal objectification of a given occasion reproduces the said occasion's internal modal structure, the objectification necessarily exhibits the same position as the superject it reproduces. Thus, the superject and any one of its causal objectifications are *two and the same* particular. The superject embodies the occasion's formative locus, whereas each causal objectification embodies a proper subregion within the standpoint of some later occasion or other. This is why the completed occasion's *extensive location is not simple but multiple*. But wherever it is thus located, it exhibits one selfsame position and is thereby one selfsame particular. Thus, according to Whitehead, the "oneness of the universe, and the oneness of each element in the universe, repeat themselves to the crack of doom in the creative advance from creature to creature, each creature including in itself the whole of history and exemplifying the self-identity of things and their mutual diversities" (*PR*, 228).

Causal objectification cannot be construed as the transmission of a signal precisely because a completed actuality exhibits one selfsame position in its formative locus and in the respective extensive loci of each of its causal objectifications in its metaphysical future. Except that there may be some abstraction from its subjective definiteness, the total quantum of information that is C embodies the selfsame *positional locus* that is exhibited alike by $C*$, $c*/\&**/D*$ and, say, $c*/\&**/N*$. For the same reason, at least some of the new information created by c/C at $c*/C*$ is also found at $c*/c*/\&**/D*$ and at $c*/c*/\&**/N*$. But there has been no transfer of information from one positional locus to a different positional locus. On the other hand, if the subjective form of D has been conformed to the subjective form of C, then some feature originated at $c*/C*$ and causally objectified at $c*/c*/\&**/D*$ has been reiterated at $d*/D*$. Since $c*/c*/\&**/D*$ and $d*/D*$ exhibit different positional loci, the feature in question may be construed as a signal "transmitted" from one occasion to another. The feature is a signal precisely because, as in its new positional locus, it necessarily refers to its old positional locus: it signifies its source. Thus, the conformal prehensions of earlier occasions by later occasions account for the transmission of signals; but it is causal objectification that makes conformation possible by *universalizing* each completed occasion into an objective component of all occasions in the said occasion's metaphysical future.

The modal scheme, immanent reference, transcendent reference, and the objectification of transcendent reference cooperate with yet another property of extension—its *prehensive* property—to account for the integral unity of every proper region and thus of the occasion embodying it. As a local manifestation of the prehensiveness of extension, each proper region is the integral and indissoluble union of the modal presences within itself of all extensive regions, actual or potential (*SMW*, 64–65). By reason of ontic reference, the proper region is the intraexperiential unity of its own self-aspect with its own other-aspect, and of all more specialized modal presences within itself. This integral unity of the occasion's formative locus is the reflection, into the locus, of the eternal unity of the universe. In this manner, the proper region is, from the onset of its existence, the real potentiality for the individual integrity of the several prehensions that will arise in response to its objective content, and for the integration of all such prehensions into the one final prehension that is the superjective satisfaction. In other words, the integrity of the formative locus grounds the integrity of the experiencing embodying that locus; and the integrity of the experiencing grounds the referential integration of the various objectifications in the other-aspect through the integration of the corresponding subjective forms in the self-aspect (*PR*, 308). On the other hand, the teleological integrity of the subjective process of self-completion is grounded in its autonomously modified subjective aim.

The real potentiality harbored by *C** at its inception is itself the realization and particularization of the eternal potentiality harbored by *C***. This realization requires a reference to the actual world correlative with the initiation of *C*'s becoming and a reference to the eternal universe *qua* envisaging creativity. The correlative actual world, as just this nexus of just these attained actualities, comes into existence with the completion of *C*'s immediate ancestor, say *B*. The completion of *B*'s becoming is the efficient cause of the immediate initiation of *C*'s becoming, but what determines which potential region is to serve as *C*'s formative locus is *!c**/fu**/&**/B**, the objective datum of *B*'s physical anticipation of *C*. However, without the envisaging creativity of the universe, the becoming of *C* could not even begin; for what does not exist cannot bring itself into existence. Hence, *C*, as in its first subphase of partially determinate existence, is *begotten* by the envisaging creativity of its correlative universe. The individualizing activity is then said to be functioning as a *vehicle* for a *transcendent decision* made by *B* in respect to *C*. In effect, the individualizing activity is adapting the new to the old. Such an adaptation would not be possible if *C***, or any subregion of it, were already actualized by another occasion. Not all physical anticipations are successful. We may speculate that their failures were common in the very early universe as each occasion physically anticipated a very large number of immediate descendants, but that habits of orderly prolongation for genetic nexūs gradually evolved and thereby increased the probability of successful physical anticipations. Perhaps some laws of conservation are manifestations of such habits.[14]

Once it is individualized, the envisaging creativity is owned by *C qua* self-determining subject; but *C*'s phases of self-determination are preceded by three *individualizing* phases in which *C* is both begotten and partially determined by the state of the universe correlative with the initiation of *C*'s becoming. Aside from the fact that they beget a new occasion, these phases are *repetitive* in nature. The first creative subphase produces the first created subphase, or *dative phase*, which harbors the reproduction of all past occasions. The second creative subphase produces a *subphase of conformal physical prehensions*, that is, of prehensions whose respective objective data are causal objectifications, and whose respective subjective forms repeat some of the eternal objects exemplified in the subjective forms of the respective objective data. The third creative subphase produces a *subphase of conformal conceptual prehensions*, that is, of evaluative prehensions whose objective data are the eternal objects exemplified in the subjective forms of the conformal physical prehensions. In this manner, the individualizing phases beget not only the occasion's initial objective content, but also its initial conformal subjectivity.

In the individualizing phases, *C* is *other-realized* and thus not yet *actual* in Whitehead's technical sense of that term: *C* is *merely real*. But once the universe has been individualized in *C*'s formative region, *C*'s subjective self-realization begins. The envisaging creativity has become the nascent subject's experiencing activity: *c*/*C* now owns the experiencing process and autonomously completes itself, progressively fashioning the final definiteness of its subjective form and thereby referentially integrating the objectifications given for it. The autonomous subjective phases are *originative* in nature. They begin with a phase of novel conceptual prehensions whose data are eternal objects whose germaneness to *C*'s inherited character is a function of God's primordial evaluation and ordering of all eternal objects; for *c*/*C* prehends the causal objectification of God and is thereby influenced by God's primordial aim that any set of eternal objects already ingressed in an occasion shall constitute a *lure* for the ingression of other eternal objects that are not yet ingressed in that occasion, but which, in some abstractive hierarchy or other, are proximately relevant to those that are already ingressed. This divine influence is noncoercive and constitutes what Whitehead refers to as the "secular function of God."[15] The postulation of this influence is meant to account for the temporal world's exhibition of a tendency toward qualitative and structural order, and toward new forms of both types of order, that is not wholly explicable by the statistical probabilities of occasions making this or that type of decision (*PR*, 206–07). Abstractive hierarchies promote order without determining or prefiguring it.

The phase of novel conceptual prehensions is followed by successive phases of origination and integration until the subject completely realizes the immanent component of its subjective aim. Notice that, in this process, *C* becomes progressively more definite, but remains the selfsame occasion because its position does not change. It thus "combines self-identity and self-diversity" (*PR*, 25). This is what allows Whitehead to say that actual entities become but do not change (*PR*, 59, 79–80; *AI*, 204). *Change* is the difference between successive occasions in a social nexus of occasions (*PR*, 73, 79).

The self-realization of any occasion contrasts with its other-realization by its correlative universe. These two processes presuppose each other, but are distinct in their functions. In the becoming of every occasion, the repetitive phases effect a *transition* from a just-completed autonomous subject-superject to an incomplete autonomous subject-superject. These phases thus constitute what Whitehead refers to as the "macroscopic process of transition"—*macroscopic* because what the process must take into account, its *determining scope of envisagement*, is the world writ large, the macrocosm including and transcending the nascent occasion's formative locus. In turn, the subsequent originative phases effect a progressive growing together, or *concrescence*, of prehensions whereby the occasion attains its final unity of satisfied, or fully made, experience. These originative phases thus constitute what Whitehead refers to as the "microscopic process of concrescence"—*microscopic* because what the process must take into account, its *conditioning scope of experiencing*, is the world writ small, the microcosm immanent in the occasion's formative locus. The two processes are respectively the processes of *efficient causation* and of *final causation*, provided these are conceived as interdependent aspects of an occasion's becoming. Here is how Whitehead summarizes this important metaphysical doctrine:

> There are two species of process, macroscopic process and microscopic process. The macroscopic process is the transition from attained actuality to actuality in attainment; while the microscopic process is the conversion of conditions which are merely real into determinate actuality. The former process effects the transition from the 'actual' to the 'merely real'; and the latter process effects the growth from the real to the actual. The former process is efficient; the latter process is teleological. The future is merely real without being actual; whereas the past is a nexus of actualities. The present is the immediacy of the teleological process whereby reality becomes actual. The former process provides the conditions which really govern attainment; whereas the latter process provides the ends actually attained. (*PR*, 214)

IV

If we hypothesize that certain empirical relations of abrupt or instantaneous succession may manifest the supersessional junctures of occasions or of generations belonging to the same social nexus, then the distinction between the process of macroscopic transition and the process of microscopic concrescence, together with the distinction between causal objectifications and conformal subjective forms, may illuminate the distinction between holistic and local influences on any given occasion or event.[16] Contingently evolved, or cosmological, laws appear to restrict most occasions to conform significantly only to predecessors with which they are extensively contiguous. Since we interpret signals to be complex qualifications of subjective form, this would mean that, by some such law, or set of laws, our cosmic epoch sets an upper limit for the speed at which a signal may travel in spatio-temporalized extension. But a cosmological restriction on the transfer of information by conformation is not a restriction on the metaphysically necessary universalization of information by causal objectification. For this reason, occasions in the metaphysical past of a given occasion, but not in or on its past light cone, still provide it with information that can exert an influence on its immanent and transcendent decisions. And what is true of an occasion is also true for enduring societies. They can influence each other not only through the conformal reception of signals, if any, but also through their causal objectifications in one another.

Additional illumination for the distinction between global and local influences may be gained from the doctrine that every occasion has some information about some of its metaphysical contemporaries by reason of the physical anticipations made by the immediate ancestors of the said contemporaries. This information may be of special relevance for two or more occasions with equivalent alpha dates, whatever their space-like separation. It may be even more relevant for enduring societies with a common origin and such that each constituent occasion of one society is paired with a constituent occasion of the other society by virtue of both occasions having equivalent alpha dates and equivalent omega dates. In the latter case, one transitional process begets each pair of occasions and, as a vehicle for past transcendent decisions, may from their inception coordinate the possibilities offered for their respective autonomous decisions. Then, from the perspective of a later occasion belonging to a human stream of experience, the paired occasions, regardless of the space-like separation of their respective formative regions, would constitute one quantum event.

These suggestions as to the possible explanatory value of Whitehead's metaphysics are vague and empirically on the cheap. Whether they, and others like them, can be translated into empirically testable hypotheses remains to be seen. But the other contributions to this volume justify my optimism regarding the relevance of Whitehead's metaphysics to contemporary physics. In any case, my goal has been to present aspects of Whitehead's metaphysics in a manner suggestive of possible applications. I now return to and conclude that task.

If we assume the adequacy of Whitehead's metaphysics, the supersessional process exhibits a rhythmic alternation between efficient transition and teleological concrescence that grounds the mutual relevance of holistic influences and local self-realizations. Throughout it all, and aside from its eternal factors, the universe manifests itself as an evolving, self-forming, informational system. On the one hand, the universe is ever being informed by the becoming and being of actual occasions. On the other, the becoming and being of every actual occasion is informed by its correlative universe. The information of the universe by the becoming of occasions is the transformation of eternal potentiality into noneternal, but permanent, actuality. The information of each occasion by its correlative universe is efficient causation complemented by the teleological self-formation of each occasion's subjective self-aspect. In this manner, every quantum increment in determinate information affects the whole, and the whole affects every such quantum. Finally, each actual or potential increment is internally related to all other actual or potential increments.

In this informationally coherent universe, *knowledge* is information that has become the object of conscious self-formation, including conscious perception, conscious recollection, and conscious anticipation. But, as such, knowledge is necessarily a player in the rhythmic swing from self-formation to other-formation, and from other-formation to self-formation. Each cognizing occasion, with its complement of immanent and transcendent decisions, is a complex causal link, at once efficient and teleological, between two determinate states of the universe. Conscious decisions do make a difference in the universe.

Moreover, since all information may be thought of as protoknowledge, or as an in-principle possible terminus of knowledge, it is now evident that we can construe the actual occasion as an *onto-epistemic quantum* equally relevant to matters ontological and matters epistemological. The distinction between being and knowing has been blurred. It should also be evident that the actual occasion is both a quantum of becoming and a quantum of determinate being, a *subject-superject*; and that it is both a quantum of metaphysical extension and a quantum of metaphysical time, a metaphysical *topo-chronon*. Accordingly, Whitehead's metaphysics of actual occasions appears eminently suited for the task of generating a philosophical cosmology that can do justice to quantum physics without erecting a barrier between our cognitive experience and the real physical behavior of the entities and processes we claim to know.

Abbreviations

Books by Alfred North Whitehead

AI — *Adventures of Ideas* (1933; reprint, New York: Free Press, 1967).
CN — *The Concept of Nature* (1919; reprint, Cambridge: Cambridge University Press, 1964).
FR — *The Function of Reason* (1929; reprint, Boston: Beacon Press, 1958).
IS — *The Interpretation of Science* (Indianapolis: Bobbs-Merrill, 1961).
MT — *Modes of Thought* (1938; reprint, New York: Free Press, 1969).
PR — *Process and Reality* (1929; New York: Free Press, 1978).
RM — *Religion in the Making* (1926; reprint, New York: Meridian, 1960).
S — *Symbolism: Its Meaning and Effect* (1927; reprint, New York: Capricorn, 1959).
SMW — *Science and the Modern World* (1925; reprint, New York: Free Press, 1967).

Books by Other Authors

MMQM — Stapp, Henry P. *Mind, Matter, and Quantum Mechanics* (Berlin: Springer Verlag, 1993).
NST — Hawking, Stephen and Penrose, Roger. *The Nature of Space and Time* (Princeton: Princeton University Press, 1996).
SNWV — Shimony, Abner. *Search for a Naturalistic World View* (2 vols. Cambridge: Cambridge University Press, 1993).

WIO Bohm, David. *Wholeness and the Implicate Order* (London: Routledge and Kegan Paul, 1980).

WMES Nobo, Jorge Luis. *Whitehead's Metaphysics of Extension and Solidarity* (Albany: S.U.N.Y. Press, 1986).

Notes

1. (Princeton: Princeton University Press, 1996). *NST* in subsequent citations. Citations from other books are also by abbreviations keyed to titles listed under Abbreviations.

2. The aspects in question are presented under an interpretation and development for which I have argued at length in *WMES* and in "Experience, Eternity, and Primordiality: Steps Towards a Metaphysics of Creative Solidarity," *Process Studies* 26 (1997): 171–204.

3. An eighth category of existents—*multiplicities*—is said to be *improper*. Of significant interest for logic, the category of multiplicities is not germane to the limited concerns of this paper (*PR*, *30*).

4. *PR*, *66*, 90–91, 288–89, and 304–05. See also *WMES*, 214–22 and 246–48.

5. God's unending becoming yields completed divine stages that are each a determinate synthesis of all the data available at the initiation of the stage in question. Data are entities. Eternal objects are the only data for the primordial stage; but all subsequent divine stages have as data some set or other of completed occasions. These subsequent stages constitute what Whitehead calls the "consequent nature of God."

6. *RM*, 89–90; *IS*, 243; *PR*, 149, 154, 214–15, 236; *AI*, 194–95.

7. *SMW*, 42, 69; *RM*, 111–12; *IS*, 241–42; *S*, 39, 58; *PR*, 56, 154–55; *AI*, 252–55, 280.

8. *PR*, 7, 50–51, 148–49, 165–66, 220; *AI*, 192–93, 197; *IS*, 243–44.

9. The term *ingredient subject* is not Whitehead's. Whitehead uses *subject* to refer to the total occasion. I believe this is a mistake that hides Whitehead's most important insights concerning the nature and structure of experience. I have argued as much in *WMES*, 351–54, 364–65, 385–87, 390–91. In that book, however, I referred to the ingredient subject as the "empirical subject," and to the total occasion as the "metaphysical subject."

10. If the distinction between enduring society and enduring object is missed, very different conclusions will be reached on these matters. Accordingly, compare my conclusions with Shimony's in *SNWV*, vol. 2, 294, 298, 303.

11. Recent experiments involving Bose-Einstein condensates [http://jilawww.colorado.edu/bec/] show that if one concentrates a large number of identical bosons in a small region, their wave functions can overlap so much that the bosons lose their identity. Like electrons, bosons are enduring objects and can be exactly alike in all respects, except for their spatio-temporal situation. However, any two bosonic societies will remain perfectly distinct. Moreover, if those two societies participate in a Bose-Einstein

condensate, their existence comes to an end. When a boson as enduring object emerges from the condensate, that boson marks the beginning of a new bosonic society.

12. The individualization of this history may be relevant to the theoretical concerns pursued by G. Chew in his contribution to this volume (chapter 8).

13. Modality is the main reason for Whitehead's rejection of simple location. By reason of modality, every occasion pervades the continuum. Also by reason of it, every occasion has an intrinsic reality at its formative locus, and an extrinsic reality at every locus in which it is objectified. Cf. *SNWV* vol. 2, 293.

14. That the laws of nature reflect the habits of the entities making up nature is a view shared alike by Peirce and Whitehead. The view is also held by David Finklestein, as evident in his contribution to this collection (chapter 14).

15. In my development of Whitehead's metaphysics, another secular function of God is to physically anticipate the set of occasions constituting the initiation of the temporal world, or nondivine, noneternal universe. We may surmise that each member of this set physically anticipated a large number of immediate descendants. We may surmise also that, in the earliest stages of the temporal universe, each occasion physically anticipated a large number of immediate descendants, so that the ratio of immediate descendants to common immediate ancestor was unimaginably high. We may also assume that, with the gradual evolution of contingent forms of orderly prolongation for genetic nexūs, primitive enduring societies and their associated enduring objects emerged, struggled for survival, proliferated, and evolved into the familiar physical entities of our cosmic epoch. With these or similar assumptions, the origination and rapid expansion of the temporal universe would receive its required metaphysical explanation.

16. Notice that Whitehead's process of macroscopic transition is analogous to David Bohm's *holomovement*, a fact not lost on C. Papatheodorou and B. Hiley in their jointly authored "Process, Temporality, and Space-Time" (*Process Studies* 26, 247–78). The difference between the enfoldment of information throughout all space and time, and the transmission of a signal corresponds to the difference between global causal objectification and local physical conformation. See Bohm, *WIO*, 168.

18

Dialogue for Part IV

Chew: I sometimes like to say that reality is "electromagnetic." A particular feature of electromagnetism which has often impressed me is the role of electric screening in creating boundaries for objects. For example, the fact that I can think of this soda can as a separate object is due to the fact that the can has a huge amount of electric charge but, because of the almost exactly equal and opposite amounts of charge, the electric fields don't leak out very much. So when I approach that soda can I don't feel it until I touch it, and then there's the boundary. So that boundary effect—which is so much at the root of the way we look at the world and how we see the world as made up of separate objects—is a basic notion rooted in electromagnetism.

Fagg: That's a very good point. Because of this screening, that's why electromagnetism isn't quite as influential cosmologically as gravitation.

Eastman: In high-density, collisional neutral environments like the soda can, or the air in this room, the scale of the screening is very small and boundaries are sharp. Some of my research concerns plasma domains located halfway to the moon within the Earth's magnetotail where the Debye length or screening distance within the collisionless space plasma is about one kilometer. The plasmas there are highly interconnected, and that gets multiplied because of the overlapping of Debye spheres. In this way, you have long-range interactions going on from earth radii scales to hundreds of earth radii in which you have rather substantial coupling. Thus, the spatial scales of both electrical screening and electromagnetic coupling between different systems can vary dramatically.

Klein: I would like to get back to the topic of seeing what we can do to connect Whitehead's process thinking to quantum thinking. One way of approaching it is to read Whitehead's text carefully and try to figure out what his words meant and what they might mean in the modern age. This approach of starting with Whitehead gets us into lots of linguistic battles about what Whitehead meant when he said this or that. The other approach is to take Whitehead's writing loosely and to understand the spirit of Whitehead. Since quantum mechanics, as we have seen, has many ingredients in common with process thinking, why don't we just take quantum mechanics as the basis for process thinking rather than the Whitehead text, and let's try to discuss what needs to be added to present quantum mechanics. I personally think that Henry Stapp's version of quantum mechanics is sufficiently rich that maybe nothing extra is needed. Let's discuss what extra ingredients might need to be added to the quantum framework to bring it to the richness that Whitehead's metaphysics brought to process philosophy.

In order to make progress on extending quantum mechanics to encompass Whitehead's worldview, it would be good to compare lists summarizing the essential ingredients of the following four worldviews: The first one is the Whitehead list, the second one is the quantum list [see the introduction in Paul A. M. Dirac, *The Principles of Quantum Mechanics*, Vol. 27, Oxford: Oxford University Press, 1982], the third one is the classical list from Rosen [*Process Studies* 26/3–4: 328–330, 1997] and the fourth is from the paper by Finkelstein and Kallfelz [*Process Studies* 26/3–4: 279–292, 1997]. Here is the parallel list by Rosen for the topic of classical evolution:

1. Classical evolution is, in general, effectively unpredictable.
2. Under classical evolution, a system may continually pass through a limited range of states in an almost periodic manner, and a tiny change of conditions can cause the system to change abruptly its behavior and settle into a single state, just like quantum collapse.
3. States of complex classical systems can involve long-range patterns and order, as emphasized, for example, by those of us doing brain modeling.

Eastman: Clearly, the panel and others here are open to modifications of Whitehead's metaphysics. We're not just into what has sometimes been referred to as Whiteheadian scholasticism. Using the best of contemporary physics and philosophy, we can be carrying out the cutting edge of such a program. So what's missing?

Stapp: The outline that I gave of quantum mechanics in my talk left open two questions, and I took it right out of Bohr's discussion of Dirac's idea that nature chooses the answers, and Heisenberg's idea that we choose the questions. This means that quantum mechanics in its present form is incomplete. So let's say the project is to start with present quantum mechanics, and then use Whiteheadian ideas to add whatever more is needed. By beginning in this way, you have at least a solid foundation that is based on scientific evidence. You will start with a coherent, logical structure grounded on scientific evidence, and then use Whitehead to enrich it to the extent that's needed to bring it up to Whiteheadian standards of completeness.

What appears to be missing at the moment in the quantum description is how these two questions get answered. How does nature answer the question of what's actually going to happen, and how does the sequence of questions put to nature get selected? Whitehead is basically suggesting, at least the way I read him, that the processes by which those questions get answered are basically psychological in nature, not local mechanistic. So I think there will be something essential missing from quantum mechanics until the basically experiential process is added.

Finkelstein: One of the elements in Whitehead is a uniform way of looking at organization at all levels. Nowadays it's called systems theory. I think that we can say that Whitehead anticipated some elements of systems theory by a few decades. And we don't have a systems theory that's founded on quantum theory. It seems to us today that systems behave very differently at the quantum level, at the macroscopic level, and so on. I suppose there's a uniform way of looking at all these levels, but I haven't seen it, and I certainly wouldn't swear to produce one, but I am interested in setting about trying.

Chew: Do we all share the belief that the observer has to be recognized explicitly in any statement of what quantum mechanics amounts to?

Stapp: Certainly, at the practical level that's what quantum theory is.

Chew: Now do you agree that this condemns our formulations to be approximate?

Finkelstein: Yes, definitely—as any other. Any statement usually is stated by somebody. Usually there's a speaker, and a speaker isn't able ever to have maximum information about the speaker. Since all the little details about where my left toe is right now have some impact on distant stars, there's no way that I can make an exact statement about the world. The bigger the speaker and the smaller the system, the more chance there is of making reasonable statements about the system, but as the world gets bigger, eventually I get to look pretty small, and the idea that I can make a perfectly accurate statement about the rest of it is lunacy, it's futile.

Eastman: For a decade, if not centuries, the general public has thought about physics as being the exemplar of exact understanding in the details, and now we have some of the most eminent physicists of the country agreeing that it's necessarily approximate.

Finkelstein: I think that it was an hallucinatory phase, roughly from 1750 to somewhere in the last part of the nineteenth century, and now we have reached the end of this little island, we're getting into our boats and joining the rest of humanity.

Eastman: And perhaps with that, getting into some of the murkiness of the real world and not just the thin ether of our concepts. There is that essential interplay.

Finkelstein: That blooming confusion.

Chew: Would you include an estimate of the accuracy of the formulation in the formulation?

Finkelstein: I think that could be done actually, but it won't be an exact estimate.

Valenza: Do you gentlemen have any first impressions on whether or not the "eternal objects" that play an essential role in Whiteheadian metaphysics are required in what you propose?

Eastman: In Whitehead there are real potentialities. The process of becoming involves an incorporation of both previous actualities and potentialities, or eternal objects in Whitehead's scheme, so there is both the incorporation of previous actualities and a range of possibilities that are part of the prehensive unification that constitutes any actual occasion.

Klein: How necessary is that to process philosophy and theology?

Finkelstein: I think it's important. We're speaking of real potentialities, which are crucial in the Whitehead system and also in quantum theory. Heisenberg spoke about this quite early on in the game. Wave functions, Ψ vectors, are examples of potentialities that may or may not become actualized.

Klein: That's a wonderful matching of quantum to Whitehead, but then I thought there's this new thing called an eternal potentiality.

Finkelstein: Remember, when we're doing quantum theory we don't ask where wave functions come from, we treat the observer as eternal. We always have these possibilities available to him. That's part of the approximation for us. We don't look at those changes, so we might pretend they don't happen.

Eastman: Some physicists might wish to entirely avoid real potentialities, and one example of this is in the many worlds interpretation. In that view, the collapse of the wave function and its possible alternatives are all actualized in a multiple of realized worlds so that there is no real potentiality. In contrast, Whitehead would say that there is real potentiality where things may become either A versus B.

Stapp: I think that's the sort of quantum mechanics we're talking about—with real potentialities. David said the other day that the other one is nonsense. Didn't you say something to that effect?

Finkelstein: I probably did—I get carried away.

Bracken: I found a real connection that wasn't noted between this notion of real potentiality and finality because efficient causality organized into an overall system tends to reduce everything to actuality. If there's a system that is somehow actual, you can go backwards and forwards in time because it already has a certain actuality as a system of thought. Only when you introduce final causality and subjectivity do you really get real potentiality—the opportunity for real creativity and novelty to emerge. And that's where there really is a clash between classical science and Whitehead because he's going back to a premodern understanding of science where final causality played a very important role, but he's not doing it in the relatively unsophisticated way that was practiced by Aristotle and Aquinas.

Nobo: On the issue of an actual entity deciding where it comes to be, that may be Hartshorne but it's not Whitehead. Where the occasion comes to be is determined by the past and specifically by at least one actual entity in the immediate past. Whitehead, as I interpret him, has a cumulative theory of actuality. Once a region of the continuum becomes determinate, it remains determinate. Because of that, it can function in later occasions where the holistic process projects the information of that entity into later entities. It's a very redundant system because the information that has been accumulated is constantly being projected into every new actual occasion, which then completes itself. But part of its process of becoming involves a structural determination that has a causal effect on where in extension, and when in the supersessional order, its immediate descendent will come to be.

Stapp: So an event itself has no input into where it's going to be; it's already fixed where it's going to be?

Nobo: That is correct. The event is begotten by the universe with a partial determination, and an incomplete occasion, in the act of completing itself, anticipates at least one successor. But it can anticipate more than one, so you can have a splitting of world lines and you can also have coalescences of world lines. I would like you to consider the possibility that there is such a thing as a metaphysical double cone of which the light cone is a subset and that, until quantum physics, the metaphysical cone has been irrelevant. But when you have to deal with quantum physics, there's a possibility that now the metaphysical cone becomes important. It's just like the light cone. An occasion has a past, a future, and an elsewhere region for contemporaries. The occasion is the frame of reference. Relative to that frame of reference, there will be events such that one is earlier than the other, but both are contemporaries with the frame of reference. These relations don't depend on light, which is a special case depending on local causation. I do agree with what you said about modifying Whitehead, and this is one of the modifications that we need to make. There's a difference between causal objectification and conformation that tends to be ignored in the interpretation of Whitehead, but going beyond that I think Whitehead goofed when he said that an actual entity has to conform to every entity in the past. If it only conforms to some, then that conformation relation is what is going to give you local causality requiring contiguity and so forth. But the relation of causal objectification is going to give you what appears to be, what is, in a sense, instantaneous communication. And there is a way in which we can have some information about contemporaries metaphysically because they are anticipated by their antecedent events. I'm very excited about this but now have to formalize it.

Barbour: I was once a physicist and became a theologian, and what I would like to ask is from a more theological perspective. Physicists want to start from and stay as close to physics as possible. Whitehead, in his more metaphysical stage particularly, was very concerned about a system that would include religious sensitivities as well as scientific ones and, if one looks at that side of Whitehead, I'm wondering if one isn't pushed to extend Henry Stapp's position to ask, "Isn't God asking the question in some situations?" It seems to me that this opens another dimension. I don't think that Whitehead would have approved of those theologians today who are exploiting quantum mechanics by saying that God intervenes at the quantum level to determine one of the potentialities. That's certainly a possible way to go, and it doesn't violate the laws of physics, but God is not pushing electrons around. He would be actualizing one potentiality among other potentialities. If every action is God's doing, then there are all kinds of problems theologically, especially the problem of evil (see Bibliography: Griffin, 1976, 1991). I think that most contemporary theologians who make use of quantum indeterminacy want to say that it must be a rare kind of event and the processes must be subject to the statistical distributions that we know are not violated.

There is one other problem that Hartshorne was quite concerned about, and I am wondering if anybody would want to comment on it, namely, what is God's relation to the system (see Bibliography: Hartshorne, 1970)? And I would be interested whether Bell's theorem in your eyes throws any new light on this. The problem is that for us it's fine to say that E1 and E2 are events that are out of communication with each other. The separation interval is longer than a signal at the velocity of light could be communicated. At the practical level, we can remain agnostic about which event is first and accept the destruction of absolute simultaneity for human observers. But if God is related to all of this, presumably God's omniscience isn't subject to that kind of limitation of the transmission of information at the velocity of light. Either one has to get beyond Hartshorne and say that God is presenting initial aims relevant to the particular event that God is relating to, so that for God himself, with the relativistic time effect and the destruction of simultaneity, you could say that God presents an absolute frame. However, most physicists would resist that, but I think that it's an issue. Following Hartshorne's having raised it, several people have tried to reply to it. What is the implication of the Bell's theorem result of the nonlocality feature (see Bibliography: Cushing and McMullin, 1989)? As human observers we can step to one side and remain agnostic about it since no causal influence can go between events outside of each other's light cone, but they're both included in God's omniscience. Does that throw any light on that aspect of Whitehead's attempt to relate God to each event? I guess what I'm essentially asking, if anyone wants to do it, is obviously more speculative and won't come out of the physics alone. It will come out of a concern for relating a God who may be known in other ways, known for Whitehead in other avenues of human experience, and relating that to the findings of either quantum mechanics or relativity. This is obviously a very large agenda. Are there any particular points that anybody would want to respond to? Whether what you have been saying throws any light on God's interaction with the world that is clearly a part of the process conceptuality?

Stapp: Well, first let me talk about things in physics before getting to God. I realize that your emphasis is on God but, first, there is a small point, but maybe an important one. You said no physicist would prefer a special frame. But you must remember that the basic thing about the theory of relativity is that the laws are supposed to be invariant under certain transformations: the general laws are supposed to be frame independent. That statement is very different from saying that the world itself is frame independent: the world itself is definitely not frame independent. The world itself has one particular structure, which is not invariant under Lorentz transformations. So a big distinction has to be drawn between the nature of the world and the nature of the general laws.

As far as the world itself is concerned, there was a big bang apparently, and our experiments are done in our one world that was created with a particular structure. Amazingly enough, if you look back in all directions it seems that there was a preferred frame. The parts moved out from the big bang in all directions with various velocities, and by looking at the light coming from these various parts, it looks like there was a preferred rest frame in which the big bang occurred. This is the rest frame of the background radiation, and it is measured with great accuracy, 1 in 10^5. That is, when you look out in all directions, you find that there's a common rest frame of the background electromagnetic radiation. So in the actual world in which we live, there is a preferred frame.

There's another result along the same lines. In the Schilpp volume on Einstein, there is a chapter by Kurt Godel, of Godel's theorem fame [*Albert Einstein: Philosopher-Scientist*, Vol. 7, Paul A. Schilpp, ed., Fine Communcations, 2001]. He points out (I don't know if it's still true but it probably still is true) that in every known cosmological model there is, in fact, a natural sequence of "nows." A natural sequence of preferred instantaneous nows is built into each of the cosmological models. Thus there are two reasons for saying that we don't necessarily have to think the order of coming into being has to be relativistically invariant. We have two good reasons for saying that maybe it's not that way: that maybe there is a preferred order for coming into being.

There's nothing really contrary to relativity theory about that. Relativity theory says two things: (1) the general laws that control causal evolution are invariant under relativistic transformation; and (2) no "signal" can be transmitted faster than light. Those requirements are both satisfied in quantum theory; there's no problem with either of them. If the laws are correct, there's no possibility of sending a signal faster than light, and the laws of evolution of the (Heisenberg picture) operators are relativistically invariant. But the boundary conditions are not relativistically invariant! The actual evolution of the actual world itself depends on the actual world itself, so the actual unfolding process itself is certainly not relativistically invariant. There's no real logical contradiction between the principles of relativity and the possibility that in the actual process of unfolding of the actual world there is a preferred rest frame. The actual world apparently started out with a preferred frame, and this frame will certainly be preserved by general laws that are relativistically invariant. The actual process of unfolding of the actual world must necessarily depend on the actual world, and hence on any preferred frame defined by that world. The effort to make the process of the unfolding of the actual world relativistically invariant is irrational.

Bell's theorem, on the other hand, says that if you believe that the experimenters have a free choice (and you can't get anywhere unless you are willing at least to imagine that the experimenters have a choice to do this experiment or that one), then you cannot impose the condition that their choices can have no effects outside the future light cone of the region where the choice is implemented. If you try to impose that condition, then a logical contradiction ensues. This result seems to be telling us that there are faster-than-light "influences," in spite of the fact that there can be no faster-than-light signals. This is what quantum theory seems to be saying. So, if you want to say this is God's doing, then you can say that God has his own view of the universe, and that he has no problem with having faster-than-light influences but no faster-than-light signals. He merely needs to work in the preferred frame of the universe, allow influences to act only into the future, and follow the quantum rules.

Klein: I think you get into a major problem that is the one Ian Barbour actually mentioned: namely, the problem of evil. For the same reason that you don't want God to have omniscience in collapsing wave packets, in general you sure don't want God to have the omniscience to do these little micro things, because then you'll have the problem of evil. Most theologies that I know of don't give God that kind of influence.

Tanaka: I would like to comment on the theological implications of Bell's theorem. One difference between Whitehead and Hartshorne is that, for Hartshorne, God is a society of divine occasions with personal order, so Hartshorne needs the cosmological now. Hartshorne is annoyed by the general theory of relativity, in the usual understanding where we say "here-now" instead of a cosmological now, which does not have an objective counterpart in the general theory of relativity. We can accept Professor Stapp's model and say that there is something like a cosmological now in the causal structure without violating relativity theory and its idea of spacetime. In my Whiteheadian model, the theoretical relation between God and the world is something like this. Max Jammer, the Hebrew historian of science, wrote in his *Concepts of Space* (3rd ed., Mineola, NY: Dover Publ., 1992) that the Newtonian concept of absolute space has an origin in the Hebrew idea of God as a person. Newton says that absolute space is a *sensorum dei*, sense organ of God, so absolute space is something like the space we experience. One possible reading of Whitehead is that the extensive continuum is something like an absolute space in the Newtonian sense, so that God is omnipresent with everything in the extensive continuum.

Klein: How does Whitehead deal with evil?

Tanaka: Whitehead distinguished creativity and God. There is a metaphysical concept called creativity, which is more fundamental than God as an actual entity. If God is omnipotent being, Whitehead says that then there is a very serious problem with the actual existence of evil in this world.

Forizs: I would like to comment on the problem of evil. In my view, Whitehead solved that problem. The solution is connected with the problem of objective immortality. There is no such thing in Whitehead's philosophy as God isolated from the metaphysical scheme which defines a primordial nature of God, and a consequent nature of God. The primordial nature is actually deficient, conceptually abundant, and unconscious; on the other hand, the consequent nature is actually abundant and in a sense conscious. There is a beautiful sentence in Whitehead's *Process and Reality* that God is in a sense a sufferer, so he suffers, and his suffering converts in a sense the ruin. It uses the ruin, losses in the temporary world, and converts this ruin into his own nature, so in a sense even the Holocaust is a part of the consequent nature of God. Contributing to loss is often all too easy for humans. What's difficult is to work through the suffering and go beyond it. The Holocaust shows the real power of the Whiteheadian scheme, the categoreal scheme. Objective immortality means that every single decision is part of the consequent nature of God. You cannot explain away the Holocaust. All you can do is to consider how and why it could have happened, and how to remember it fully yet go beyond it. The Holocaust is not a shame on God; it's a shame on us.

Klein: From what I've heard of process theology, God is more of a positive force. How does that mesh with the notion that God is always present on these process decisions in creating evil?

Forizs: That is God's function in the old-fashioned sense. For Whitehead, God is a process. God is an actual entity in the making.

Fagg: Whitehead commented somewhere in *Religion in the Making* that the limitation of God is of goodness and the strength of God is of an ideal.

Barbour: I think the process answer to the problem of evil involves a number of elements. David Griffin has written two very good books on it [Griffin, 1976, 1991] and it involves the whole idea of divine self-limitation. The fact is that in the Whiteheadian system, no event is entirely deprived of God's action, and God is not the God who coerces anything to happen. In fact, God can't make anything happen alone. God introduces elements into what's already there and is bound by those structures and usually respects the lawful character of things, and no act is purely the act of God. God is the fellow sufferer who understands and participates in the world. There tends to be a problem with this in traditional theology. Whitehead strongly reacted to the monarchial God, the sovereignty, the omnipotent God. God doesn't know the future because the future can't be known. Omniscience is very restricted and omnipotence, in particular the traditional notion of omnipotence, is very strongly rejected in Whitehead's writings. While there is a problem of evil, I don't think it's a problem with a God as lure; a God of persuasion if you want to use an anthropomorphic term. Not a sovereign God in the traditional sense. Even creation is never out of nothing.

Bracken: I would like to introduce one distinction that Whitehead does not always make, and that is, when you talk about evil, you need to distinguish between deliberate attempts to produce negative results and what I would call tragedy, that is the confluence of freedom with nobody making a decision on this. I think it's important to make that distinction when we talk about evil, because in a world in which you have free creatures, even with good intentions and with as much knowledge as you possibly could have, the future is somewhat open. You can then have destructive events happening with no bad intent, whereas evil would be the bad intent to produce negative results.

Eastman: Earlier, we have noted how Whitehead needs to be modernized from a scientific point of view. Can someone comment on the need for such modernization from the point of view of developments in philosophy and metaphysics?

John Wygant: Whitehead does need to be modernized in the light of what has happened in the past 70 or more years. In doing that, I think that it is important to distinguish between his metaphysical scheme that's outlined in the second chapter of *Process and Reality* and the applications that he makes of it in Part II. The more important question is, "How do the metaphysical categories need to be modernized?" Those categories are very, very abstract and general and compatible with a lot of different scientific theories. I think they're compatible with action-at-a-distance theories, they're compatible with contact theories, and they're compatible with different versions of quantum theory. The point is that the nature and method of metaphysics is really very different from the nature and method of science, so in revising I think one also has to look at what's happened in the last 70 years in metaphysics as well. Now, as to the problem of universals which was raised by the topic of eternal objects as a point in case, a lot has been said about the problem of universals and the issue of eternal objects in the intervening 70 years. No one here has mentioned the topic of societies, but that's how Whitehead makes the connection between his categories and ordinary objects, and his concept of societies is an essential concept. A society perpetuates a defining characteristic from occasion to occasion, and that defining characteristic is an eternal object, a kind of Aristotelian potentiality. What's most relevant for that are the many developments in metaphysics, especially modal logic, since the 1960s. There has been extensive discussion of substantialism in metaphysics, and that's relevant to a reevaluation and interpretation of what Whitehead says about societies, eternal object, and so on. As you engage in your project, I think you have got to think through what the nature of metaphysics is and how it relates to physics (it's not as obvious as some of the things you've suggested), but also to take into account what has happened during the intervening 70 years in metaphysics as well [see Bibliography: Kim and Sosa, 1995].

This dialogue section provides concluding remarks by the panelists.

Valenza: I have always accepted Whitehead as a working hypothesis and have had a lot of intellectual fun trying to chase out the consequences of it. I've always been astounded by the man's depth and intelligence, but have always felt ill at ease with the grounding of process metaphysics, which either isn't well grounded or threatens an infinite regress. It disturbs me that that happened in these meetings to the extent that we get into trouble with language crossing categories when something is physical or protophysical. That still hasn't been sorted out. Insofar as our discussions highlight that issue, they're useful, but there are some very deep questions, both physical and metaphysical, that need to be asked about the whole process system. Unless they're asked or at least a coherent and systematic conversation emerges about these issues, then we are in each other's way in terms of making progress.

Chew: One of the questions I was hoping to get an answer to at this meeting was whether or not Whitehead's cosmological scheme was based in some way on a notion of matter. I know that it's based on a notion of process, there's no question about that, but does Whitehead's notion of process carry with it some implied, a priori meaning for matter? I will proceed in a Whiteheadian spirit and, as Phil Clayton indicated, I have been inspired by this discussion. I don't know why, but I feel more motivated, even more than I was, to try to develop this idea that a cosmology can be based on a notion of history where the history doesn't start with some notion of matter, but the notion of matter is emergent from certain patterns of history. There also is a much larger component of history which is nonmaterial, and I have heard here all sorts of possible relevance for this idea about nonmaterial history.

Tanaka: I think that there is no "matter" in Whitehead's metaphysics. Whitehead criticized the concept of matter in *Science and the Modern World.* In the conceptual scheme of *Process and Reality* there is nothing like matter; instead Whitehead proposes creativity as an ultimate category. Whitehead says that there are enduring objects. He did use that word, but there is no "matter" in the conventional sense in Whitehead's metaphysics, and to Whitehead the historical reality is most basic, with no substantial substratum.

Eastman: Whitehead carries out a critique of the notion of self-identical substance, the classical type of substance or the philosophical concept of substance. What we refer to as matter, as substance, are things that emerge in the ongoing process of becoming and being. It's in this dialectic of the prehensive unification of things into actual entities and then into sequences of occasions (the being of any actual entity is constituted by its process of becoming; its prehensive unification of past particles, fields, and its own immediate past self). From that whole process you get what we refer to as substances, the physical world, the table and everything; so there is a notion of substance, it's just not the classical notion.

Jungerman: I think that the fact that we're meeting here exemplifies that even after 70 years Whitehead has a lot to offer in inspiring us and, at the same time, it seems clear from our discussions that he certainly isn't the last word; he needs to be updated. I'm very encouraged and excited by physicists here who might actually develop some models based on physics that could provide a firm foundation for the metaphysics. It's a great start, but it seems to be terribly ambitious when you think about religious experience or feelings and emotions, and trying to incorporate those things into quantum mechanics. I'm glad I'm not doing it.

Fagg: The discussion of quantum problems has really stimulated a lot of thought in my mind and I really appreciate that. Also, I'm especially interested in what Geoff Chew is trying to do with his model and even more interested in his statements about how quantum theory must come to grips with electromagnetism, especially in the problem of measurement. He's absolutely right. It's very rarely ever mentioned whenever any discussion of quantum theory and its interpretation comes into play.

Tanaka: Metaphysics originally comes from Aristotle with his book written after (meta-)physics, thus the term *metaphysics*. So Aristotle's metaphysics presupposes the physics. Similarly, Whitehead wrote books about physics [*Principle of Natural Knowledge*, *Concepts of Nature and Theory of Relativity*]. He was well versed in relativity physics and its vital significance for contemporary physical processes before he began to write his own metaphysics. There is another aspect of metaphysics; that it must be the science of the most concrete elements of our experience. We must go deeper than usual, so he tried to do a phenomenological analysis of our own experience; the depth structure of experience in the world. Experience is very important in Whitehead's metaphysics. In one sense, Whitehead is a radical empiricist—that is my understanding.

Eastman: As one tries to work with these metaphysical propositions, as Whitehead says, you start in the ground of immediate experience and you fly off, as in an airplane, to work with conceptualities and models. But you necessarily must land again and be grounded in immediate experience and test your hypotheses or concepts both in metaphysics and in the sciences.

Klein: For me our discussions were both inspiring and stimulating. I was stimulated from engaging with many people on these deep metaphysical issues. I enjoyed the episode where David Finkelstein, Henry Stapp, and I got together for lunch and discussed the possibility of coming up with a minimalist set of quantum postulates on which we felt everyone would agree. The next step would be to compare this structure with Whitehead's metaphysics to determine what, if anything, needs to be added or subtracted from Whitehead's framework. I doubt if the three of us will carry out this project, but the project still seems reasonable to me and I do hope a Whitehead scholar will be inspired to carry it out. Quantum physics is presently our best description of reality, and any metaphysics without a quantum basis will be flawed.

The most inspiring portions of our discussions were the aspects where we were discussing the connections between quantum mechanics, Whitehead, philosophy, and theology. The central question is where will future societies get grounding for their values. It disturbs me to think that physicists, biologists, and other scientists might become arbiters of values. Plenty of other people, including theologians, are doing that quite successfully. The problem isn't how to come up with values; the problem is how to develop a comprehensive framework, a metaphysics, that provides a home both for the scientific world and the world of values. From our discussions I have gained confidence that the triple meshing of Whitehead's process thinking with the detailed structure of quantum mechanics and with the world of values and meaning is a feasible enterprise. That's a pretty good outcome from a wonderful conference.

Stapp: Our experience here has been an 'event' in the adventure of ideas. Specifically I think that all of us here are, in our own way, trying to advance human understanding of this world that we're living in, and it's obviously a meeting of different currents, of different ways of understanding this world. This cross filtration of ideas is a very good thing to happen. We physicists come here concerned with a certain part of human experience about the world. But as physicists trying to push the frontier a little bit, we are always coming up against philosophical questions of one sort or another. So I think it's very good for us to have an opportunity to come in contact with, and exchange ideas with, people who are working in a philosophical tradition that is close enough so that we can really exchange ideas. Because the ideas from the two sides are close enough, I would hope that maybe some of the ideas coming from physics could help supply what is needed to advance Whiteheadian metaphysics, and vice versa. This cross-fertilization of ideas could be a significant event in the advancement of our ideas about how the world works.

Finkelstein: First of all, it's inspiring to be among so many seekers for the truth or a better way of looking at nature and life. The idea which has attracted me for decades has been that of a quantum view of nature since for me, quantum theory is a process theory. Whitehead's attempt at a process theory of nature is particularly important for me as an indication of how such an enterprise might proceed. I certainly didn't imagine getting involved in a search for quantum theology and I feel that I don't have much to contribute to such a search, but the search for a quantum philosophy strikes me as worthwhile and if God happens to drop in, I'm open. I really did come with some basic questions. I'm trying to get myself to give up the search for "the" law of nature which has driven me for decades and so I'm delighted to get reinforcement, from Whitehead himself apparently.

Let me spell out my full belief, which is based on the precedent of general relativity. Once we used to think that there was a "right" geometry, and then for a century people began waffling and looking around for "the" right geometry. Maybe Euclid was wrong. Then along came Riemann and Einstein and they say no, geometry itself is a variable, and then Einstein comes along and says, no, geometry is the only variable in the unified field theory. So, nowadays we do these things more quickly and so, from the suggestion that the law is a variable, I go immediately to the inference that it's a quantum variable, everything is quantum. Superposition of the different laws are therefore as good as the laws themselves and, finally, with a variable this complex, who needs any other?

So probably if the law is a variable, it's the only one. That's somehow different than the search for "the" law of nature. I look around—there it is. I've been looking all my life for it and look, there it is. Right in front of me all this time. But something is still missing. There is this remarkable tendency for events to take a certain course, again and again, and it's not just enough to say look around and say that's the way things are. We need more explanation, and

I'm convinced that this has to do with the structure of the vacuum. I'm searching for those features which make a vacuum a vacuum, a fit substratum to serve as a carrier of the "law" of nature. The general impression this meeting makes for me is one of great encouragement. There are lots of people looking in this general direction.

Nobo: As you can judge from my demeanor, I have been extremely excited by the meeting, all three sessions. I have been tremendously stimulated, even challenged. I'm going to go home and start working on some of this. I'm very pleased to see that some ideas of Whitehead that I always maintain have been neglected are precisely the ones that are making an impact on or have some use for physicists. Again, that's very encouraging. Papatheodorou unfortunately didn't make it, but while reading his article with Basil Hiley [*Process Studies*, 26/3–4: 247–278, 1997; electronically reprinted in Eastman and Keeton, PSS 2003 (see Bibliography/Internet Resources)], I noticed that he pays more attention to part IV of *Process and Reality*. It strikes me that most people who call themselves Whiteheadian seldom look at part IV of *Process and Reality*, perhaps because it is too technical. Part IV is about extension and things like that, and the kinds of things that some Whiteheadians are interested in, experience or aesthetics, they can find in the other parts. But it's there and also in *Science and the Modern World* and in earlier books that you get some of the ideas that are most useful for physics.

I'm thrilled that these ideas are being looked at, and perhaps some of you will be interested in exploring them further. I'm also very thrilled by the fact that here we have a group of eminent scientists, and not one of them suffers from what we sometimes call "scientism," the belief that you can base a worldview solely on science. It's exhilarating for me because even in philosophy we have so many thinkers that are slaves to science, what Whitehead called the "obscurantists of the modern age." They laugh ideas out of court because they don't fit the current fashionable way of talking or the current fashionable way of thinking. Some of you expressed uneasiness about talking philosophy among philosophers. You need not fear—you're all philosophers.

Barbour: Very briefly, I've appreciated being able to attend most of your sessions. I only wish I could have attended all of them. I respect the careful grounding in the evidence that physical scientists provide, and I agree with the reluctance to move rapidly towards some natural theology. I am concerned as a theologian that the theological community makes use of science in a way that respects its integrity. I don't think that one can derive theology from a scientific picture alone. I am surprised that biological scientists were not represented in this workshop, not one biologist (for relevant works in biology, see for example Birch, 1990, 1995). I think that if we want to make any continuity with the biological world we have got to further broaden our categories, but I do greatly respect the sensitivity to the limitations of science that have been expressed by several of you. I do think that the connections with theology tend to be more general ones, the kind of elements of holism, and recognition of the interplay of law and chance. These must be taken seriously in any worldview, including that of theologians, so I have greatly benefited from what I have heard and am eager to see what further developments might come out of this without expecting that one will be able to develop a metaphysics purely out of quantum mechanics and relativity.

Glossary

action principle: Action is the integral of the Lagrangian of a physical system with respect to time. Of all possible motions, the dynamical motion of a system of particles and fields is one for which the action is stationary (not always a minimum).

actual entity: In Whitehead's metaphysical system, the category of actual entity—also termed *actual occasion* or *occasion of experience*—is the basic metaphysical genus of existents. The other metaphysical genera of the system consist of existents that are either generic operations, features, or relationships of actual entities (see *contrast*, *nexus*, *prehension*, *proposition*, and *subjective form*) or are uncreated ontological presuppositions of all actual entities (see *creativity*, *envisagement*, *eternal object*, and *extension*). Actual entities are the final real constituents of the actual world. They are discrete and interconnected, and their generic properties are those Whitehead deemed essential to any discrete moment of experience, human or nonhuman. Actual entities are "drops of experience, complex and interdependent" (see Bibliography: Whitehead, 1929, *Process and Reality*, 18).

actual occasion: See *actual entity.*

asymmetry: Vulnerability to a possible change.

atomism: The view that there are discrete irreducible elements of finite spatial or temporal extent. For propositions, the view that relations are external and that some true propositions are irreducible (logical atomism).

autocatalytic (chemical) reaction: A (chemical) process that increases in speed as it progresses. Non-autocatalytic reactions generally become slower as time passes. However, if a chemical substance produced by a reaction happens to increase the speed (rate) of the chemical reaction that produces it, then that product is said to be an autocatalyst—and the reaction rate increases without limit, unless some necessary reactant other than the autocatalyst becomes depleted. This sequence of events (mechanism) is called "product activation" or "direct autocatalysis." Several other relatively simple chemical mechanisms (e.g., inhibition by a reactant) also yield autocatalytic behavior (indirect autocatalysis).

baryon: An elementary particle formed of three quarks, for example, a proton or a neutron.

Bell's theorem: A mathematical proof, assuming locality, that determinate projections for the spins of electrons are incompatible with spin correlations predicted by quantum theory. Observational tests supporting Bell's result suggest that reality must be nonlocal. These results appear to resolve Einstein's EPR argument, which indicated a necessary incompleteness to quantum mechanics (derived from John Bell, 1928–1990).

Boolean algebra: An algebraic system that treats variables, such as computer language elements or propositions, with the operators AND, OR, IF, THEN, NOT, and EXCEPT (derived from George Boole, 1815–1864).

bootstrap: The notion that the nature of matter reflects self-consistency rather than arbitrary "elementary" constituents. See *S-matrix approach.*

Bose-Einstein statistics: See *quantum statistics.*

causality: The relationship between a cause and its effects. There are determinate, statistical, and other types of causal relations. The causal principle, roughly that the same cause always produces the same effects, is essential to science.

change: In considerations of symmetry or asymmetry, the production of something different.

classical logic: See *logic, quantum and classical.*

closure: Completion of a collection of items or relationships in such a way as to yield a set that is, in some sense, complete.

collapse of a quantum state: The change in a quantum state as a quantum system undergoes a transition from the potential to the actual during a measurement.

concrescence: In Whitehead, the process of concrescence—or the microscopic process—is the second stage in the becoming of an actual entity. The first stage is the process of transition. In the stage of concrescence, the process of becoming, though conditioned by the determinations of transition, is self-determining. The process of concrescence is thus a process of self-causation. Since concrescence is guided by the actuality's subjective aim, it is also a process of final causation.

contrast: In Whitehead, contrast is the most encompassing category of existence, excluding only existents belonging to the category of eternal object. A contrast is any synthesis of items in an actual entity.

Copenhagen interpretation of quantum theory: A synthesis of Bohr's complementarity interpretation and Heisenberg's ideas on the uncertainty principle. Sometimes referred to as the "orthodox" interpretation of quantum mechanics.

cosmic background radiation (CBR): The total radiation from outer space that is not associated with specific, identifiable sources (considering photons but not particles). The equivalent blackbody temperature of the CBR is approximately 3°K.

covariance: See *relativistic invariance.*

creativity: In Whitehead, creativity is that eternal aspect of the universe by reason of which there is an endless becoming of actual entities. It individualizes itself, and thus manifests itself, in the becoming of every actual entity. But it is never exhausted by such individualized manifestations.

Debye screening: Within a plasma there is a redistribution of space charge to prevent penetration by external electrostatic fields. The length scale associated with such shielding is called the Debye length, and shielding clouds of this dimension are called Debye spheres. See *plasmas*.

decoherence: Environment-induced loss of interference terms in quantum solutions that can contribute to resolving some problems in quantum measurement theory.

diachronic: Covering events at different times.

dielectric: Material placed between plates at different electrical potentials (capacitor).

dissipative structure: An arrangement of processes, taking place in an open system (a defined spatial region that can exchange material and/or energy with its environment), such that any alteration in that arrangement (whether internally or externally generated) engenders a response that tends to restore the original arrangement (see *limit cycle*). Dissipative structures contrast with equilibrium structures—self-restoring arrangements of items that persist in closed (isolated from the environment) systems. Interaction with the outside world tends to break up equilibrium structures (such as an ice crystal), but such interaction is necessary for persistence of dissipative structures (such as a flame in the tip of a gas jet).

dualism: For metaphysics, the view that for any given domain there are two independent and mutually irreducible substances (e.g., the Cartesian dualism of matter and mind). For epistemology, the view that there is a duality of the content immediately present to the mind (e.g., sense datum) and the perceived object.

dynamic system: In chemistry, a collection of interrelated processes—states of affairs that change as time progresses. Such combined interactions may lead to a unique equilibrium state—a condition that maintains itself indefinitely (without changing the surroundings) by means of the balance of countervailing changes. Alternatively, such combinations of changes may produce one or more nonequilibrium steady states—conditions that maintain a particular set of properties, while exchanging material and/or energy with the environment. Nonequilibrium steady states may be either stable (self-restoring after disturbance) or unstable (destroyed by any disturbance). In the condition known as bistability, two stable nonequilibrium steady states exist, as well as a third, unstable, steady state. The condition of the overall system will correspond to one or the other of the two stable steady states, depending on the past history of the system.

dynamics: The analysis of energy, force and associated motions. See *kinematics*.

electromagnetism: See *force fields*.

electron: A pointlike particle that contains one unit of negative charge and has a mass of 0.5 Mev/c^2. Electrons form the component of atoms outside the nucleus.

entelechy: In Aristotle's philosophy, the mode of being of a thing whose essence is completely realized or actualized, in contrast to potentiality of the form.

envisagement: In Whitehead, envisagement is that eternal aspect of the universe by reason of which every individualization of creativity is in part a function of the state of the universe correlative with the initiation of that individualization.

epistemology: One of the two main branches of philosophy, which is devoted to studies of the origin, methods, structure, and validity of knowledge. See *metaphysics (or ontology).*

epochal theory of time: Whitehead's theory that temporal process is a discrete succession of epochs, each having the duration needed for the emergence and completion of an actual occasion; also called a cell theory or atomic theory of time.

EPR argument: An argument by Einstein, Podolsky, and Rosen (EPR) that quantum mechanics is necessarily incomplete. It is based on two metaphysical principles, the separability principle and a locality principle, and derives an inconsistency between these realist principles and predictions of quantum mechanics. Einstein appears to have lost this argument (see *Bell's theorem, nonseparability*).

eternal object: In Whitehead, the only category of existence whose members are not created by the becoming of actual entities. A geometrical form, a shade of blue, an emotion, and a scalar form of energy are all examples of eternal objects. Basically they are qualia and patterns whose reproduction within, or ingression into, an actual entity render determinate the latter's objective and subjective content. Thus, relative to actual entities, they are said to be forms of definiteness or pure potentials for the specific determination of fact. But, in themselves, eternal objects are each an existent with a unique individual essence conjoined with a nonunique relational essence that it shares with some other eternal objects. Apart from their joint ingression into an actual entity, eternal objects are isolated from one another.

Euclidean geometry: See *geometry, Euclidean and Riemannian.*

experience: In Whitehead, experience refers both to an experiencing process and to the experience product generated by that process. The experiencing process is one and the same with the becoming of an actual entity; and the experience product is that same actual entity as a being, or as already become. The essential features of experiencing do not include consciousness or sense perception, but do include receiving, taking account of, and responding to, data that are primarily actual entities already become and eternal objects. The receiving of data constitutes the objective side of experiencing, whereas the responding to data constitutes the subjective side. A completed actual entity is a determinate synthesis of objective data and subjective responses. Such a synthesis is both the aim and product of the actual entity's process of becoming, or experiencing.

extension: In Whitehead, extension (or extensive continuum) is a technical term with a variety of related meanings. In the metaphysical theory, extension is an eternal aspect of the universe: it is an infinite, indeterminate continuum differentiable into potential finite regions that only become actual when embodied by the becoming and being of individual actual entities. The finite region embodied by an actual entity is its extensive standpoint, is uniquely its own, and does not overlap with the extensive standpoint of any other actual entity. Extension, in this sense, is neither physical space nor physical time; rather, the becoming of actual entities effects the spatialization and temporalization of extension. In the cosmological theory, under the influence of relativity physics, it is assumed that actual extensive standpoints contingently constitute an ever-expanding four-dimensional continuum of spatiotemporalized extension. This assumption is in no way necessitated by Whitehead's metaphysics.

extensionality: In the logic of classes, there is an axiom of extensionality which effectively assumes that everything is a class, and treats the words *set* and *class* as synonymous.

extensive connection: See *topology* and *extension.*

Fermi-Dirac statistics: *See quantum statistics.*

Feynman diagram: In quantum field theory and especially for quantum electrodynamics, simple diagrams can effectively replace complex mathematical terms of the field equations.

Feynman paths: Classical action-carrying "potential" paths in space-time whose aggregation determines wave-function propagation.

final causation: See *concrescence.*

fine structure constant: A fundamental dimensionless constant of physics, the fine structure constant α (= 1/137) derives from studies of closely spaced groups of lines observed in the spectra of the lightest elements.

Fock space: The fundamental representation of the infinite-dimensional Heisenberg algebra. See *Hilbert space.*

force fields: Four fundamental interactions are distinguished in modern physics: gravitational, electromagnetic, weak nuclear, strong nuclear. The characteristic strengths of these force fields are 10^{-39}, 10^{-2}, 10^{-5}, and 1, respectively. Both gravity and electromagnetism are long-range force fields, whereas both nuclear forces are very short range (10^{-13} cm or less). See *gauge theory.*

gauge theory: Gauge invariance is central to current theories of the fundamental interactions. Derived from Weyl's work in relating scale changes and the equations of electrodynamics, gauge theories constructed to embody various symmetry principles have been very successful in representing the fundamental interactions. See *force fields.*

geometry, Euclidean and Riemannian: Geometry is the mathematics of the properties and relationships of points, lines, surfaces, and solids. The system of geometry that dominates our practical affairs is a modified version of the assumptions of Euclid of Alexandria (325–265 B.C.). Georg Riemann (1826–1866) introduced a new system of geometry to handle curved surfaces and curved spaces. Einstein used this Riemannian geometry to develop his general relativity theory, providing a quantitative representation of both gravity and accelerated reference frames.

gluon: A virtual particle that is exchanged between quarks that constitutes the strong force. Gluons not only interact with quarks, but with each other.

grand unification theory (GUT): A theory that seeks to provide a common derivation for the color and electroweak forces and for quarks and leptons.

hermeneutics: The science and methodology of interpretation; a discipline initiated through biblical analysis in the nineteenth century.

Hilbert space: A vector representation space for the properties of quantum systems. A finite-dimensional Hilbert space is a finite-dimensional Euclidean space in which vectors are represented as complex numbers instead of real numbers.

holism: The theory that the world is composed of organic or interrelated wholes which are more than their constituent parts. See *particularism.*

hyperfine structure: A slight shift in the frequency of radiation from atoms due to the interaction with the atom's nucleus.

hypothetico-deductive method: This method is used in factual sciences as well as in metaphysics. It is constituted, according to Whitehead, by a balanced mutual relationship between three elements: (1) the creative element, which accounts for an imaginative construction of hypotheses and theories and as such is free of all kinds of restrictions; (2) the rational element, which secures the coherence and internal consistency of the hypotheses forming the theory and the possibility of deduction of consequences from this axiomatic base; and (3) finally, the empirical element, which calls for at least the possibility of indirect empirical testing of metaphysical theories through the integration of mediating scientific schemes.

idealism: Any philosophical framework whose basic interpretive principle is that of Idea, Mind, or Spirit. Process philosophers generally avoid traditional dualisms such as idealism/materialism and mind/body. See *realism.*

implicate order: Physicist David Bohm introduced a distinction between an explicate order, comprising the given world of experience, and a holistic implicate order, which is a fundamental causal order in parallel with and underlying the explicate order.

indeterminacy principle: For particular observable pairs (e.g., position and momentum; energy and time), precise measurement of one observable necessarily causes uncertainties in possible knowledge of the complementary observable.

kinematics: The study of motion excluding the effects of mass and force. See *dynamics.*

K-meson: One of the strongly interacting elementary particles with baryon number 0. Observations of K-mesons first indicated an intrinsic time asymmetry in high-energy physics interactions.

Lagrangian: The difference between total kinetic energy and potential energy for a dynamic system of particles expressed as a function of generalized coordinates and their time derivative.

Lamb shift: A small change in the frequency of the light emitted by a hydrogen atom due to the existence of virtual pairs.

lattice field theories: Theories where spacetime, rather than being continuous, consists of a discrete set of points at which fields are defined.

leptons: Elemental fermions, which have spin 1 statistics (see *quantum statistics*), electrons, muons, neutrinos, and tauons are all leptons. See also *quarks.*

limit cycle: A unique sequence of states of a dynamic system, each described by two or more variables (say, x and y), such that the sequence traces out a closed curve in the x, y plane which encloses a single point (x', y') that corresponds to an unstable steady state of the system—and furthermore, that any deviation from that sequence of states (the limit cycle) engenders a response that tends to restore the system to one of the states of the limit cycle. All trajectories of the system eventually reach the limit cycle, irrespective of their starting conditions. Autocatalysis is necessary (but not sufficient) for existence of a limit cycle.

linguistic turn: The point of view that meaning is constituted by linguistic practice of a community and that there is no external standard for such practice.

logic, quantum and classical: Logic investigates the structure of propositions and deductive reasoning by focusing on the form of propositions instead of their content. Classical logic derives from Aristotle and is captured in modern Boolean logic. In contrast, some physicists argue that quantum theory requires a non-Boolean or quantum logic. The basic logical structure for deductive reasoning is the syllogism wherein one can infer, for example, from major premise "If *A* then *B*" and minor premise *A* that *B* is true. Similar logical implication is obtained through disjunctions. For example, one can infer the truth of *A* from combining the major premise "*A* or *B*" and minor premise "not *B*."

logical conjunction, disjunction, and implication: See *logic, quantum and classical.*

logicist: The claim that all mathematics can be derived from logic. Implementing this reduction was the goal of Whitehead and Russell's *Principia Mathematica.* This goal proved untenable by 1931 with the discovery of Gödel's incompleteness theorems for axiomatic systems.

metaphysics and sciences, relations between: See *sciences and metaphysics*, relations between.

metaphysics (or ontology): One of the two main branches of philosophy, which is devoted to studies of the nature of being or ontology. See *epistemology. Plain*: A theory of the most general features of reality and real entities. *Exact*: A theory of the most general features of reality and real entities which makes explicit use of formal sciences (logic, mathematics, game theory) in theory construction (reducing the ambiguities of ordinary language by defining the basic terms explicitly and making explicit the relations between the basic notions and axioms) and theory criticism (testing internal and external consistency). *Scientific*: An exact metaphysical theory which further takes into account the (main) results of contemporary scientific research.

metasystem level: Level of description embracing both experimenter and system under study.

Neoplatonism: Arising in the second century A.D. as extensions of Platonism, the view that ideal patterns or universals are existent substances and that body and soul are independent substances.

nexus: In Whitehead, any set of actual entities is a nexus. The term "does not presuppose any special type of order, nor does it presuppose any order at all pervading its members other than the general metaphysical obligation of mutual immanence" (see Bibliography: Whitehead, 1933, 201). The mutual immanence of discrete actual entities is an apparently paradoxical doctrine that receives consistent explanation in terms of Whitehead's theory of metaphysical extension (see chapter 17). Nexūs are classified into various main types according to the specific contingent forms of order they exhibit. A social nexus, or society, is by far the most important type of nexus.

nonseparability: The supposition that Bell's theorem results on spin correlations indicate a fundamental connectedness between occasions, including some regions outside the normal causal light cone of relativity theory. However, nonlocal correlations depend on a common initial event at the source of the associated particles or photons.

objectification: In Whitehead, objectification refers to the manner in which one actual entity is immanent in another actual entity. There are three modes of objectification: causal, presentational, and anticipational. The different natures of the three modes cannot be understood apart from the properties of metaphysical extension (See chapter 17).

occasion of experience: See *actual entity.*

ontology (or metaphysics): The theory of being qua (as) being. For Aristotle, ontology is the science of the essence of things. See *metaphysics.*

particularism: The view that all apparent wholes are mere aggregates of discrete, separable parts. See *holism.*

phenomenology: The descriptive analysis of subjective processes. Roughly, pure phenomenology and pure logic are mutually independent disciplines. Modern usage of the term derives from Husserl (1859–1938).

photon: The particle aspect of light, or more generally, electromagnetic radiation. Photons have no rest mass and carry one unit of angular momentum in units of h-bar. Each photon has an energy which is Planck's constant, h, multiplied by its frequency.

Planck scale: An exceedingly short distance, equal to 1.6×10^{-35} m, based on combining Planck's constant, the speed of light, and the gravitational constant.

plasmas: Plasmas are an electrically conducting interactive mix of uncharged particles, positively charged particles, negative electrons, electric fields, and magnetic fields. The fraction of uncharged particles in a plasma varies dramatically, from more than 95% in the lower ionosphere to less than 1% in the solar wind, the continuous stream of plasma from the sun. In contrast with neutral gases and liquids, plasmas are noted for their highly interactive properties and collective effects. They make up more than 99% of the visible universe. See *Debye screening* and the plasma Web site at http://www.plasmas.org.

prehension: In Whitehead, the activity of prehending is the most concrete creative operation involved in the becoming of an actual entity. Its created product is the most concrete component of an actual entity already become. Each prehension is analyzable into a subject, an objective datum, and a subjective form. The subject is the actual entity insofar as it autonomously decides its own final subjective definiteness. The objective datum is any entity—such as another actual entity, an eternal object, or a *nexus*—that is taken into account and responded to by the subject in its process of self-formation. And the subjective form is the definiteness with which the subject clothes itself in response to that datum. The subjective form is how the subject defines itself in response to that datum.

proposition: In Whitehead, a proposition is the prehensible contrast of an actual entity or nexus with an eternal object, or a set of eternal objects, expressing a possible determination of that actual entity or nexus. A proposition may serve as the datum for a propositional prehension. The prehension of a proposition does not require consciousness. However, in a high-grade actual entity, a few propositions may be consciously prehended against a vast background of unconsciously prehended ones. Self-referential propositional prehensions are possible, and one kind, termed the subjective aim, is generic to all actual entities.

pulsar: A rapidly rotating neutron star that emits radiation in pulses linked to its rotational period.

quantum logic: See *logic, quantum and classical.*

quantum measurement theory: The analysis, for different interpretations of quantum mechanics, of what observables can have determinate values in a given quantum state.

quantum state: A mathematical entity that contains all the available information about the probabilities of all possible measurements on a quantum system.

quantum statistics: Statistical physics endeavors to deduce information about macroscopic properties of a system based on analyzing statistics of its microscopic constituents. In application to quantum systems, a basic distinction is found between particles having half-integer spin or Fermi-Dirac statistics and integer spin systems with symmetric quantum wave functions, which display Bose-Einstein statistics. Since electrons have spin ½, Fermi-Dirac statistics apply and no two electrons can simultaneously occupy the same quantum state. The resulting Pauli exclusion principle forms the basis for generating the periodic table of the elements.

quark: A pointlike constituent of neutrons, protons, and mesons. The quarks contain all the nucleon mass and are indivisible as far as we know. Three quarks form a baryon, such as a proton, neutron, and other short-lived particles. Two quarks form mesons. See also *leptons.*

realism: For ontology, the theory of the reality of abstract terms or universals in which universals exist before things, in contrast to nominalism, for which universals have a being only after things. In epistemology, realism holds that it is possible to have faithful and direct knowledge of the actual world. Process philosophers generally avoid traditional dualisms such as idealism/realism and mind/body. If one insists on the idealism/

realism pairing, it can be argued that Whitehead was both an idealist and a realist. See *idealist*.

reductionism: Reduction is the subsumption of one conceptual scheme by another. In metaphysics, reductionism holds that there are systematic identities between entities of a higher level and those in a lower, reducing level. In epistemology, reductionists typically point to semantic equivalences between propositions in the higher level and those in a lower, reducing level. Reduction in science is the effort to systematically explain one scientific theory by laws and phenomena in another lower-level theory.

quantum mechanics, field theory: The physical theory of matter, electromagnetic radiation, the force fields, and their interactions. See *force fields*.

relativistic invariance: Properties of a system are invariant if they are unchanged during a change in the frame of reference. The Lorentz transformation needs to be applied to length and time measurements to achieve relativistic invariance. In comparison, the classical Galilean transformation is a good approximation at low speeds but differs noticeably from the Lorentz transformation at high speeds, especially those approaching the speed of light. Applying relativistic invariance enables the laws of physics to have the same form for any system of coordinates; this is called the principle of covariance.

Riemannian geometry: See *geometry, Euclidean and Riemannian*.

S-matrix approach: Theory based on the general principles governing particle collisions (such as frame independence, causality, and probability conservation), rather than on arbitrarily specified field equations.

science and metaphysics, relations between: *Stimulation*: On the one hand, a scientific metaphysics provides a most general scheme of reality from which (with the help of additional specifying hypotheses and limiting conditions) specific notions can be logically derived, which may function as basic concepts in a science and thereby throw new light on old scientific problems, transform them, and initiate new scientific research strategies and experiments. *Criticism*: Science is a critic of metaphysics. The more special schemes of the sciences mediate between the abstract and general metaphysical scheme and empirical fact as disclosed by experimentally guided observation, the greater is the possibility that the metaphysical system receives indirect empirical support (confirmation) or criticism (falsification). If a scientific approach that is initiated by a metaphysics does not lead to fruitful new insights issuing in new experiments yielding positive, i.e., confirming results, not only this scientific approach must be modified or even abolished, but also the initiating metaphysics has to be reconsidered.

screening: See *Debye screening* and *plasmas*.

semantics: The study of the meaning of signs and symbols.

society: In Whitehead, a society is a nexus made up of successive generations of actual entities exhibiting a common form of order that the members of each generation inherit from the members of the preceding generation. The common form of order is the society's defining characteristic. Societies, or their defining characteristics, are the enduring objects of nature.

space plasma: See *plasmas*.

spin network: A quantum-theoretic network of coupled spins used by Roger Penrose as an atomistic geometry of the surface of the sphere.

spinor: A vector describing spin of the electron.

state vector: Polarization and other quantum states are represented by matrices called state vectors. See *wave function.*

steady state model: A model of the cosmos in which hydrogen is steadily created so that the cosmos will continue in a steady state.

subjective aim: In Whitehead, the subjective aim of an actual entity is what guides its process of becoming. It involves a proposition prehended by the actual entity with the subjective form of purpose to realize its process of becoming. The actual entity is itself the logical subject of the proposition, and an eternal object, or set of eternal objects, is the logical predicate of the proposition.

subjective form: In Whitehead, a particular subjective definiteness gained by an actual entity by reason of its prehension of some datum or other. The definiteness of each subjective form is due to the ingression of an eternal object or set of eternal objects.

substantialism: The view that all reality is basically composed of substances whose change of configuration constitutes change without reference to any fundamental process as in process thought.

supersession: In Whitehead, supersession is a technical term standing both for the process constituting the becoming of an actual entity and for the chronological relations generated by that process. The relata of supersession are either actual entities or phases in the becoming of each actual entity.

supervenience: The relationship between the characteristics of a collection and the properties of the components that compose that collection. For ordinary macroscopic objects (shoes, ships, sealing wax), the mass of a collection is simply the sum of the masses of the components. However, much of the energy produced by the Sun derives from the circumstance that the mass of the nucleus of the helium atom is somewhat less than the mass of four hydrogen nuclei (protons) that may be taken to compose it. The standard philosophical treatment of wholes and parts ("mereology") treats parts as being unaffected by their aggregation into larger units. This approach may not apply to all composites of philosophic and/or scientific interest. To the extent that the existence of a composite changes the characteristics of the components, the properties of the aggregate may not correspond exactly to ("supervene on") the sum of the corresponding properties of the parts of that aggregate.

symmetry: Immunity to a possible change.

synchronic: Covering all events at one time.

syntactics: Formal (content-free) analysis of the linguistic forms of languages.

topology: The study of geometric spaces that are invariant under deformation.

transition: In Whitehead, the process of transition—or the macroscopic process—is the first stage in the becoming of an actual entity. The second stage is the process of concrescence. In the stage of transition, the process of becoming is determined by the state of the universe correlative with the initiation of the process. The process of transition is thus a process of efficient causation.

uncertainty principle: See *indeterminacy principle*.

uranium: A silvery-white metallic element of atomic number 92, which has fourteen isotopes, of which U-238 is the most abundant. The isotope U-235 is fissionable with slow neutrons and, in a critical mass, is capable of sustaining a chain reaction that can proceed explosively.

vacuum: Observationally, there is no pure vacuum anywhere in the universe, which is permeated by radiation, particles and plasmas even if extremely tenuous in most space plasma regions. See *plasmas*. In quantum field theory, ground states are referred to as a 'vacuum.' The electromagnetic, weak, and strong fields have fluctuations about a vanishing expectation value in such a vacuum. See *virtual particle pair*.

vector: A quantity having both magnitude and direction. See *state vector*.

Venn diagram: Classes can be diagrammed as overlapping circles, which can help test the logical consequences of given Boolean propositions. (John Venn, 1834–1923).

virtual particle: A particle that is formed for a very short period of time and then ceases to exist.

virtual particle pair: A pair of particles produced spontaneously in the vacuum as a result of the Heisenberg uncertainty principle. For example, an electron and positron can be produced and exist for a very short time.

wave function: The state of motion of a particle can be described by a complex wave function $\Psi(x, y, z; t)$. The probability of finding the particle in a volume element dV at point (x, y, z) and time t is equal to $\Psi^*\Psi$. Wave functions are commonly represented by linear operators or matrices called vectors; examples include what are termed ket, bra, and state vectors.

wave packets: Groups of waves, traveling at the group velocity, which are combined disturbances of a set of sine waves with a limited range of frequencies and wavelengths. The pure sine waves used to define phase velocity do not really exist because they would require infinite spatial extent.

Bibliography

Agar, Wilfred E. *A Contribution to the Theory of the Living Organism*. Melbourne: Melbourne University Press, 1943.

Anshen, Ruth N. *Alfred North Whitehead: His Reflection on Man and Nature*. New York: Harper & Brothers, 1961.

Athearn, Daniel. *Scientific Nihilism: On the Loss and Recovery of Physical Explanation*. Albany: State University of New York Press, 1994.

Barbour, Ian. *Religion in an Age of Science: The Gifford Lectures 1989–1991*, Volume 1. New York: HarperCollins Publishers, 1990.

Barbour, Ian. *Ethics in an Age of Technology: The Gifford Lectures 1989–1991*, Volume 2. New York: HarperCollins Publishers, 1993.

Barbour, Ian. *When Science Meets Religion*. New York: HarperCollins Publishers, 2000.

Barbour, Ian, and Robert J. Russell, eds. "David Bohm's Implicate Order: Physics, Philosophy, and Theology." *Zygon* 20 (2) (1985).

Berman, Morris. *The Reenchantment of the World*. Ithaca: Cornell University Press, 1981.

Birch, Charles. *Feelings*. Sydney: University of New South Wales Press, 1995.

Birch, Charles, Jay B. McDaniel, and William R. Eakin. *Liberating Life: Contemporary Approaches to Ecological Theology*. Maryknoll, NY: Orbis Books, 1990.

Bohm, David. "The Implicate Order: A New Order for Physics," edited by D. Fowler. *Process Studies* 8 (2): 73–102 (1978).

Bohm, David, and F. David Peat, *Science, Order, and Creativity*, 2nd ed. New York: Routledge, 2000.

Bradley, James. "The Speculative Generalization of the Function: A Key to Whitehead." *Tijdschrift voor Filosofie* 64, 253–271, 2002.

Bright, Laurence. *Whitehead's Philosophy of Physics*. New York: Sheed and Ward, 1958.

Brody, Thomas. *The Philosophy Behind Physics*. Berlin: Springer-Verlag, 1993.

Browning, Douglas, and William T. Myers, eds. *Philosophers of Process*. 2nd ed. New York: Fordham University Press, 1998.

Bub, Jeffrey. *Interpreting the Quantum World*. Cambridge: Cambridge University Press, 1997.

Burgers, J. M. *Experience and Conceptual Activity: A Philosophical Essay Based upon the Writings of A. N. Whitehead*. Cambridge: MIT Press, 1965.

Cahill, Reginal T. "Process Physics: Modeling Reality as Self-Organising Information." *The Physicist* 37(6): 191–195 (2000).

Čapek, Milič. *Philosophical Impact of Contemporary Physics*. Princeton: D. Van Nostrand, 1961.

Čapek, Milič. *Bergson and Modern Physics: A Reinterpretation and Re-evaluation*. Dordrecht: D. Reidel, 1971.

Čapek, Milič. *New Aspects of Time: Its Continuity and Novelties*. Dordrecht, Boston: Kluwer Academic, 1991.

Chew, Geoffrey. "Historical Quantum Cosmology." Lawrence Berkeley Laboratory preprint 42946 (May 1999).

Clayton, Philip. *God and Contemporary Science*. Edinburgh: Edinburgh University Press, 1997.

Clayton, Philip. *The Problem of God in Modern Thought*. Grand Rapids: W.B. Eerdmans, 2000.

Code, Murray. *Order and Organism: Steps to a Whiteheadian Philosophy of Mathematics and the Natural Sciences*. Albany: State University of New York Press, 1985.

Cushing, James T. and Ernan McMullin, eds. *Philosophical Consequences of Quantum Theory: Reflections on Bell's Theorem*. Notre Dame: University of Notre Dame Press, 1989.

Davies, Paul. *The Cosmic Blueprint: New Discoveries in Nature's Creative Ability to Order the Universe*. New York: Simon & Schuster, 1988.

Earley, Joseph E., ed. *Individuality & Cooperative Action*. Georgetown University Press, Washington, DC, 1981.

Eastman, Timothy E., ed. "Process Thought and Natural Science." *Process Studies* 26 (3–4) (1997); 27 (3–4) (1998).

Emmet, Dorothy. *Whitehead's Philosophy of Organism*, 2nd ed. London: Macmillan, 1966.

Emmet, Dorothy. *The Effectiveness of Causes*. Albany: State University of New York Press, 1985.

Emmet, Dorothy. *The Passage of Nature*. Philadelphia: Temple University Press, 1992.

Fagg, Lawrence W. *The Becoming of Time: Integrating Physical and Religious Time*. Atlanta: Scholars Press, 1995.

Fagg, Lawrence W. *Electromagnetism and the Sacred: At the Frontier of Spirit and Matter*. New York: Continuum, 1999.

Ferre, Frederick. *Being and Value: Toward a Constructive Postmodern Metaphysics*. Albany: State University of New York Press, 1996.

Finkelstein, David R. "A Process Conception of Nature." In *The Physicist's Conception of Nature*, ed. J. Mehra, 709–713. Dordrecht: D. Reidel, 1973.

Finkelstein, David R. "All Is Flux." In *Quantum Implications: Essays in Honour of David Bohm*, eds. B. J. Hiley and F. D. Peat, 289–294. London: Routledge, 1987.

Finkelstein, David R. *Quantum Relativity: A Synthesis of the Ideas of Einstein and Heisenberg*. Berlin: Springer-Verlag, 1996.

Finkelstein, David R. "Organism and Physics." *Process Studies* 26 (3–4): 279–292 (1997).

Finkelstein, David R. "Relativity and Interactivity." In *Buddhism & Science: Breaking New Ground*, ed. B. Alan Wallace, 365–384. New York: Columbia University Press, 2003.

Fitzgerald, Janet A. *Alfred North Whitehead's Early Philosophy of Space and Time*. Washington, DC: University Press of America, 1979.

Folse, H. J. "Complementarity, Bell's Theorem, and the Framework of Process Metaphysics." *Process Studies* 11 (4): 259–273 (1981).

Fowler, Dean R. *Relativity Physics and the Doctrine of God: A Comparative Study of Whitehead and Einstein*. PhD dissertation. Claremont: Claremont Graduate University, 1975.

Fowler, Dean R. "Whitehead's Theory of Relativity." *Process Studies* 5(3): 159–174 (1975).

Fowler, Dean R., ed. "Whitehead and Natural Science." *Process Studies* 11 (4) (1981).

Griffin, David Ray. *God, Power and Evil: A Process Theodicy*, Philadelphia: Westminster Press, 1976.

Griffin, David Ray, ed. *Physics and the Ultimate Significance of Time: Bohm, Prigogine, and Process Philosophy*. Albany: State University of New York Press, 1986.

Griffin, David Ray, ed. *The Reenchantment of Science: Postmodern Proposals*. Albany: State University of New York Press, 1988.

Griffin, David Ray. *Evil Revisited: Responses & Reconsiderations*. Albany: State University of New York Press, 1991.

Griffin, David Ray. *Unsnarling the World-Knot: Consciousness, Freedom, and the Mind-Body Problem*. Berkeley, CA: University of California Press, 1998.

Griffin, David Ray. *Religion and Scientific Naturalism: Overcoming the Conflicts*. Albany: State University of New York Press, 2000.

Grunbaum, Adolf. "Whitehead's Method of Extensive Abstraction." *British Journal for the Philosophy of Science* 4: 215–226 (1953).

Grunbaum, Adolf. "Whitehead's Philosophy of Science." *The Philosophical Review* 71: 218–229 (1962).

Hahn, L., ed. *The Philosophy of Charles Hartshorne*. La Salle, IL: Open Court, 1991.

Hahn, L., ed. *The Philosophy of Paul Weiss*. Peru, IL: Open Court, 1995.

Hammerschmidt, William W. *Whitehead's Philosophy of Time*. New York: King's Crown Press, 1947.

Harris, Errol E. *The Foundations of Metaphysics in Science*. Lanham, MD: University Press of America, 1983 (originally published by George Allen and Unwin Ltd., 1965).

Harris, Errol E. *The Reality of Time*. Albany: State University of New York Press, 1988.

Hartshorne, Charles. *Creative Synthesis and Philosophic Method*. La Salle, IL: Open Court, 1970.

Hartshorne, Charles. *Whitehead's Philosophy: Selected Essays, 1935–1970*. Lincoln: University of Nebraska Press, 1972.

Hartshorne, Charles, and Paul Weiss, eds. *Collected Papers of Charles Sanders Peirce, Vol. VI. Scientific Metaphysics*. Cambridge: Harvard University Press, 1935.

Haught, John F. *Science & Religion: From Conflict to Conversation*. New York: Paulist Press, 1995.

Haught, John F., and F. J. Ayala, eds. *Science and Religion in Search of Cosmic Purpose*. Washington, DC: Georgetown University Press, 2000.

Hawking, Stephen, and Roger Penrose. *The Nature of Space and Time*. Princeton: Princeton University Press, 1996.

Henry, Granville C. *Forms of Concrescence: Alfred North Whitehead's Philosophy and Computer Programming Structures*. Cranbury, NJ: Associated University Presses, 1993.

Henry, Granville C., and Robert J. Valenza. "The Preprojective and the Postprojective: A New Perspective on Causal Efficacy and Presentational Immediacy." *Process Studies* 26 (1–2): 33–56 (1997).

Henry, Granville C., and Robert J. Valenza. "Eternal Objects at Sea." *Process Studies* 30 (1): 55–77 (2001).

Hiley, Basil J., and F. David Peat. *Quantum Implications: Essays in Honour of David Bohm* London: Routledge, 1987.

Holzey, Helmut, Alois Rust, and Reiner Wiehl, eds. *Natur, Subjektivitat, Gott: zur Prozessphilosophie Alfred N. Whitehead*. Frankfurt am Main: Suhrkamp, 1990.

Hurley, Patrick J. "Whitehead's 'Relational Theory of Space'; Text, Translation and Commentary." #1259 in *Philosophy Research Archives*, 4 (1978).

Hurley, Patrick J. "Russell, Poincaré, and Whitehead's 'Relational Theory of Space'." *Process Studies* 9(1–2): 14–21 (1979).

Hyman, A. T. "A New Interpretation of Whitehead's Theory." *Il Nuovo Cimento* 104B: 387–398 (1989).

Johnson, A. H. *Whitehead's Theory of Reality.* Boston: The Beacon Press, 1952.

Johnson, A. H. *Experiential Realism.* London: George Allen and Unwin, 1973.

Jungerman, John A. *World in Process: Creativity and Interconnection in the New Physics.* Albany: State University of New York Press, 2000.

Kafatos, Menas, and Robert Nadeau. *The Conscious Universe: Part and Whole in Modern Physical Theory.* Berlin: Springer-Verlag, 1990.

Kauffman, Stuart A. *The Origin of Order: Self-Organization and Selection in Evolution.* Oxford: Oxford University Press, 1993.

Kauffman, Stuart A. *At Home in the Universe: The Search for Laws of Self-Organization and Complexity.* Oxford: Oxford University Press, 1995.

Keeton, Hank. *The Topology of Feeling: Extensive Connection in the Thought of A. N. Whitehead, Its Development and Implications.* PhD dissertation. Berkeley: Graduate Theological Union, 1984.

Kim, Jaegwon, and Ernest Sosa, eds. *A Companion to Metaphysics.* Oxford: Blackwell Publishers, 1995.

Kirk, James. *Organicism as Reenchantment: Whitehead, Prigogine, and Barth.* Frankfurt: Peter Lang, 1993.

Kitchener, Richard F. *The World View of Contemporary Physics: Does It Need a New Metaphysics?* Albany: State University of New York Press, 1988.

Klein, Etienne, and Marc Lachieze-Rey. *The Quest for Unity: The Adventure of Physics.* Oxford: Oxford University Press, 1999.

Klein, George. *Alfred North Whitehead: Essays on His Philosophy.* Englewood Cliffs, NJ: Prentice-Hall, 1963. Reprint, Lanham, MD: University Press of America, 1989.

Kraus, Elizabeth M. *The Metaphysics of Experience: A Companion to Whitehead's Process and Reality.* New York: Fordham University Press, 1979.

Kreek, Michael J. *Whitehead, Von Weizsacker and the Problems of Time and Experience in Contemporary Cosmology.* MA thesis. Nashville: Vanderbilt University, 1983.

Krips, Henry. *The Metaphysics of Quantum Theory.* Oxford: Clarendon Press, 1987.

Lango, John W. *Whitehead's Ontology.* Albany: State University of New York Press, 1972.

Lawrence, Nathaniel. "Whitehead's Method of Extensive Abstraction." *Philosophy of Science* 17: 142–163 (1950).

Leclerc, Ivor. *The Relevance of Whitehead.* London: Allen and Unwin, 1961.

Leclerc, Ivor. *The Philosophy of Nature.* Washington, DC: The Catholic University of America Press, 1986.

Lillie, Ralph. *General Biology and the Philosophy of Organism*. Chicago: University of Chicago Press, 1945.

Lobl, Michael. *Wissenschaftliche Naturerkenntnis und Ontologie der Welterfahrung: Zu A. N. Whiteheads Kosmologiemodell im Horizont von Relativitatstheorie und Quantentheorie*. Frankfurt: Peter Lang, 1996.

Longair, Malcolm, ed. *The Large, the Small and the Human Mind: Roger Penrose with Abner Shimony, Nancy Cartwright and Stephen Hawking*. Cambridge: Cambridge University Press, 1997.

Lowe, Victor. *Understanding Whitehead*. Baltimore: John Hopkins Press, 1962.

Lowe, Victor. *Alfred North Whitehead: The Man and His Work, Vol. I: 1861–1910*, ed. J. B. Schneewind. Baltimore: Johns Hopkins Press, 1985.

Lowe, Victor. *Alfred North Whitehead: The Man and His Work, Vol. II: 1910–1947*, ed. J. B. Schneewind. Baltimore: Johns Hopkins Press, 1990.

Lucas, George R. Jr. *The Rehabilitation of Whitehead: An Analytic and Historical Assessment of Process Philosophy*. Albany: State University of New York Press, 1989.

Madigan, Patrick S. "Space in Leibniz and Whitehead." *Tulane Studies in Philosophy* 24: 48–57 (1975).

Malin, Shimon. *Nature Loves to Hide: Quantum Physics and Reality, a Western Perspective*. Oxford: Oxford University Press, 2001.

Margenau, Henry, and Roy A. Varghese, eds. *Cosmos, Bios, Theos: Scientists Reflect on Science, God, and the Origins of the Universe, Life, and Homo Sapiens*. La Salle, IL: Open Court, 1992.

Martin, Richard M. *Events, Reference, and Logical Form*. Washington, DC: The Catholic University of America Press, 1978.

Martin, Richard M. *Primordiality, Science, and Value*. Albany: State University of New York Press, 1980.

Mays, Wolfe. *Whitehead's Philosophy of Science and Metaphysics*. The Hague: Martinus Nijhoff, 1977.

McHenry, Leemon. "Quine and Whitehead: Ontology and Methodology." *Process Studies* 26 (1–2): 2–12 (1997).

Murphy, Nancey, and George F. R. Ellis. *On the Moral Nature of the Universe: Theology, Cosmology, and Ethics*. Minneapolis: Fortress Press, 1996.

Nobo, Jorge Luis. *Whitehead's Metaphysics of Extension and Solidarity*. Albany: State University of New York Press, 1986.

Northrop, Filmer S. C. "Whitehead's Philosophy of Science." In *The Philosophy of Alfred North Whitehead*, 2nd ed., edited by Paul A. Schilpp, 167–207. New York: Tudor, 1951.

Palter, Robert M. *Whitehead's Philosophy of Science*. Chicago: The University of Chicago Press, 1960. [Note: App. IV, Survey of Writings on Whitehead's Theory of Relativity]

Papatheodorou, C., and Basil Hiley. "Process, Temporality and Space-Time." *Process Studies* 26(3–4): 247–278 (1997).

Peat, E. David. *Einstein's Moon: Bell's Theorem and the Curious Quest for Quantum Reality*, Los Angeles: NTC Publishing Group, 1990.

Peat, E. David. *Infinite Potential: The Life and Times of David Bohm*, Cambridge: Perseus Publishing, 1997.

Penrose, Roger. *Shadows of the Mind: A Search for the Missing Science of Consciousness*. Oxford: Oxford University Press, 1994.

Peters, Ted, ed. *Cosmos as Creation: Theology and Science in Consonance*. Nashville: Abingdon Press, 1989.

Pirsig, Robert M. *Zen and the Art of Motorcycle Maintenance: An Inquiry into Values*. New York: Morrow Press, 1974.

Plamondon, Ann L. *Whitehead's Organic Philosophy of Science*. Albany: State University of New York Press, 1979.

Prigogine, Ilya. *The End of Certainty: Time, Chaos, and the New Laws of Nature*. New York: Free Press, 1998.

Prigogine, Ilya, and Isabelle Stengers. *Order Out of Chaos: Man's New Dialogue with Nature*. New York: Bantam Books, 1984.

Ranke, Oliver von. *Whiteheads Relativitatstheorie*. Regensburg, Germany: Roderer Verlag, 1997.

Rapp, Friedrich, and Reiner Wiehl. *Whitehead's Metaphysics of Creativity*. Albany: State University of New York Press, 1990.

Rayner, C. B. *Foundations and Applications of Whitehead's Theory of Relativity*. PhD dissertation. London: University of London, 1953.

Reinhardt, Michael, and A. Rosenblum. "Whitehead Contra Einstein." *Physics Letters*, 48A: 115–116 (1974).

Rescher, Nicholas. *Process Metaphysics: An Introduction to Process Philosophy*. Albany: State University of New York Press, 1996.

Rescher, Nicholas. "On Situating Process Philosophy." *Process Studies* 28 1–2: 37–42 (1999).

Rosen, Joe. *The Capricious Cosmos: Universe Beyond Law*. New York: Macmillan, 1991.

Rosen, Joe. *Symmetry in Science: An Introduction to the General Theory*. Berlin: Springer-Verlag, 1995.

Russell, Bertrand. "Whitehead and Principia Mathematica." *Mind* 57: 137–138 (1948).

Russell, Bertrand. *My Philosophical Development.* New York: Simon and Schuster, 1959.

Russell, Robert J., and Christoph Wassermann. "Kerr Solution of Whitehead's Theory of Gravity." *Bulletin of the American Physical Society* 32: 90 (1987).

Russell, Robert J. "Whitehead, Einstein, and the Newtonian Legacy." In *Newton and the New Direction in Science*, eds. G. V. Coyne et al., 175–192. Vatican City: Specola Vaticana, 1988.

Russell, Robert J., Nancey Murphy, and A. R. Peacocke. *Chaos and Complexity: Scientific Perspectives on Divine Action.* Vatican City: Vatican Observatory Foundation, 1995.

Russell, Robert J., Philip Clayton, Kirk Wegter-McNelly, and John Polkinghorne, eds. *Quantum Mechanics.* Ed. Notre Dame, IN: University of Notre Dame Press, 2001.

Rust, Alois. "Die Organismische Kosmologie von Alfred North Whitehead." *Monographien zur philosophischen Forschung* 248, Frankfurt am Main: Athenäum 1987.

Schild, Alfred. "Conservative Gravitational Theories of Whitehead's Type." In *Recent Developments in General Relativity*, eds. S. Bazanski et al., 409–413. Oxford: Oxford University Press, 1962.

Schilpp, Paul A., ed. *The Philosophy of Alfred North Whitehead.* Evanston, IL: Northwestern University Press, 1941. Reprints Cambridge: Cambridge University Press, 1943; New York: Tudor Publishing Co., 1951.

Schlegel, Richard. *Superposition & Interaction: Coherence in Physics.* Chicago: University of Chicago Press, 1980.

Schmidt, Paul F. *Perception and Cosmology in Whitehead's Philosophy.* New Brunswick, NJ: Rutgers University Press, 1967.

Seaman, Francis C. "Whitehead and Relativity." *Philosophy of Science* 22: 222–226 (1955).

Shields, George W. *Process and Analysis: Whitehead, Hartshorne, and the Analytic Tradition.* Albany: State University of New York Press, 2002.

Shimony, Abner. "The Methodology of Synthesis: Parts and Wholes in Low-Energy Physics." In *Kelvin's Baltimore Lectures and Modern Theoretical Physics: Historical and Philosophical Perspectives*, eds. Robert Kargon and Peter Achinstein. Cambridge: MIT Press, 1987.

Shimony, Abner. "Metaphysical Problems in the Foundations of Quantum Mechanics." In *The Philosophy of Science*, eds. Richard Boyd, Philip Gasper, and J. D. Trout. 517–528. Cambridge: MIT Press, 1991.

Shimony, Abner. *Search for a Naturalistic World View, Vol. II: Natural Science and Metaphysics.* Cambridge: Cambridge University Press, 1993.

Shimony, Abner. "On Mentality, Quantum Mechanics and the Actualization of Potentialities." In *The Large, the Small and the Human Mind*, ed. M. Longair. Cambridge: Cambridge University Press, 1997.

Shimony, Abner. "Can the fundamental laws of nature be the results of evolution?" In *From Physics to Philosophy*, eds. Jeremy Butterfield and Constantine Pagonis. 208–223. Cambridge: Cambridge University Press, 1999.

Sklar, Lawrence. *Philosophy of Physics*. Boulder: Westview Press, 1992.

Smith, Raymond. *Whitehead's Concept of Logic*. Westminster, MD: The Newman Press, 1953.

Smith, Vincent Edward. *Philosophical Physics*. New York: Harper, 1950.

Smolin, Lee. *The Life of the Cosmos*. Oxford: Oxford University Press, 1997.

Stapp, Henry P. "Whiteheadian Approach to Quantum Theory and the Generalized Bell's Theorem." *Foundations of Physics* 9 (1–2): 1–25 (1979).

Stapp, Henry P. *Mind, Matter, and Quantum Mechanics*. Berlin: Springer-Verlag, 1993.

Stengers, Isabelle. "Whitehead and the Laws of Nature." *Salzburger Theologische Zeitschrift* 3 (2): 193–206 (1999).

Stolz, Joachim. *Whitehead und Einstein: Wissenschaftsgeschichtliche Studien in Naturphilosophischer Absicht*. Frankfurt: Peter Lang, 1995.

Stolz, Joachim. "The Research Program in Natural Philosophy from Gauss and Riemann to Einstein and Whitehead: From the Theory of Invariants to the Idea of a Field Theory of Cognition." *Prima-Philosophia* 10 (2): 157–164 (1997).

Swimme, Brian, and Thomas Berry. *The Universe Story: From the Primordial Flaring Forth to the Ecozoic Era—a Celebration of the Unfolding of the Cosmos*. San Francisco: Harper, 1992.

Synge, John L. "The Relativity Theory of A. N. Whitehead." *Fluid Dynamics and Applied Math. Institute*, Lecture Series 5. College Park: University of Maryland, 1951.

Tanaka, Yutaka. "Einstein and Whitehead: The Principle of Relativity Reconsidered." *Historia Scientiarum* 32: 45–61 (1987).

Tanaka, Yutaka. "Bell's Theorem and the Theory of Relativity: An Interpretation of Quantum Correlation at a Distance Based on the Philosophy of Organism." *Annals of the Japan Association for Philosophy of Science* 8 (2): 49–67 (1992).

Toulmin, Stephen. *The Return to Cosmology: Postmodern Science and the Theology of Nature*. Berkeley: University of California Press, 1982.

Toulmin, Stephen. *Cosmopolis: The Hidden Agenda of Modernity*. Chicago: University of Chicago Press, 1990.

Uschenko, Andrew P. "A Note on Whitehead and Relativity." *The Journal of Philosophy* 47: 100–102 (1950).

Van der Merwe, Alwyn, ed. *Old and New Questions in Physics, Cosmology, Philosophy, and Theoretical Biology: Essays in Honor of Wolfgang Yourgrau*. New York: Plenum Press, 1983.

von Ranke, Oliver. *Whitehead's Relativitatstheorie*. Regensburg: Roderer Verlag, 1997.

Wassermann, Christoph. "Mathematical Grundlagen von Whitehead's Religions Philosophie." In *Natur Subjektivitat Gott*, eds. Helmut Holzhey, Alois Rust and Reiner Wiehl, 240–261. Frankfurt: Suhrkamp, 1990.

Wassermann, Christoph. *Struktur und Ereignis*. Geneva: Faculte de Theologie, 1991.

Weiss, Paul. *Creative Ventures*. Carbondale, IL: Southern Illinois University Press, 1992.

Weissman, David. *Hypothesis and the Spiral of Reflection*. Albany: State University of New York Press, 1989.

Weizsäcker, Carl Friedrich von. *Aufbau der Physik*. Munich: Carl Hanser Verlag, 1985.

Weizsäcker, Carl Friedrich von. *Zeit und Wissen*. Munich: Carl Hanser Verlag, 1992.

Welton, Willibrord. "Whitehead, Einstein et la relativite: l'uniformite de l'espacetemps." *Gregorianum* 61: 77–95 (1980).

Whipple, Elden C. Jr. "Events as Fundamental Entities in Physics." *Il Nuovo Cimento* 923A (11): 309-327 (1986).

Whittemore, Robert C., ed. *Studies in Process Philosophy*. New Orleans: Tulane University Press, 1974.

Whittemore, Robert C., ed. *Studies in Process Philosophy II*. New Orleans: Tulane University Press, 1975.

Whittemore, Robert C., ed. *Studies in Process Philosophy III*. New Orleans: Tulane University Press, 1976.

Wiener, Norbert. *A Comparison Between the Treatment of the Algebra of Relations by Schroeder and That by Whitehead and Russell*. PhD dissertation. Cambridge, MA: Harvard University, 1913.

Wilber, Ken, ed. *Quantum Questions: Mystical Writings of the World's Great Physicists*. Boulder, CO: Shambala, 1984.

Woodbridge, B., ed. *A.N. Whitehead: A Primary-Secondary Bibliography*. Bowling Green, OH: State University, 1977.

Selected Writings of Alfred North Whitehead

A Treatise on Universal Algebra, with Applications, Vol. I. Cambridge: Cambridge University Press, 1898.

"On Mathematical Concepts of the Material World." *Philosophical Trans. Royal Society of London*, Ser. A (205): 465–525 (1906).

The Axioms of Projective Geometry. Cambridge: Cambridge University Press, 1906.

The Axioms of Descriptive Geometry. Cambridge: Cambridge University Press, 1907.

Principia Mathematica (with Bertrand Russell). Vols. I–III. Cambridge: Cambridge University Press, 1910–1913; 2nd ed., 1925–1927. Abridged as *Principia Mathematica to *56*. Cambridge: Cambridge University Press, 1962.

An Introduction to Mathematics. London: Williams & Norgate, 1911.

An Enquiry Concerning the Principles of Natural Knowledge. Cambridge: Cambridge University Press, 1919.

The Concept of Nature. Cambridge: Cambridge University Press, 1920.

The Principle of Relativity, with Applications to Physical Science. Cambridge: Cambridge University Press, 1922.

Science and the Modern World. New York: Macmillan, 1925; Cambridge: Cambridge University Press, 1926.

Religion in the Making. New York: Macmillan; Cambridge: Cambridge University Press, 1926.

Symbolism: Its Meaning and Effect. New York: Macmillan, 1927; Cambridge: Cambridge University Press, 1928.

The Aims of Education and Other Essays. New York: Macmillan; London: Williams & Norgate, 1929.

The Function of Reason. Princeton: Princeton University Press, 1929.

Process and Reality: An Essay in Cosmology. New York: Macmillan; Cambridge: Cambridge University Press, 1929. Critical edition by D. R. Griffin and D. W. Sherbourne. New York: Macmillan, 1978.

Adventures of Ideas. New York: Macmillan; Cambridge: Cambridge University Press, 1933.

Nature and Life. Chicago: University of Chicago Press, 1934.

Modes of Thought. New York: Macmillan; Cambridge: Cambridge University Press, 1938.

Essays in Science and Philosophy. New York: Philosophical Library, 1947; London: Rider & Co., 1948.

The Wit and Wisdom of Whitehead, ed. A. H. Johnson. Boston: The Beacon Press, 1947.

Alfred North Whitehead: An Anthology, eds. F. S. C. Northrop and Mason W. Gross. New York: Macmillan, 1953.

Internet Resources

Resource Guide to Physics and Whitehead, eds. Timothy E. Eastman and Hank Keeton, Process Studies Electronic Supplement 2003, http://www.ctr4process.org/publications/pss/.

Contributors

Timothy E. Eastman, editor, has carried out basic research in plasma physics and space physics for over 25 years and has been national coordinator for space plasma research for six years while at NSF and NASA headquarters. Working with QSS Group, Inc., Dr. Eastman is group manager for space science support at the Space Science Data Operations Office located at NASA's Goddard Space Flight Center in addition to being a consultant in plasma science and applications. He is well known for discovering the low-latitude boundary layer of Earth's magnetosphere and other research results important to current work in solar-terrestrial relations and space weather. In addition to space physics research, Eastman has pursued philosophical interests through extensive reading and graduate level studies, conferences and publications, especially in the area of process philosophy. He has given formal responses in conferences to Hilary Putnam and Henry Stapp, and edited special issues on process thought and natural science in two issues of the Process Studies journal (*Process Studies*. 26/3–4 (1997); 27/3–4 (1998)). In collaboration with Dr. Keeton, he created a comprehensive guide to physics and Whitehead published in *Process Studies Electronic Supplement* ("Resource Guide to Physics and Whitehead," *PSS* 2003; http://www.ctr4process/publications/pss/)

Hank Keeton, editor, became interested in Whitehead's thought after deciding to do graduate work in philosophy, rather than physics. But his inclination toward particle physics led him to work on the Alvarez experiment at the Lawrence Berkeley Laboratory in the early 1970s. This combination of interests led him to complete his Ph.D. under Bernard M. Loomer at the Graduate Theological Union in Berkeley, concentrating on Whitehead's evolution from theoretical physics into philosophical metaphysics. Keeton explored the mathematical development of Whitehead's theory of extensive connection in his doctoral work. Since then, he has continued to be active at the interface between process thought and the philosophy of science, with an emphasis on current developments in theoretical physics. His greatest hope is that the evolving ideas of process philosophy will assist current developments in relativity and quantum mechanics.

Other Contributors (alphabetical ordering)

Geoffrey F. Chew of Lawrence Berkeley National Laboratory is a leading theorist in fundamental physics. He originated the famous S-matrix theory, which has been very influential in high energy physics. The book of essays related to his work *A Passion for Physics* (C. De Tar, C. Tan, and J. Finkelstein, editors, World Science, 1985) was prepared in Dr. Chew's honor. Most recently, he has been developing an 'historical quantum cosmology.'

Philip Clayton is Ingraham Chair at the Claremont School of Theology and holds doctoral degrees in both philosophy and religious studies from Yale University. He is the author of *Explanation from Physics to Theology* [Yale Press, 1989], *God and Contemporary Science* [Eerdmans, 1998] and *The Problem of God in Modern Thought* (Eerdmans, 2000), and co-editor of seven volumes, including *Quantum Mechanics: Scientific Perspectives on Divine Action* [Notre Dame, 2001] *In Whom We Live and Move and Have Our Being: Panentheism and Science* [Eerdmans, 2003], *Evolutionary Ethics: Human Morality in Biological and Religious Perspective* [Eerdmans, 2004], and *Science and the Spiritual Quest: New Essays by Leading Scientists* [Routledge, 2002]. Professor Clayton is a leading scholar in the science and religion field and is principal investigator of the *Science and the Spiritual Quest* project at the Center for Theology and the Natural Sciences in Berkeley, California.

Joseph E. Earley, Sr. is a professor emeritus of chemistry at Georgetown University who has specialized in far-from-equilibrium chemical systems. In addition to his forefront work in modern chemistry, Professor Earley is active in the emerging field of philosophy of chemistry. He is the editor of *Individuality and Cooperative Action* (Georgetown University Press, Washington, D.C., 1991) and the editor of *Chemical Explanation: Characteristics, Development, Autonomy* (New York Academy of Sciences, New York, 2003).

David Ritz Finkelstein of the Georgia Institute of Technology is a leading theorist in fundamental physics. He is the author of *Quantum Relativity: A Synthesis of the Ideas of Einstein and Heisenberg* (Springer-Verlag, Berlin, 1996). Professor Finkelstein is currently developing a quantum theory based on elementary processes rather than elementary particles.

Niels Viggo Hansen is now at the Danish Ministry of Health, Knowledge and Research Center for Alternative Medicine, engaged in the development of scientific approaches and methods for the understanding and assessment of alternative and complementary medicine. Before this, he worked at the Department of Philosophy, University of Aarhus, Denmark, with metaphysical, epistemological, and sociological dimensions of the philosophy of science. In his philosophical work, he has been particularly interested in exploring and articulating a line of thought, radical constructivism, which claims to overcome the classical realism/idealism dichotomy and is a common legacy of a few radically processual thinkers including Whitehead, Deleuze, and the second century Buddhist saint-philosopher, Nagarjuna.

John A. Jungerman is Professor Emeritus of the University of California at Davis. His research has been primarily in nuclear physics. He was Founding Director of the Crocker Nuclear Laboratory there. He studied process philosophy with Dr. Rebecca Parker, President of the Starr King School for the Ministry. Most recently he has authored *The World in Process, Creativity and Connection in the New Physics* (State University of New York Press, Albany, 2000).

Shimon Malin teaches at Colgate University in Hamilton, New York. His research is in various aspects of quantum field theory, relativity, and cosmology and has authored two books, *Representations of the Rotation and Lorentz Groups* with M. Carmeli [Dekker Inc., New York, 1976] and *Nature Loves to Hide: Quantum Physics and Reality* (Oxford University Press, Oxford, 2001).

Jorge Luis Nobo is professor of philosophy at Washburn University in Topeka, Kansas. He is a leading scholar of metaphysics, process philosophy, and Whitehead, and is the author of *Whitehead's Metaphysics of Extension and Solidarity* (State University of New York Press, Albany, 1986).

Franz G. Riffert is a philosopher and psychotherapist with the Institute for Education at the University of Salzburg, Austria. He has written three books, including *Whitehead und Piaget: Zur interdisziplinären Relevanz der Prozeßphilosophie* (Peter-Langer, Vienna, 1994), and a forthcoming edited volume, (together with Michel Weber) soon to be submitted, entitled *Searching for New Contrasts: Whiteheadian Contributions to Contemporary Challenges in Psychology, Neurophysiology and the Philosophy of Mind* (Peter-Lang, Vienna, 2003, in press).

Joe Rosen is former chair of the Department of Physics and Astronomy at the University of Central Arkansas. A distinguished contributor to the study of symmetry, he is the author or editor of ten books, including *Symmetry in Science: An Introduction to the General Theory* [Springer, 1995] and *The Capricious Cosmos: Universe Beyond Law* [Macmillan, 1991]. Dr. Rosen is now a consultant and resides in Rockville, Maryland.

Henry P. Stapp of Lawrence Berkeley National Laboratory is a leading theorist in fundamental physics, especially the quantum theory of measurement. His most recent work *Mind, Matter, and Quantum Mechanics* [Springer, 1993] has moved into issues of the origin of mind and consciousness, always with solid roots in the physics.

Yutaka Tanaka is professor in the department of philosophy at Sophia University in Tokyo. He is director of the Japan Internet Center for Process Studies, editor of *Process Thoughts* (Japanese philosophical journal), and director of Touri-Kadan (forum of classical Japanese poetry). Dr. Tanaka has written *From Paradox to Reality Kouro-sha* (Kyoto, 1993, in Japanese) and *Whitehead Koudan-sha* (Tokyo, 1998, in Japanese).

Note on Supporting Center

This series is published under the auspices of the Center for Process Studies, a research organization affiliated with the Claremont School of Theology and Claremont Graduate University. It was founded in 1973 by John B. Cobb Jr., Founding Director, and David Ray Griffin, Executive Director; Marjorie Suchocki is now also a Co-director. It encourages research and reflection on the process philosophy of Alfred North Whitehead, Charles Hartshorne, and related thinkers, and on the application and testing of this viewpoint in all areas of thought and practice. The center sponsors conferences, welcomes visiting scholars to use its library, and publishes a scholarly journal, *Process Studies*, and a newsletter, *Process Perspectives*. Located at 1325 North College, Claremont, CA 91711, it gratefully accepts (tax-deductible) contributions to support its work.

Index

SUNY series in Constructive Postmodern Thought
David Ray Griffin, series editor

David Ray Griffin, editor, *The Reenchantment of Science: Postmodern Proposals*

David Ray Griffin, editor, *Spirituality and Society: Postmodern Visions*

David Ray Griffin, *God and Religion in the Postmodern World: Essays in Postmodern Theology*

David Ray Griffin, William A. Beardslee, and Joe Holland, *Varieties of Postmodern Theology*

David Ray Griffin and Huston Smith, *Primordial Truth and Postmodern Theology*

David Ray Griffin, editor, *Sacred Interconnections: Postmodern Spirituality, Political Economy, and Art*

Robert Inchausti, *The Ignorant Perfection of Ordinary People*

David W. Orr, *Ecological Literacy: Education and the Transition to a Postmodern World*

David Ray Griffin, John B. Cobb Jr., Marcus P. Ford, Pete A. Y. Gunter, and Peter Ochs, *Founders of Constructive Postmodern Philosophy: Peirce, James, Bergson, Whitehead, and Hartshorne*

David Ray Griffin and Richard A. Falk, editors, *Postmodern Politics for a Planet in Crisis: Policy, Process, and Presidential Vision*

Steve Odin, *The Social Self in Zen and American Pragmatism*

Frederick Ferré, *Being and Value: Toward a Constructive Postmodern Metaphysics*

Sandra B. Lubarsky and David Ray Griffin, editors, *Jewish Theology and Process Thought*

J. Baird Callicott and Fernando J. R. da Rocha, editors, *Earth Summit Ethics: Toward a Reconstructive Postmodern Philosophy of Environmental Education*

David Ray Griffin, *Parapsychology, Philosophy, and Spirituality: A Postmodern Exploration*

Jay Earley, *Transforming Human Culture: Social Evolution and the Planetary Crisis*

Daniel A. Dombrowski, *Kazantzakis and God*

E. M. Adams, *A Society Fit for Human Beings*

Frederick Ferré, *Knowing and Value: Toward a Constructive Postmodern Epistemology*

Jerry H. Gill, *The Tacit Mode: Michael Polanyi's Postmodern Philosophy*

Nicholas F. Gier, *Spiritual Titanism: Indian, Chinese, and Western Perspectives*

David Ray Griffin, *Religion and Scientific Naturalism: Overcoming the Conflicts*

John A. Jungerman, *World in Process: Creativity and Interconnection in the New Physics*

Frederick Ferré, *Living and Value: Toward a Constructive Postmodern Ethics*

Laurence Foss, *The End of Modern Medicine: Biomedical Science Under a Microscope*

John B. Cobb Jr., *Postmodern and Public Policy: Reframing Religion, Culture, Education, Sexuality, Class, Race, Politics, and the Economy*

Catherine Keller and Anne Daniell, editors, *Process and Difference: Between Cosmological and Poststructuralist Postmodernisms*

Nicholas F. Gier, *The Virtue of Nonviolence: From Gautama to Gandhi*